RED VIENNA

RED VIENNA

Experiment in
Working-Class Culture
1919–1934

Helmut Gruber

New York • Oxford
OXFORD UNIVERSITY PRESS
1991

Oxford University Press

Oxford   New York   Toronto
Delhi   Bombay   Calcutta   Madras   Karachi
Petaling Jaya   Singapore   Hong Kong   Tokyo
Nairobi   Dar es Salaam   Cape Town
Melbourne   Auckland

and associated companies in
Berlin   Ibadan

Copyright © 1991 by Oxford University Press, Inc.

Published by Oxford University Press, Inc.,
200 Madison Avenue, New York, New York 10016

Oxford is a registered trademark of Oxford University Press

All rights reserved. No part of this publication may be reproduced,
stored in a retrieval system, or transmitted, in any form or by any means,
electronic, mechanical, photocopying, recording, or otherwise,
without the prior permission of Oxford University Press.

Library of Congress Cataloging-in-Publication Data
Gruber, Helmut, 1928–
Red Vienna : experiment in working-class culture, 1919–1934 /
Helmut Gruber.
p.   cm.
Includes index.
ISBN 0-19-506914-5
1. Vienna (Austria)—Social conditions.   2. Vienna (Austria)—
Social policy.   3. Working class—Austria—Vienna—History—
20th century.   4. Vienna (Austria)—Popular culture—History—
20th century.   5. Austro-Marxist school—History—20th century.
6. Vienna (Austria)—History—1918–   I. Title.
HN418.V5G78   1991
306'.09436'13—dc20   90-24065   CIP

9 8 7 6 5 4 3 2 1

Printed in the United States of America
on acid-free paper

For Käthe Leichter (1895–1942)
Sociologist, Socialist, Feminist,
and Courageous Human Being

Preface

In carrying out this study and critique of the Viennese socialists' attempt to create a working-class culture, I have confronted the symbolic and mythic aspects of the subject, which have distorted it. These stem in part from the public representations of socialist leaders at the time. They also have been revived in recent decades by Austrian historians of the working class in shaping a heroic past to serve as a tradition usable by the postwar Socialist party (SPÖ). In the process of demythologizing the cultural experiment, its Austromarxist leadership has unavoidably lost its heroic sheen. However limited the success of these leaders or contradictory their perception of both workers and culture, their commitment to socialist ideals and dedication to the class in whose name they claimed to speak cannot be challenged. Although their daring vision overshadowed and distorted the actual accomplishments of their experiment, "red Vienna" remains a model of cultural experimentation in the socialist movement. In that guise it provides us with fresh insights into the practices of all such experiments directed from above, and the resistance of structures, customs, and actors they are bound to encounter.

All books are collective efforts in the sense that authors depend on a group of advisors and critics who constitute a miniature audience. I have been very fortunate in receiving the supportive and challenging reading of versions and sections of the developing manuscript. I am especially indebted to Anson Rabinbach, Geoffrey Field, and Adelheid von Saldern, who pointed out exuberances and excesses that needed rethinking and modification and who encouraged my attempt to view the Vienna experiment in a larger, international context. Several draft chapters also had the benefit of crtitical readings by Istvan Deak, Joan Scott, and Felix F. Strauss, for which I would like to express my gratitude.

My perception of the book's orientation, sense of the myriad details, and access to invaluable oral history collections were enhanced by contacts and conversations over several years with the following younger historians of the Austrian working class: Reinhard Sieder, Joseph Ehmer, Alfred Pfoser, Theo Venus, Gottfried Pirhofer, Siegfried Mattl, Friedrich Stadler, Gerhard Meisl, Gerhard Steger, Karl Fallend, Hans Safrian, and Joseph Weidenholzer. Although they may not agree with my critical perspective, they nevertheless helped to shape it.

The friendly reception and personal assistance I received at various Viennese archives and university institutes—which gave me unlimited access to all of their material and allowed me to read theses and dissertations before being catalogued, and whose directors offered helpful hints—were indispensable to my work. My thanks for this high degree of scholarly cooperation go to the following: Institute für Wirtschafts- und Sozialgeschichte, Institut für Zeitgeschichte, and Institut für Volkskunde—all of the University of Vienna; Kammer für Arbeiter und Angestellte, Dr. Eckhart Früh and Dr. Karl Stubenvoll; Dokumentationsarchiv des Österreichischen Widerstandes, Dr. Herbert Steiner and Dr. Wolfgang Neugebauer; Österreichisches Circus- und Clownmuseum, Mr. Berthold Lang; Bezirksmuseum Rudolfsheim-Fünfhaus, Dr. Joseph Ehmer; Institut für Geschichte der Medizin am Josefineum, Dr. Karl Sablik; Allgemeines Verwaltungsarchiv des Österreichischen Staatsarchiv, Dr. Isabelle Ackerl; Archiv der Stadt und Land Wien; Archiv des Österreichischen Gewerkschaftsbundes; Film Laden, Dr. Franz Grafl; Archiv der *Volksstimme;* Institut für Wissenschaft und Kunst.

The center of my research in Vienna for five summers was the Verein für Geschichte der Arbeiterbewegung. Its director, Dr. Wolfgang Maderthaner, a colleague in every sense of the term, provided an atmosphere of scholarly conviviality, served as a source of information and contacts to persons and places vital to my work, and supplied most of the photographs for the book. I am most grateful for his generosity.

Discussions following the public presentation of portions of the work in progress were both stimulating and useful. These included: the 17th International Conference of Labour Historians at Linz in 1981; the Harvard University Center for European Studies at Cambridge, Massachusetts, in 1984; the Second UNESCO International Forum on the History of the Working Class at Paris in 1985; and the University Seminar in the History of the Working Class at Columbia University in New York City in 1986. At Oxford University Press, Nancy Lane and David Roll's enthusiasm for my book from the very beginning was translated into a caring guidance of the manuscript in its various transformations.

The origins of this book are difficult to trace. No doubt the vivid memory of eighteen months spent as a boy in Vienna in 1938–39 before I was forced to emigrate—with the possibility, in those troubled times, of exploring the city from end to end in the company of friends and without adult supervision—was one of the impulses. I am certain that my wife, Françoise Jouven, will appreciate that I have resisted bowing to the convention of reciting platitudes of gratitude on her behalf.

Paris H. G.
August 1990

Contents

1. **Introduction** 3

    Obituary for Austrian Socialism 3
    A Model of Proletarian Culture 5

2. **Vienna as Socialist Laboratory** 12

    Vienna, 1919–1921: A Montage 13
    Austromarxism: A Theory for Practice? 29

3. **Municipal Socialism** 45

    Public Housing: Environment for "neue Menschen" 46
    Public Health and Social Welfare: Shaping the "Orderly" Worker Family 65
    Public Education: Equality for Workers and Rising Expectations 73

4. **Socialist Party Culture** 81

    Elite Culture Rejected and Desired 83
    Magical Powers of the Word 87
    Enrichments of Taste: Music, Theater, and the Fine Arts 96
    To Culture Through Action: Sports and Festivals as Symbols of Power 102

5. **Worker Leisure: Commercial and Mass Culture** 114

    Commercial Culture 116
    Noncommercial Leisure-Time Activities 120
    Popular Culture Condemned 123
    The Cinema: A Dream Factory? 126
    Radio: Pulpit of the People? 135
    Spectator Sports: Gladiators of Capitalism? 141

6. The Worker Family: Invasions of the Private Sphere 146

The "New Woman" and the "Triple Burden" 147
Sexuality: Repression and Expression 155
Population Politics 157
Youth: Abstinence, Discipline, and Sublimation 165
Puritanism and Sexual Realities 170

7. Conclusion 180

Political Limits 181
Cultural Limits 184

Notes 187

Index 257

RED VIENNA

CHAPTER 1

Introduction

Obituary for Austrian Socialism

Between the twelfth and seventeenth of February 1934 the international press gave front-page coverage to the uprising of Austrian workers against the repressive, corporatist government of Engelbert Dollfuss. These early, confused dispatches pictured a civil war between armed workers of Vienna and industrial cities in the provinces determined to defend democracy, on one side, and government troops and elements of the Heimwehr[1] bent on destroying the republican essence of the Austrian state, on the other. Socialism, in this small central European country, appeared to be fighting for its life.

The foreign labor and socialist press greeted the first reports with messages of solidarity. Léon Blum, head of the French Socialist party (SFIO), paid homage to the Austrian workers for their defense of liberty and pledged solidarity with their cause.[2] Andrew Conley, chairman of the British Trade Union Congress (TUC), declared that "for many months the people of Vienna have been living under a dictatorship with the constitution suspended and political liberties a mockery"; they therefore had the right to defend the "institutions of free citizenship."[3]

As the actual situation in Vienna became clearer in the next few days—only a few thousand workers had risen to engage the full force of the state's repression, and the call for a general strike had failed—the earlier optimism gave way to heroic epitaphs. Blum now compared the Austrian workers to the Communards before the *mur des fédérés* and ended his peroration by declaring, "The revolutionary commune of Vienna has been crushed; long live the commune!"[4] Similar sentiments were expressed by the Labour party intellectual Harold Laski, under the headline "Salute to the Viennese Martyrs," in comparing the Vienna uprising to the Paris Commune and the Russian Revolution of 1905.[5] Assessments in the communist movement consisted mainly of recriminations. *L'Humanité* accused the Austrian socialist leaders of having brought on the debacle by putting "their faith in bourgeois democracy and bourgeois governments" instead of "accepting a united front with the communists."[6] In the Communist International's own vitriolic condemnation, the "Austrian workers had been brought under the yoke of fascism" by their leaders.[7]

As might be expected, the foreign conservative press interpreted the ongoing events differently. *Le Temps* accused the socialists of having "obstinately refused to support the government in its heroic struggle against external forces threatening the independence of Austria. By not having compromised with reaction, they have strengthened the hand of national socialism."[8] *Le Figaro* suggested that "with courage but also with foolish carelessness, socialism has played its last card [to rally its constituents] and has lost, thereby playing into the hands of Hitler's followers."[9] The London *Times* was more concerned with the foreign political implications than with the fate of the Austrian socialists. It concluded that Chancellor Dollfuss had been fighting on two fronts and had strengthened his foreign political position by eliminating the left at home.[10] Interestingly, the *New York Times* offered the most detailed and balanced coverage: daily dispatches, analysis, and large photo essays. Though various feature articles expressed great sympathy for casualties among women and children and even commended the female workers for fighting "like the old pioneer women of the American prairies," the underlying concern was with the foreign political implications.[11]

Such were the obituaries for Austrian socialism and the republic which disappeared with the storming of the worker enclaves in Vienna. There had been no civil war.[12] A tiny minority of workers had risen spontaneously by disregarding the cautious wait-and-see policy of their leaders, who clung desperately to constitutional safeguards which had ceased to exist when Chancellor Dollfuss suspended parliament in March 1933. For more than a year the leaders of the Austrian Socialist party (SDAP) had suffered the same paralysis of will which had immobilized the German left as Hitler rose to power. They repeatedly postponed the use of force to prevent the overthrow of the republic, although force had been a measure of self-defense in the SDAP program since 1926. This strategy of passivity in the face of the gradual demolition of republican safeguards was in effect a capitulation to the corporate state publicly projected by Dollfuss and the Christian Social party (Christian Socials). Ultimately the insurrection of the few was only a desperate act in the face of a defeat that was already apparent; it was also directed against those socialist leaders who had cautioned and counseled against self-defense by the workers, and who were absent during the crucial hours from February 12 to 14.[13]

The subjugation of the Austrian working class was viewed in virtually all the press accounts as increasing the danger of fascist expansion in Europe: conservatives feared that Austria would fall prey to Italian or German aggression; socialists saw in the suppression of the SDAP by clerical fascism an ominous repetition of the destruction of German socialism a year earlier—the loss of two pillars of the Labor and Socialist International, which weakened the working-class movement. Despite the fact that passing homage was paid to the accomplishments of Viennese municipal socialism in both the labor and middle-class press,[14] strategic geopolitical or institutional concerns were paramount.

Understandably, that perspective—the destruction of a republic and the liquidation of a party of 660,000—was put before all other considerations in the crisis atmosphere of that time. It is unfortunate, however, that until very recently the most significant casualty in February 1934, namely, the experiment to create a working-class culture in the socialist enclave of Vienna, has not received the attention it deserves. That attempt to develop a comprehensive proletarian counterculture, going beyond piecemeal cultural reform efforts of socialist parties in other countries and serving as an alternate model to the Bolshevik's experiment in Russia, is the subject of this book. In the Bolshevik example popular enlightenment ideals, cultural liberalism, and utopian visions had been at risk virtually from the beginning in the struggle for order and control. The Bolshevik party's "vanguard" position determined the controlled Soviet culture for the masses emerging at the end of the 1920s.[15] The Austromarxists, and particularly Otto Bauer, distanced themselves from what they considered the dictatorship of a caste over the masses.[16] Their cultural experiment was to be predicated on democracy in a dual sense: relying on the political guarantees of a republican government and on the SDAP's relation to the rank and file of the party.[17] What follows is neither an obituary nor a testimonial for this unique cultural landmark but an endeavor to study it within its context and to assess its wider significance.[18]

A Model of Proletarian Culture

The Socialist party's attempt to create a comprehensive proletarian counterculture was not an experiment in a formal, methodological sense of positing a hypothesis, extending and testing it in practice, and evaluating its results. Its experimental quality lay in the daring attempt to explore the unknown—the blending of culture and politics through a complicated network of organizations aimed at transforming the working class. Even though the cultural project did not follow a blueprint but rather evolved on the basis of experience in daily practice, it flowed from a central belief in Austromarxist theory that culture could play a significant role in the class struggle. If, as I argue, a main strength of this theory was its flexibility, allowing socialist leaders of the republic to regard themselves as always acting within its compass, at the popular level available to SDAP members and unaffiliated workers it also had an emblematic and confidence-inspiring quality.

Unlike other versions of Marxism, it promised a foretaste of the socialist utopia of the future in the present by locating the beginning of the great transformation leading to a new socialist humanity within capitalist society itself, before the ultimate revolution. For the younger generation of Austromarxists, engaged in realizing the socialist project in Vienna, a boundless optimism about attaining their goal in the here and now assumed the character of an illusion. Marie Jahoda, one of the authors of the internationally famous sociological study of unemployment, *Die Arbeitslosen von Marienthal*,[19] recalled: "In Vienna we lived with the great illusion that we would be the generation of fulfillment, that our generation would establish democratic socialism in Austria. Our whole lives were based on this fundamental idea. Today there is no doubt that this was an illusion, but it is also doubtless that this illusion was constructive and enriching to life."[20]

During the early years of the cultural project it was infused with the grand idealism, as party leader Otto Bauer put it, of "creating a revolution of souls." Translated into practical measures, this meant educating workers, improving their environment, shaping their behavior, and turning them into conscious and self-confident actors.

The cornerstone of the SDAP's experiment was the municipal socialism the party was able to implant in Vienna. Historical circumstances made that city receptive to innovations that would enhance the lives of its working class. Once the sparkling cultural capital of central Europe, outshining Budapest and Prague as well as Berlin in the brilliance of its intellectual and artistic innovations and practices, Vienna saw its luster vanish with the onset of war and the consequent collapse of the monarchy. During the 1920s talent migrated from Vienna to Berlin, which became the new center of intellectual and artistic ferment.[21] The stage was set for the working class, largely invisible during the heyday of bourgeois culture, to leave its imprint on the city. Universal suffrage in 1919 gave the socialists absolute control of the municipal government, and Vienna's unique power to raise substantial taxes allowed the city fathers to carry out extensive social reforms. An ambitious housing program, coupled with the extension of public health and social welfare services and the radical reform of education, was attempted with a view toward improving the environment of the working population.

Had the SDAP's cultural project rested with the municipal socialist program in Vienna, it would have been a unique accomplishment in and of itself. Although similar social reforms were carried out under socialist initiative in Berlin, Frankfurt, Hannover, Brussels, Paris, Lyons, London, Stockholm, and other major cities, nowhere else did the socialists aspire to as comprehensive a goal of transforming the municipal environment or succeed in making so many starts toward its realization as in Vienna. But the Austrian socialists were even more daring. Success in municipal reform encouraged them to undertake the far more comprehensive cultural transformation of the workers' lives implicit in their Austromarxist perspective. A "revolution in the soul of man," after all, conjured up a delving into the innermost reaches of life in the private sphere—an expansion of the notion of culture to encompass the workers' total life, from the political arena and workplace to the most personal and intimate settings.

These interventions included prescriptions for an orderly family life and a new definition of a woman's role in society; lectures, a vast array of publications, and libraries to stimulate and elevate the mind; associations for the enrichment of artistic taste; encouragements for abstinence from tobacco and alcohol, and admonitions about sexuality; organizations to instill socialist ethics and community spirit in children and young adults; a virtual sports empire to create healthy proletarian bodies, to steel them for coming struggles, and to project a new symbol of power; and mass festivals to demonstrate solidarity, discipline, and collective strength.

The point of departure as well as the destination of this monumental task was the party, operating through a dense network of special, often overlapping, institutions. Its leaders were to provide the content and form of this civilizing mission to alter the workers' total environment. If their main method was to educate by providing knowledge, their desire to transform the workers into higher beings required the shaping and altering of behavior above all. The need to intervene in order to integrate the workers into party life, taking "the place of everything else," was fraught with numerous contradictions.

Repeatedly Austrian socialist leaders, and Otto Bauer in particular, proclaimed their rejection of Bolshevik vanguard elitism and avowed their democratic socialism.[22] Their party, they claimed, was a mass party of members which was not simply directed from the top but equally animated by its rank and file. This view may have expressed the hopes of Bauer and other SDAP leaders, but it was not reflected in the life of the party, where a stable oligarchy dominated the pyramidal organizational structure and warded off factions or grass-roots initiatives which challenged its supremacy. This failing was not particular to the SDAP but prevailed in all the parties of the Labor and Socialist International, which found it impossible to cope with masses of workers seeking membership or direction. The much-vaunted "democracy" by which they distinguished themselves from bourgeois parties as well as communists, contradicted practices within the Socialist parties themselves.[23]

Glaring differences in perception, needs, and priorities between socialist-leadership oligarchies and the masses of workers within and outside their parties were a major problem in interwar Europe. It was doubly so in the SDAP, which was committed to blending culture with politics on a hitherto unknown scale. The principal directors of the SDAP were not "organic leaders" formed by an existing working-class culture, but largely autonomous middle-class intellectuals drawn to the workers as the "historically progressive class." Despite the undoubted strength of their conviction in making this choice, they had experienced a socialization and imbibed a cultural value system far removed from that of ordinary workers. Consequently these leaders cast their lot with the working class, to which they dedicated their intellectual abilities, but also retained the psychological imprints of their prior development. This inner duality was reflected in their outward relations with the workers: they could be for them but not of them; even superficial aspects of behavior—speech, dress, comportment—marked the boundaries of understanding and empathy.[24]

But even those members of the SDAP elite who had risen from working-class origins to prominence in the party imbibed the values of their bourgeois/socialist confreres and exhibited the same features of social and cultural distance from the rank and file. Organic leaders such as the doyen Karl Renner or the rising young socialist Joseph Buttinger acquired tastes and values in the service of the party that distanced them from their origins.[25] It would seem that the very act of leadership conferred a differentiating character on those assuming such positions. One problem faced neither then nor now was the tendency for the bond between leaders and followers to be one of domination/submission rather than democracy/consensus: both intellectual and organic leaders responded to psychologically distancing mechanisms embedded in their roles.

This dichotomy between leaders and followers was not resolved by the Austrian socialist leaders during the course of their cultural experiment. By a strange historical coincidence, while the SDAP leaders were putting their cultural innovations into practice, the Italian communist leader Antonio Gramsci was grappling with the theoretical problem of proletarian cultural power and the role of leaders, while incarcerated in a fascist prison cell. Gramsci argued that, in order to free the working class from the intricate network (consensual, but backed by the power to coerce) by which the bourgeoisie exercised a cultural and ideological hegemony over society, it was absolutely necessary for the workers to establish a counterhegemony over civil society before attempting to capture state power. Gramsci too could not resolve the problematic position of the intellectual/leader in the process: even if such individuals cast their lot with the working class because they viewed it as "historically progressive," he argued, they remained autonomous and their loyalty was uncertain.[26]

In the Vienna experiment the problem was not the loyalty of leaders but the cultural distance that separated them from the rank and file. This distance reinforced the paternalism of leaders toward the workers they hoped to transform and liberate. The tendency to infantilize is explicit and implicit in the preconceptions about the workers which underlay the diverse party organizations and programs. Paramount among these was the total denigration of the workers' existing subcultures. On the one hand, the workers were viewed as aping the worst aspects of petty bourgeois cultural forms and aspirations. On the other, they were regarded as uncivilized: disorderly, undisciplined, and even brutish in their daily lives. To call this a very distorted image is not to argue that worker subcultures were not weighed down by a variety of social and psychological deficits (alcoholism, poor hygiene, male chauvinism, intrafamily violence). But these were judged by the workers themselves to be outside the pale of respectability within their communities.[27]

The Viennese working-class subcultures were complex. No doubt they included features of the dominant value system as well as aspects of social dissonance; but they were also enriched by community standards and norms of behavior created during more than two generations of urban life. Far from existing in a state of nature, as the socialists tended to imply, the workers were disciplined and observed social codes, engaged in hard work and experienced an all too frequent delayed gratification of basic needs made necessary by a low standard of living. All these forms of self-management imparted a sense of dignity that helped them cope with the difficult circumstances of everyday life. The SDAP leaders' inability or unwillingness to appreciate existing subcultural forms and networks partly stemmed from their unfamiliarity with life at the bottom. Such social blindness also served their need to view the worker as malformed, or at least unformed, and therefore in need of and ready for the transformational program offered by the party. For the socialist leaders to have given some credence to the positive values of indigenous worker cultures would have implied a willingness to negotiate, to adapt the cultural project to them, and to replace paternalism with more flexible forms of organization and action.

The command structure of socialist party culture was of little use in dealing with fundamental questions arising from the cultural work itself. Was all of bourgeois elite culture to be rejected, or were the workers to be given their share of what was considered a national heritage? In the latter case, how were elite forms to be given a socialist interpretation in order to make them appropriate for working-class appreciation? And what was the heritage of elite culture to which the workers were entitled: the classics, or the modern and avant-garde as well? Whose canon of taste would be used to make selections, or was a consensus of taste among leaders presumed? And how could they square the circle of denouncing the worldview of the bourgeoisie while at the same time educating workers to appreciate the historical treasures of that milieu?

How to deal with commercial and mass culture was even more problematic. From the *Gasthaus* to the *Varieté*, from radio and cinema to spectator sports, these forms of amusement competed for the workers' leisure time. They could not simply be called trash or kitsch and condemned out of existence, though such denunciations were certainly made. Attempts to compete with mass culture on its own terms failed for want of finances and determination. The SDAP cultural establishment's final effort took the form of an attempt to ennoble mass cultural projects. Unfortunately, such "upgrading" of value mainly consisted of introducing elements of the classic bourgeois culture to the new media. Virtually no one in the upper echelons of the SDAP understood these media or considered them as anything but degradations of previous art forms and conveyors of a debased culture in general. This inability to see radio and film as new art forms, or at least as unique forms of entertainment, was widespread among intellectuals outside the socialist camp at the time.[28] But for the socialists in particular, entertainment for its own sake and as a form of relaxation for the workers was rejected puritanically as incompatible with the program of cultural uplift to which their transformational experiment was committed.

The Viennese cultural experiment unfolded in a complicated political and economic climate. The SDAP had the advantage of not having to compete with a significant Communist party, as did the socialists in Weimar Germany and in France during the Popular Front. The party was also firmly in control of the municipal and provincial government as a result of its significant majorities, allowing it to initiate programs which could not be controverted locally. The SDAP and trade unions together commanded an extremely large and loyal membership. But this numerical, electoral, and governmental strength in Vienna was deceptive, for the national government remained firmly in the hands of their political opponents, the Christian Social party and its allies. From the beginning the country was divided into two camps—Vienna and scattered industrial enclaves against the largely agrarian and Catholic provinces—whose hostility was not simply political. It expressed itself in hate mongering by the Catholic church, the far-from-silent partner of political reaction, for whom socialism was the Antichrist.[29]

The state was not the neutral, republican foundation the socialists imagined it to be, but an instrument of their increasingly antirepublican opponents. The socialists' belief was shaken when politically frustrated masses in Vienna stormed the Palace of Justice on July 15, 1927, and set it ablaze; the police fired point-blank into the crowds, leaving eight-five workers dead. These events represented a turning point in the fate of the republic. They also signaled a shift in the SDAP's cultural program: earlier it had been an instrument in the class struggle; now it increasingly became a surrogate for politics, the arena of which shifted from electoral contests to force and violence in the streets.[30]

Economic conditions in postwar Austria were far from encouraging for social and cultural programs.[31] The fragmented economy of the small state created a sense of continual instability. Unemployment was extremely high even in the period of recovery (1925–30) and soared during the depression, affecting one third of the labor force by 1933.[32] It would not be an exaggeration to say that during most of the republic's fifteen years a significant section of the working class lived on the edge of poverty. Neither the SDAP nor the trade unions found the means to alter that harsh economic reality. Both were largely reactive to capitalist pressures to intensify production and keep wages from increasing in real terms. The staggering decline of over 40 percent in trade union membership was indicative of the weakness of the party and the trade unions in this contested terrain.[33] Small wonder that the workplace, so central to the everyday life of workers, was largely left out of the socialist's cultural programs.

The Austrian socialists' cultural experiment offers an excellent demonstration of the relationship between the much-sought-after proletarian culture, and the elite and subcultural forms it attempted to eradicate from the lives of workers. It exposes all the limitations of such a quest, arising from a paternalistic leadership tied to inherited values, the complexity and resistance to change of worker life-styles, and the seductive competition of commercial and mass-culture leisure activities. Though fragmented and falling short of permeating the workers' public and private sphere, the experiment's real accomplishments acted as a powerful symbolic force far greater than the sum of its achievements. It signaled strength and accorded dignity, a sense of worth, and confidence to the workers, because "red" Vienna was somehow theirs. As I shall demonstrate, this symbolic strength was also deceptive, in that the cultural program attempted to compensate for the workers' economic deprivation and the increasing political powerlessness of the SDAP after 1927. Thus the Viennese experiment presents itself as a model for studying the dynamics of comprehensive projects involving cultural transformation. As such it offers a striking image of idealist intentions, presents some significant accomplishments, reflects the inevitable contradictions resulting from actual practice, and carries somber warnings about the danger of substituting the symbolic for the real.

What follows is clearly an interpretation rather than a comprehensive history of the subject. This study seeks to examine the major components of the SDAP's cultural project in Vienna, from the reforms of municipal socialism to the ambitious goals of party culture, touching on the latter's relationship to elite, commercial, and mass culture as well as to the workers' domestic world. A secondary goal of this study is to utilize the experiences in the Viennese "laboratory" to raise more general questions about efforts to fashion and implement comprehensive cultures from above. Failed attempts in various countries during the past forty-five years make it all the more intriguing to lay bare and assess the Viennese model, developed with the best motives and the highest ideals.

CHAPTER 2

Vienna as Socialist Laboratory

Since the publication of Carl Schorske's *Fin-de-siècle Vienna* in 1980, Vienna has conjured up images of modernism in a setting of decadence.[1] The book has stimulated great interest in the high culture and politics of the multinational capital. Its powerful evocation of the intellectual and artistic climate, in which Freud, Schnitzler and Hofmannsthal, Lueger and Herzl, Klimt and Kokoschka, and Wagner and Schönberg reflected and challenged a very special bourgeois culture, unfortunately also has become an obstacle to seeing Vienna as a thriving metropolis in which social, political, and cultural experiments of great scope and originality were attempted in the period following the belle époque. The Vienna of Schorske's making is like a Chirico painting—symbolic and frozen in time and space. The golden age of high culture he re-creates so artfully does not appear to have roots or resonance in the complex experience of two million Viennese and therefore seems to be a dead end. Indeed, *Bildung* and culture as a substitute for politics among the bourgeoisie come to an end in the republican Vienna emerging after 1918.

But the ability of the city to harbor experiments of international importance continued. The organized working class, incubating during the belle époque, emerged as a powerful force which under the leadership of the Socialist party (SDAP) attempted to transform Austrian society. The socialists' experiment concentrated on Vienna, where it encountered and confronted all aspects of postwar turbulence in seeking new ways to survive. There are few connections between the hermetically sealed world of bourgeois high culture Schorske has depicted and the "red Vienna" the socialists sought to create—between elite and popular experiences and expectations of the good life. There is one interesting parallel which needs to be mentioned now and explored later: culture for the fin-de-siècle bourgeoisie was a surrogate for the politics from which it was excluded; the creation of a proletarian counterculture became paramount for the socialists, because they were unable to shift the balance in the national political arena in their favor.

My concern in this chapter is with Vienna as city: one of the four principal metropolises of Europe,[2] experiencing a critical transformation at the end of a lost war and collapsed old regime. In *Die freudlose Gasse*, G. W. Pabst's brilliant film of 1925, postwar Vienna is portrayed with an unvarnished realism as a city in crisis: inflation is rampant; tough profiteers lord it over a declassed bourgeoisie; the poor hunger to satisfy their most basic wants; and a moral decay hangs heavily over all.[3] This is but one of the many realities of postwar Vienna. My aim is to throw a brief, sharp light on the many others: revolutionary ferment and republican reform; Germans, Jews, Czechs, and others; domiciles, workplaces, and infrastructures; the new state's uncertain viability; political camps and trade union loyalties; titles, parades, uniforms, hatreds, and other residues of the old regime; the Catholic church, anti-Semitism, and Germanic christianity; the identity of the Viennese; and psychological shocks to the metropolitan ego caused by the city's newly diminished status as a capital of alpine yokels. Like a film camera panning across an urban landscape, I hope to illuminate a great number of the ingredients which constituted Vienna in ferment from 1919 to 1921— the period during which the stage was set for the performance of the socialists' cultural experiment.[4]

Hopefully, such a kaleidoscopic presentation of Viennese realities will not obscure the serious problems experienced by Vienna's citizens. It will be instructive to list these, so as to appreciate the actual challenges facing the socialists as practical politicians. From their commanding position in the municipality, the socialists dedicated themselves to solving these with reforms which became the foundation of their effort to create a proletarian counterculture. In both their municipal reforms and their larger cultural aspirations the socialists claimed to be guided by Austromarxism. In reviewing the main tenets of this theory it will be interesting to see whether it was attuned both to the situation of Vienna during this painful postwar transition, and to the comprehensive cultural project on which the socialists were embarking.

Vienna, 1919–1921: A Montage

The republic of Austria, with Vienna remaining as capital, was finally proclaimed on November 12, 1918.[5] There was little enthusiasm for the new republic. The Christian Social and Pan-German parties had only shortly before declared themselves committed to monarchy; the SDAP hesitated. All three conceived of the new state as being "German Austria." The Entente powers preparing for a "Carthaginian peace" in Paris had insisted on the contours of the republic under the simple name "Austria." None of the political forces represented by the Provisional Assembly meeting in Vienna were satisfied with the minuscule state, pasted together from the leavings of a dismembered monarchy: the Christian Socials favored a central European empire under Habsburg leadership; the left socialists as well as

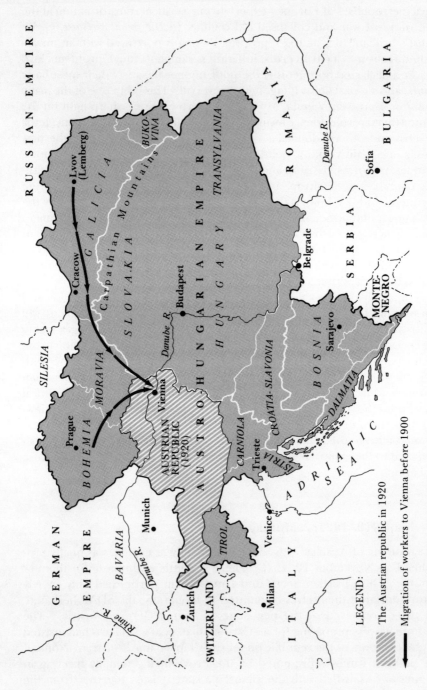

The Dual Monarchy of Austria-Hungary in 1914 and the Austrian republic of 1920

the Pan-Germans demanded *Anschluss* with Germany (though for opposite reasons); and the socialist leader Karl Renner dreamed of a Danubian confederation.[6]

But the Entente powers had other plans for the reconstruction of central Europe and these did not include either a Germany enlarged by Austrian annexation or the re-creation of a multinational operetta monarchy. No doubt the Provisional Assembly, aware of this reality and faced with Emperor Karl's apparent abdication on November 11, was forced to proclaim the republic, for which none of the political players was very eager.[7] With tens of thousands of soldiers returning or at least crisscrossing Vienna and the provinces, many of whom had experienced the Russian Revolution at first hand, a republic based on a constitution was a far lesser evil for the politicos gathered in Vienna than the anarchy or revolution conjured up by angry, frustrated, and demanding soldiers operating in a power vacuum.[8]

What about Vienna in these months following the end of the war, during which political formalism ground out a republican solution? The much-celebrated, joyous musical capital of 2.1 million, setting the tone in an empire of 52 million people of assorted nationalities, was no more. Vienna emerged from wartime with only 1.8 million inhabitants in a newly crafted state of 6.4 million. In the immediate postwar months the city still harbored many demobilized soldiers of the new succession states in transit to their homes and still retained some of its prewar flavor of ethnic diversity. These transients were replaced by civil servants of the old regime—displaced persons returning from all the corners of the old empire. They had no great impact on the character of the Viennese population. Only two significant minorities remained in Vienna by 1921: about 120,000 to 150,000 Czechs (about 6–8% of the population) and a little over 200,000 Jews (about 10.8% of the population).[9]

Physically Vienna's urban landscape remained much the same as it had been before the war. Between 1890 and 1904 most of the suburbs had been incorporated into the city, with a gradual extension of the infrastructure to all twenty-one districts.[10] But whereas it had been situated in the center of the former monarchy, it now virtually marked the boundary with Czechoslovakia to the northeast and with Hungary to the southeast. The western reaches of Vienna, including the Vienna Woods, extensive vineyards, and countless small garden plots, humanized the city and somewhat disguised its commercial, financial, and industrial character. Although Vienna had become a modern industrial metropolis before the war, with the exception of the electrical industry, light forms of manufacturing (clothing, paper, comestibles, and graphics) predominated. Nonetheless, by 1913 some sixteen industrial plants employed more than one thousand workers each.[11] By the end of 1919, however, a decline in trade with the succession states, the end of war production, and a grave shortage of coal closed half of these large plants.[12]

The political uncertainty created by the collapse of the monarchy and the less-than-enthusiastic proclamation of the Republic was exacerbated by the displacement of women from the labor force,[13] returning soldiers, the growing number of homeless caused by the critical housing shortage, and the appearance of refugees from the four corners of the former monarchy, who joined the already existing hordes of panhandlers and beggars, indigent and unemployed. Most visible among these—in long black coats *(kapottes)* and broad-brimmed black hats—were Jews who had fled from Galicia during the war. Although the total number of these Jewish refugees did not exceed 25,000 by 1919, their presence in Vienna was exaggerated in the upsurge of an already well-established anti-Semitism.[14]

To these sources of social tension must be added the hardships of the Viennese population, wearied by four years of wartime privations, in finding the food and fuel necessary for daily existence.[15] The sense of insecurity and crisis among the Viennese was increased during the first peacetime winter by an epidemic of the "Spanish grippe" which killed thousands. Other, less sensational illnesses abounded among the Viennese. Prostitution and concomitant venereal disease had increased markedly during the war and continued at a high level during the first peacetime years. Most threatening because most constant was tuberculosis, called the "Viennese disease," which accounted for one-quarter of all deaths in the city, and nearly half in working-class neighborhoods.[16]

Despite the proclamation of the republic on November 12, 1918, followed by elections and the creation of a coalition government by the SDAP and Christian Socials in February 1919, a political vacuum continued to threaten the new republic until the end of that year. Austria was not involved in a parliamentary or constitutional revolution, as has frequently been suggested.[17] It faced a real revolution at the hands of soldiers' and workers' councils, influenced by the revolutionary soviet models in Bavaria and Hungary. The socialists clearly did not want that kind of revolution, so heavily influenced by the Bolshevik example, and very adroitly outmaneuvered the revolutionary element among the workers in the councils themselves.[18] Although the socialists mastered the threat from the left in the spring and summer of 1919, they neither fully exploited this threat to strengthen their own political position in relation to the Christian Socials, nor used it to carry out far-reaching reforms to secularize and securely establish the republican form of the state. Thus the socialists' retrospective reference to the events of 1918–19 as a revolution was a dangerous delusion about the power relations in Austria. These had not been altered sufficiently to make the republic secure.

To what extent was Vienna a workers' city observable in real and symbolic forms in its workaday, public, and private life? Already, by 1914, 70 percent of all wage earners were workers of whom four-fifths toiled in workshops and one-fifth in industry proper. By 1919, the number and proportion of middle-sized industrial plants had increased, at the expense of small workshops rendered obsolete during wartime.[19] If the oldest historical parts of the inner city were identified with the nobility, and the elegant Ringstrasse boulevard with the affluent bourgeoisie, an outer belt of linked

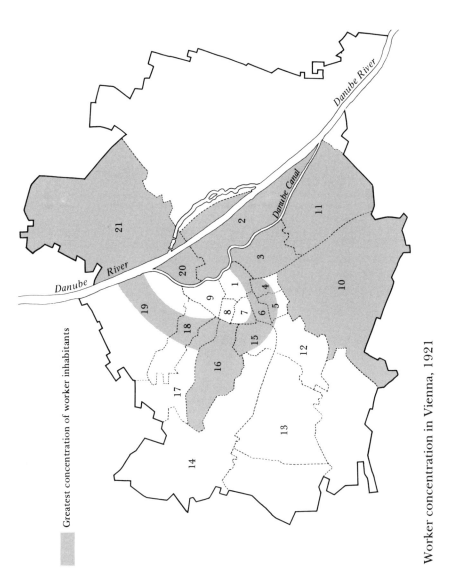

Worker concentration in Vienna, 1921

avenues, the Gürtel, and its numerous side streets harbored the tenements *(Zinskasernen)* housing Viennese workers. In the early period of industrialization, workers arriving in Vienna from Bohemia or Galicia frequently had settled in existing ethnic enclaves: Czechs, for instance, in the 10th, 16th, and 20th districts, and Jews in the 2nd.[20]

The location of larger factories in the outlying newer districts and near railway stations and yards, as well as the absence of an adequate transportation system, led to the growth of worker settlements in proximity to the new plants. No matter whether closer (Leopoldstadt) or farther (Floridsdorf) from the city center, worker neighborhoods were easily recognizable by the dreariness and dirtiness of the streets, the decaying façades of the tenements with bed linens airing on the windowsills, the children at play in the streets or in empty lots, the shoddy display in local shops, and the presence of pawnshops and smoke-filled, crudely furnished *Gasthäuser*.[21]

Throughout 1919 and into 1920, Viennese workers engaged in street actions protesting food and housing shortages and high prices, which frequently led to violence. No doubt some of these incidents stemmed from initiatives of soldiers' and workers' councils and their leaders.[22] But a good number were spontaneous acts of frustration on the part of a working population which had borne the hardships of war (on both the battlefield and the home front) and continued to experience privation, food and fuel shortages, speculation, overcrowding, and the mounting pressure of an inflation that reached astronomic and incomprehensible dimensions.[23] Such frustration over real grievances—food prices or housing needs—also led to the wrecking and looting of shops and cafés in the inner city, where the architectural monuments of traditional power—the goal of direct actions—were located. It should surprise no one that the Viennese middle class and its

Demobilized soldiers in 1919, a source of both the soldiers' councils and the Volkswehr (Verein für Geschichte der Arbeiterbewegung [VGA])

Organized workers' demonstration. The banner proclaims: "Long Live the International World Revolution." (VGA)

political spokesmen viewed these violent outbursts as the end of civilization and the prelude to Armageddon. They had not yet assimilated the collapse of the old regime; they could not comprehend or accept the fact that the lower orders had shaken off the previous invisible bonds of authority sufficiently to rise up in anger and desperation. The massive strike waves in January 1918 and following the mutiny of Austrian sailors at Cattaro that June were political as well as economic and signaled a new militancy and confidence.[24]

It is remarkable that the socialists also should have viewed with suspicion the workers' newly discovered ability to act in their own interest en masse. No doubt the socialists feared that the workers would be swayed by communists in the workers' and soldiers' councils to undertake reckless actions, or that the revolutionary élan of the Bavarian or Hungarian Soviet Republics would spill over into Austria, leading the workers in a dangerous radical direction.[25] To forestall such eventualities, the SDAP assumed a tight rein over the councils and at the same time used the Volkswehr, the newly established volunteer army under decisive socialist influence if not outright control, to keep street politics in check with calibrated measures of force.[26]

It is a sign of the uncertainty of those postwar years that the socialists overreacted to spontaneous expressions of worker militancy. The impact of the Bolshevik Revolution on Austrian workers was considerable, especially in Vienna, where firsthand accounts enjoyed currency in the workers' and soldiers' councils. It is also clear that no person of prominence in the SDAP wanted to follow in the Russian Bolsheviks' footsteps.[27] Friedrich Alder, who enjoyed the greatest popularity among Viennese workers at the time, succeeded in neutralizing the newly formed Communist party (November 1918) by persuading it to join the national organization of workers' councils. In this "parliament of the working class" the communists were outnumbered and outvoted, and their plans for a soviet republic foiled.[28] Leaders of the SDAP from Renner to Bauer tended to view pressure from below, no matter how immediately and practically related to the turbulence of the moment, as endangering plans made and initiatives taken at the top of the party.

A crisis in the relationship between leaders and masses arose from the fact that the leaders had been completely overwhelmed by the imperial collapse. There had been no preparation for such an eventuality: no discussion of republican versus other political forms; no consideration of popular participation; no agenda of practical measures for the transformation toward socialism. The socialist leaders' decisions in those crucial years were improvised without real discussion or debate at the grass roots. Only later did party theoreticians like Bauer provide theoretical respectability to acts of crisis management.[29] None of the prominent socialists at the time seemed to be troubled by their oligarchical party's increasing unresponsiveness to the rank and file, even though they made a point of distinguishing their innerparty democracy from the Bolsheviks' dictatorship. The more the SDAP's program steered away from the Bolshevik revolutionary example, the more the party resembled its leadership structure and its belief in the necessity of totally shaping and controlling the rank and file.[30]

The socialist leaders did not appreciate the high level of organization and discipline of the Viennese and Austrian workers. The danger to working-class interests did not lie in their rare anarchic outbursts but in their excessive regimentation, adherence to formal structures, and conformism to routines and empty rituals. By 1920, 458,635 Viennese workers were dues-paying members of the (socialist) Free Trade Unions, representing roughly 50 percent of the total Austrian membership.[31] Although the size of Viennese membership in the SDAP was less spectacular, having reached 188,379 or about 38 percent of national membership by 1921, it grew rapidly to 418,000 or about 58 percent of national membership by 1929.[32] The social composition of the Viennese SDAP was unusual in two respects: in 1921 women members accounted for 26 percent of the total, rising to 36 percent by 1929; the number of younger members, from twenty to forty years of age, were overrepresented in the party in comparison to the age distribution of the population.[33] That the SDAP triumphed in the municipal elections of 1919 with 54 percent of the vote, thereby producing the first socialist mayor of a metropolis (Jakob Reumann) and creating the political foundations for red Vienna, should be attributed not only to the large trade union and significant socialist party membership, but also to the positive resonance of the socialists' appeal for renovation and reform in the lower middle class and among professionals.[34]

Although the municipal council election of May 1919 had demonstrated the socialists' immediate and even long-range power in Vienna, the election for the Constituent Assembly in February of that year did not permit such an optimistic reading of the results. To be sure, the SDAP had polled 40.76 percent of the national vote and obtained 43.4 percent of the seats in the Assembly.[35] But the Christian Social and Pan-German parties had polled 35.93 and 18.36 percent of the vote respectively and together captured 54.7 percent of the Assembly seats. Lacking a majority, the socialists were forced to go into coalition with the Christian Socials. The SDAP held the key positions of chancellor (Renner), foreign minister (Bauer), and war minister (Deutsch) partly because they were the dominant party in the legislature and partly because threats of revolution from below during the spring and summer of 1919 put the conservative parties on the defensive. But their timidity was short-lived. The fall of Béla Kun's Hungarian Soviet Republic on August 1, following the failure of a communist-inspired putsch in Vienna on June 15, marked the beginning of a tougher stance on the part of Christian Socials and Pan-Germans. In the parliamentary elections of 1920 and 1921 the socialists lost both votes and mandates, leveling off at about 40 percent for both.[36] The socialists refused to read these results with anything but rose-colored glasses and continued to nurture the hope that a parliamentary majority could be attained.

In the months of greatest political uncertainty, when the Viennese *Bürger* was most anxious about a possible Bolshevik Austria, the SDAP improvised and pushed through a series of reforms intended to anchor the republic and to lay the groundwork for a socialist transformation. Fundamental to the creation and functioning of the republic was the electoral law of 1918, which enfranchised adult men and women without conditions.[37] Economic legislation passed by the first coalition government included unemployment and sickness insurance, restrictions on female and child labor, and the eight-hour day.[38] The latter was offset by the increase of work tempo introduced by productivity-minded scientific management in the larger plants.[39] Two further laws were more daring. The first provided for paid vacations for workers and employees: one week for those who had worked continuously for a year, and two weeks for those with five years of continuous labor. The second established the Chamber of Workers and Employees, with the function of overseeing collective contracts between trade unions and employers, supervising the execution of labor laws, and advising the national legislature on labor legislation.[40]

By far the most radical and innovative piece of labor legislation established so-called factory councils, which on the face of it resembled instruments of codetermination or industrial democracy at the workplace. A less positive view of the factory councils might regard them as instruments of worker pacification devised by the SDAP and socialist trade unions to replace the politically highly charged and unpredictable workers' councils. However one might interpret the intended function of these councils, they were an intrinsic part of the project to socialize the Austrian economy.[41]

Putting the productive processes in the hands of the producers themselves had long been a cornerstone of Marxist socialism. But as so often happens with pillars of theory, the idea remained undeveloped in the prewar worker parties for whom the issue was a distant one at best.

When the SDAP was suddenly confronted with the possibility of some form of socialization early in 1919, there was no concrete plan to fall back on. Even Bauer, who by then had emerged as the party's spokesman on theory, sketched a program lacking in practical details. He foresaw a long transition from private to public ownership, beginning with the primary economic sector and ending with the socialization of the banks. Expropriation was to be accomplished with full compensation; socialized enterprises were to be managed by a triad of trade unions, consumers, and the state. In projecting the future management of the economy, Bauer assigned the task of preparing the workers for their responsibility to the factory councils. His ideas proved too grandiose and schematic to have any impact on the Socialization Commission over which he presided in the spring and summer of 1919.[42] When the threat of revolutionary contagion from Bavaria and Hungary ended, the Commission faded away without having made the smallest changes.[43]

No doubt the piecemeal reforms of 1919 helped improve the quality of the workers' lives; at the same time they failed to establish the basis for a more extended socialist experiment. The door to the latter was opened by the fundamental recasting of the legal political standing of Vienna itself. Even before the war the socialists in the city had expressed a desire to separate it politically from the province of Lower Austria in order to provide a just representation for Viennese workers.[44] By 1920 this socialist goal was virtually accomplished, with Vienna achieving the status of capital of the republic and separate province.[45] Henceforth the socialist majority in both the municipal and provincial Viennese government was in a position to command the power of local taxation and use it in developing a program of municipal socialism.[46] By shifting the form of taxes from indirect to direct, from necessities to luxuries, and by introducing a graduated scale that favored wage earners, the socialists created a source of revenue for public projects truly unique at the time.[47]

Less apparent than the special powers of taxation for the socialists' project of municipal reform was the conjunction of a prewar municipal rent-control law with the postwar runaway inflation. Until 1918 Viennese workers, and to a lesser extent petty bourgeois, had been at the mercy of landlords, who were permitted by the tolerant Christian Social municipal administration to raise rents arbitrarily, to refuse to make even basic repairs, and to evict at will.[48] We will look more closely at the condition of worker tenements in the next chapter. Suffice it to say here that horrible overcrowding, brought on by the financial reliance on subtenants and bed renters, as well as a kind of nomadism—an annual or even semiannual forced change of domicile—characterized life in working-class neighborhoods. In 1918, as the war was winding down, an emergency law seeking to forestall social unrest set limits on rent increases and eviction notices for both apartments and commercial establishments. The first coalition government of 1919 confirmed the law.

In the context of 1918 this law had been far from radical. It had done no more than protect tenants from the worst excesses by landlords (rapacious increases and arbitrary evictions). It had failed to lower rents or alter their proportion of worker or employee wages (between one-quarter and one-third). But the postwar inflation virtually wiped out the rents stabilized by the control law, which had made provision for only minimal increases for maintenance of properties. This aberrant legislation created a windfall for Vienna's workers, whose rent declined to 3 from 7 percent of their wages.[49] Worker families could look forward to more stable living arrangements and no longer needed subtenants and bed renters to help meet the monthly payments. The subtenants and bed renters, however, found themselves turned adrift to face a grim housing shortage for which no immediate relief was at hand. The rent-control law became a bastion for the SDAP as well as a principal subject of struggle with the Christian Social camp. It also created one of those rare instances in which lower-middle-class renters and businessmen found themselves allied with the SDAP and the workers it represented.

From November 1918 until the summer of 1920 the atmosphere in Vienna was one of turmoil and change. Socialist brooms, it appeared, were sweeping out the old order and taking measures to safeguard the new republic against desires for restoration on the right and Bolshevizing pressures on the left. Otto Bauer was not alone in calling the transition from monarchy to republic a revolution.[50] If one accepts his characterization for the sake of argument, one would have to append the reservation that revolutions leave large aspects of society, life, and values untouched.

In looking at Vienna with that caveat in mind, one finds the republican mood for change tempered by atavistic remains of the old regime in the form of hierarchy, paternalism, and authoritarianism. Aristocractic titles—not only the prefix "von" but also Baron, Graf, Fürst, Herzog—were abolished with the founding of the republic. At the same time a long list of honorific designations establishing social rank and distance were allowed to stand and proliferate. The new republic lent itself to a title mania in which "Doktor" was the least one might expect of anyone who counted for anything. Beyond that minimum lay higher ranks demanding greater social deference: Dozent, Professor, Hofrat, Diplomingineur, Sanitätsrat, Schulrat, Diplomkaufmann, Rechtsrat, Chirurg, Midizinalrat, Studienrat, Kammersänger, Kommerzialrat, Direktor, Generaldirektor. Wives of those so entitled expected to be addressed by the appropriate honorific. In the immediate postwar months Vienna swarmed with officials of the former imperial bureaucracy, now redundant with only their titles to cling to. Most bizarre was the common use of the title "Doktor" among the socialists, leading to such ambiguous appellations as "doctor colleague" or "professor comrade."[51]

Rituals of the old order—marches, parades, demonstrations, and public festivals—took on a sharpened political meaning as instruments in the struggle between the socialist and reactionary camps. And so did the accouterments of mass display and action: the crosses and red flags carried; the uniforms of athletes, youth groups, and paramilitary groups; peasant costumes and "progressive" clothing; and above all, decorations in red or black.[52]

Uniforms in particular continued to play a remarkably large role in both the political and the daily life of the Viennese: they identified the various official authorities in working-class neighborhoods for children playing in the street[53]; they set apart the worker in municipal transport or the railroads as one of the lucky ones with steady employment and prestige; and, as the trappings of large groups, uniforms served to distinguish members from nonmembers. Even in the earliest postwar years uniforms in parades and demonstrations suggested more than festivity or identity of the participants. In that context the uniform became a pseudomilitary emblem, suggesting combat without the need to negotiate or compromise—a forceful contrast of action to the endless discussions and maneuvers of parliament.[54] It was not just military-style uniforms that enjoyed a great popularity on both sides of the political divide; military terms also had wide currency, from the socialist Rote Falken (Red Falcons) to the Catholic youth, and from SDAP party platforms to bishops' pastoral letters. The list of such atavistic remnants could easily be expanded. They contributed to the uncertainty surrounding the republic and the new Austrian state in the early years.

Could Vienna live up to its glorious past in the context of a minuscule state, amputated industry and markets, and reactionary Catholic provincialism? Was Austria viable *(lebensfähig)* as an independent state?[55] The answer was "no" *(lebensunfähig)* across the political spectrum in the first postwar years. Not only did the Christian Socials yearn for a return of crown and cross, and the Pan-Germans call for a union of all racial Germans in a "greater German" state, but the socialists too demanded that the Allies' artificial Austria be incorporated into a unified German socialist state. No doubt the First Austrian Republic had serious economic problems, including a shortage of investment capital, scarce fuel resources, a weak capital goods sector, and uncertain markets. But these conditions also prevailed in other small countries and particularly among the succession states. Austria also had decided assets, which various politicians were loath to put into the balance: its lack of war damage; the quick return from the front of its soldier work force; the productivity of nearly 90 percent of its arable land; the fact that its population was more homogeneous than that of most countries; and Vienna's continued role as a commercial hub between east and west.[56]

The question of national viability lingered beyond the critical early years of restructuring, to be dusted off and posed yet again in the ongoing struggle of the two camps. In 1922, for instance, when inflation had gotten completely out of hand, Bauer proposed a customs union with Germany (as a prelude to *Anschluss*), and Seipel, as foreign minister, while negotiating the stabilization of Austria's currency by the League Powers, talked of an annexation with Germany, Czechoslovakia, or Italy.[57] For the socialists, especially, the subject was like a ghost refusing to be laid to rest until Hitler assumed power in Germany. No matter how one wants to interpret the significance of the force and persistence of the idea of *Lebensunfähigkeit*—whether as clever diplomatic ploy to wrest concessions from the Allied powers or as an idée fixe—its psychological impact was to weaken public confidence in the state, in the republic, and in a republican Vienna. By making the present seem to be only temporary and undeserving of loyalty, it added to the power of centripetal forces: the provinces against the capital; Germans against "aliens"; and the political camps against each other.

In prewar Vienna one of the less endearing qualities of daily life had been a rampant chauvinism of German Austrians against the ethnic minorities.[58] In the Viennese vernacular each one received a derogatory designation: "Krowot" for Croat, "Behm" for Czechoslovakian, "Katzelmacher" for Italian, "Saujud" for Jew, to mention but a few. With the end of the war the Viennese population suddenly became more homogenous as tens of thousands moved in and out of the city. "The Czech, Polish, and Magyar troops returning from the front, the Italian and Russian prisoners of war, were only in a hurry to get away.... The foreign element drained away quickly from the desperate city."[59] As mentioned earlier, Czechs and Jews were the two sizable minorities remaining. The former consisted of both Austrian citizens and citizens of Czechoslovakia, of those included in or excluded from the political life of the republic.[60] Two-thirds of the Czechs were workers in close relation with the SDAP and the Viennese trade unions. They enjoyed an active but separate cultural life, a form of ethnic self-segregation which also kept anti-Czech chauvinism alive. But the extent of overt chauvinism or covert hostility was somewhat less than before the war, for the simple reason that Czechs had standing as nationals or potential nationals of a neighboring state and had access to Czechoslovakian diplomats in Vienna to air grievances and make complaints.[61]

With the Czechs enjoying the status of a protected minority, that left only the Jews, neither a nation nor a people, not simply a religion but "surely a race," to serve as leading scapegoat for the discharge of popular resentment. Anti-Semitism was deeply rooted in Austrian public policy and culture, in both a religious and a racist form.[62] It had been the cornerstone of the Christian Social party, particularly under the leadership of Mayor Karl Lueger before the war.[63] It was whipped up immediately after the war by the all-too-visible presence of some 25,000 Jews who had fled from pogroms in Galicia to the asylum of Vienna. No one had any use for this army of impoverished peddlers. "For the Christian Socials they are 'Jews'; for the German nationalists they are 'Semites'; and for the socialists they are 'unproductive elements.'"[64] In 1920 Leopold Kunschak, head of the Catholic Workers' Association and a close associate of Seipel in the Christian Social party, proposed that these Jewish refugees be given the choice of leaving Vienna voluntarily or of being put into concentration camps to segregate them from the population.[65]

Clearly this handful of East European Jewish refugees were only the surrogate for the much larger, generally assimilated, Jewish population of Vienna. It is the Viennese Jews prominent in the professions and arts, in journalism and the rising mass media, in industry and high finance, but especially in the SDAP, who were the target in the hate campaigns which were a permanent fixture of the First Republic.[66] Jews in the first tier of leaders among the socialists included Otto Bauer, Max Adler, Friedrich Austerlitz, Friedrich Adler, Julius Deutsch, Julius Braunthal, Wilhelm Ellenbogen, Robert Danneberg, Julius Tandler, Therese Schlesinger (Eckstein), Emmy Freundlich, Helene Bauer (Gumplowicz), and Hugo Breitner.[67]

It was to them in particular that the Christian Social party manifesto of Christmas 1918 was addressed: "The corruption and power mania of Jewish circles, evident in the new state, forces the Christian Social party to call on the German-Austrian people for a most unrelenting defensive struggle against the Jewish peril."[68] The Pan-German party program was equally direct and vehement: "The party . . . is in favour of a campaign of enlightenment about the corrupting influence of the Jewish spirit and the racial anti-Semitism necessitated thereby. It will combat Jewish influence in all areas of public and private life."[69] Such public declarations more often than not included other code words to make the political intention of the anti-Semitic slurs clear. Thus the Jewish danger became the Jewish Bolshevik danger which, as all right-thinking "German-Austrians" knew, represented the Viennese socialist danger.[70]

No one was clearer or sharper in describing the present and long-range danger of Jews to a Christian Austria than Ignaz Seipel, who became head of the Christian Social party and chancellor in 1922. In a public analysis of the "Jewish problem" in 1919 he concluded that the Jews were not Europeans, but aliens with a merchant ethos pervading all their activities, and that the only basis on which they could exist in Austrian society was as a fully segregated minority. For a short time in the same year Seipel flirted with a Jewish exceptional law being prepared by Kunschak which would segregate Jews, annul past and prevent future assimilitation, and reduce their activity to correspond with their numerical proportion of the national population.[71] In the following years Seipel's anti-Semitism served as a flexible instrument of his politics and was used by him with skill in dealing with his socialist opponents on innumerable occasions.[72]

The socialist leaders' response to the virulent anti-Semitism directed against them and their party was either passive or perverse. All the Jewish leaders of the SDAP were fully assimilated and took pains to avoid any identification of the party with Jewry.[73] To be sure, the party formally opposed anti-Semitism, but such a passive approach simply allowed their opponents to make the wildest charges and associations—"Jewish-Bolshevik conspiracy"—without being confronted as hate mongers. When the SDAP sought to answer slanderous attacks, it played into the hands of the anti-Semites by charging that they too were under the influence of or were pawns of Jewish bankers or Jewish capital. The perversity of Jewish self-hatred among Jewish socialist leaders was expressed in the attempt to fight anti-Semitism with anti-Semitism in socialist pamphlets and broadsides.[74] But at no time did the SDAP publish a full-scale rebuttal of anti-Semitism or expose the close ties between its Pan-German and Christian Social exponents and the Catholic church and racist organizations.[75]

That the SDAP allowed such gutter politics to go essentially unchallenged from the beginning of the republic to its end, with the prominent Jews in its leadership keeping a low profile, weakened the party and undercut the republic as well. Otto Bauer and other prominent Jewish socialists lacked the political courage to answer the anti-Semitic slanderers fearlessly and powerfully in public debate. By contrast, when Léon Blum was subjected to an anti-Semitic slur in the Chamber of Deputies in 1923, he replied: "I am a Jew indeed. . . . One does not in any way insult me by recalling the race in which I was born, a race which I have never denied and towards which I retain only feelings of gratitude and pride." Answering a similar slander in 1936, Blum said that he belonged to "a race which owed to the French Revolution human liberty and equality, something that could never be forgotten."[76]

It is difficult to explain the very different response to anti-Semitism of French and Austrian Jewish socialists. Perhaps the French enjoyed the advantage of the revolutionary heritage of a nation which had also painfully experienced and risen above the Dreyfus Affair, whereas the Austrians confronted a tradition that had prided itself in resisting change.[77] Put more boldly, one might say that the difference of response lay in the difference between the two republics: the French was secular and the Austrian clerical.

Defeat on the battlefield swept away the old monarchy, but the Catholic church remained undiminished in its power. The Austrian episcopate lost no time in declaring itself to be the moral guardian of a "Christian and German nation."[78] At the same time the reigning cardinal and bishops impressed upon their flocks the need to vote for those parties representing Christian principles in the upcoming national and municipal elections. While the politicians were debating articles of the constitution and the relative powers of the provinces and national government of the federated republic, the Catholic church quietly laid claims to its enduring place in the new Austria. It retained control over secular functions exercised under the monarchy, such as compulsory religious education in the schools and religious marriage.[79]

The Catholic church was better prepared than anyone else to argue for the continuity between the old and the new, and thereby to effectively forestall a serious consideration of the separation of church and state. Thus Catholic priests were paid salaries by the state, a privilege not accorded to the officials of other religions. And most important, priests were permitted to hold public office, a situation made blatant in the person of the Jesuit Ignaz Seipel, who as leader of the Christian Social party became head of government. Furthermore, the numerous thinly disguised Catholic lay action groups to be activated at will on behalf of political campaigns by Chancellor Seipel or his party lent a particularly warped character to the Austrian republic.[80]

It should be pointed out that the Austrian church was one of the least flexible, most reactionary, and ultramontanist in Europe.[81] The policy of the Austrian church, whether regarding the working class or anti-Semitism or the role of the family, it was pointed out by the confessor to the Papal Nunzio, was made by the Pope and not in Austria. That revelation was made to Otto Bauer, leader of the Religious (Catholic) Socialists, in two separate interviews with Father Georg Bichlmaier. The latter told Bauer (known as "der kleine Bauer") that the reigning Cardinal Piffl "has nothing to say. . . . He is a zero. . . . Church policy is made in Rome."[82]

It seems that the Catholic church was much more than simply another institution in the Austrian republic, and that its influence in the political struggle between the two camps was far greater than is generally assumed. How influential was the church on the Viennese workers themselves? In working-class neighborhoods one could still see little girls dressed all in white for their confirmation or the celebration of Corpus Christi. The old sacramental practices surrounding birth, marriage, and death continued to be observed.[83] But it would appear that these were "rites de passage" in which the ceremony rather than the spiritual content continued to be attractive, and that such religiosity was nominal for the overwhelming majority of Viennese workers, who were two or three generations removed from rural life.[84] Whereas atavistic religious practices continued among Viennese workers, anticlericalism was also on the increase among them. Between 1919 and 1921 some 7,000 to 8,000 workers formally left the Catholic church each year, and beginning with the late 1920s annual defections were in the tens of thousands.[85] This becomes all the more significant in view of the fact that these were formal resignations involving a legal and official procedure. It suggests an even larger number who took no official steps but simply disregarded the church or viewed it as a hostile institution.

What about the commonly held assumption that women were the perpetuators of religious tradition and practice? One study suggests that Viennese working-class women were anticlerical on political grounds, because the church tended to make their lives more difficult with proscriptions. There seems to have been little churchgoing among them. They were pantheist rather than atheist; spontaneous prayers at home to a supreme being were adapted to their family's needs and trials. Both membership in the cremation society Die Flamme and resignations from the church were common. At the same time an adherence to some atavistic rituals appears also to have been fairly common.[86]

The mere nominal Catholicism of the overwhelming majority of the Viennese working class was hardly a closely kept secret. In view of that fact, it is difficult to explain why, during the months of the socialists' greatest power in 1919, the SDAP did not force through a clear separation of church and state in order to constitutionally secure a secular republic and curtail the power of the Christian Social party.[87] It is equally difficult to explain why the SDAP was not then or at any later time prepared to fight against the pernicious reactionary influence of the church by asking workers or at least party members to leave it en masse. In the virtual *Kulturkampf* waged by the Catholic church against the SDAP on every single aspect of the socialists' political and cultural program, the SDAP refused to use its most powerful weapon, which would have made the struggle unmistakably an open one.[88] Despite the scathing criticism of the Free Thinkers, who asked the socialists to join them in battle with the church, the SDAP stood by its position that "religion is a private matter."

The socialists shrank from taking up the open struggle with the Catholic church over a secular republic, just as they shrank from challenging Austrian capitalists on socialization. In both instances they argued that provinces would secede and Western financial credits would be cut off. One cannot prove them wrong, but at the same time one is prompted to doubt that the Allies, who had created the new Austria, would have allowed it to be altered by territorial defections. The socialists seemed to have basked in taking the morally high ground by allowing the class-conscious worker to shed all vestiges of religion on his own; but by avoiding the struggle with the church when they had the advantage, they undermined their own projects. Later, when the clerical republic was already entrenched, Bauer offered brilliant arguments for the separation of church and state.[89]

Hopefully, the preceding remarks on conditions in Vienna between 1919 and 1921 have highlighted the turmoil of those first years of the new republic. There may not have been a revolution, in the sense that property and class relations to power were not changed fundamentally, but daily life was shot through with the sense of uncertainty and the unexpected, with the improvisation and impermanence so characteristic of revolutionary experience. In the national political arena the socialists went into a traditional social democratic opposition. The two political camps, the SDAP on the one hand and a coalition of Christian Socials and Pan-Germans on the other, divided power between them. But it was not an equal division, with the former in control of the capital and province of Vienna and the latter in command of the nation. Yet the SDAP remained hopeful on the national front (it was fixated on winning 51% of the vote), and at the same time had a large metropolitan laboratory in which to attempt to implement its socialist program. The party faced this challenge with a special theoretical armory—Austromarxism. In what follows we will look at the main tenets of this theoretical framework with an eye to assessing the extent to which it was applicable to the practical municipal socialist reforms the socialists put on the agenda for Vienna.

Austromarxism: A Theory for Practice?

What was Austromarxism? Or, better put, was there an Austromarxism and, if so, what role did it play in the life of the First Republic and in the socialist cultural experiment conducted in Vienna? So tentative and questioning an introduction should alert the reader to the difficulty in answering these questions, arising from the subject's elusiveness. Virtually every book dealing with Marxist theory in the decade before World War I or with any aspect of the First Republic has had to stumble through the thorny field of definition.[90] In this reader's view, all the "Austromarxisms" which have appeared in what amounts to a considerable literature, have begged the question and failed to offer some clear shared body of theory or firm common theoretical orientation that would justify the use of the term "school."[91] Even Charles Gulick in his magisterial political history of nearly two thousand pages consigned the subject to the last quarter of his work with the comment (forty years ago) that the term had been overused.[92]

It is surely significant that Otto Bauer himself turns to explaining Austromarxism only in 1927 in an unsigned lead article in *Die Arbeiter-Zeitung*, and does so in a manner that could not have been terribly illuminating to readers of the party newspaper at that time.[93] Bauer calls it a school of prewar Marxism which took its departure from Kant and Mach and disappeared with war and revolution. Since then, he adds, it has been a pejorative term used by the opponents of Austrian social democracy, but it has also been taken up in a positive sense as summarizing the current principles of the SDAP as outlined in the Linz Program: the place of unity above all other values and the synthesis of real-political and revolutionary tendencies. The problem suggested above is especially vexing, because the term "Austromarxism" has also been attached to the socialist cultural experiment in Vienna. In what follows, I hope to clear the air sufficiently to proceed with my subject.

In answering the question "was there an Austromarxism?" I would like to suggest that there were two. The first of these consisted of a small group of Marxist theoreticians and intellectuals active in the decade before World War I: Karl Renner, Rudolf Hilferding, Max Adler, Otto Bauer, and Friedrich Adler. The second comprised the SDAP oligarchy of doers and reformers during the First Republic, including the leading figures in the Viennese municipal and provincial government, party leaders, and those responsible for the party's educational, cultural, and publication activities. Not only were these two distinctive groups composed of almost completely different individuals, but they were linked by neither a clear body of theory nor by the ties of active party or political work. The second group of party officials, who created the Viennese cultural experiment, believed themselves the heirs of the Marxist theoretical contributions made by the first. The term "Austromarxism" was thus used to cover a multitude of vague assumptions about a school of thought that depended more on inference (to lend a sense of prestige or continuity to arguments or activities) than on shared ideas. It will be necessary to take a close look at the putative "founders" of Austromarxism to determine the legacy they were presumed to have left to the socialist actors of the republic.

It is far from simple to sketch an intellectual collective portrait of this small group of Austrian Marxists who were part of what George Lichtheim calls "the revisionist generation of 1905,"[94] engaged in explicating Marx on the basis of current experience. What did they have in common?[95] Of the five, all but Renner were Jews or, in the case of Friedrich Adler, the son of a Jew who had converted to Christianity. In the common anti-Semitic parlance of Karl Lueger's Vienna and much later, "once a Jew, always a Jew," by which was meant: if a Jew converted to the Christian religion he was a dissembler boring from within; if he chose the legal status of being "without confession," it was simply to disguise Jewish plotting. Austrian Marxists' Jewishness is mentioned here to indicate that, in the popular Viennese anti-Semitic prejudice of the time, they were considered troublesome foreigners corrupting German culture and threatening the social order.

All five had doctorates and enjoyed the elevated cultural position of being considered and addressed as "Herr Doktor" in Viennese society and within the Socialist party as well. Four of the five, Renner excluded, came from middle-class families and milieus; all but Renner were born in Vienna. They were occupied either in the liberal professions or as intellectuals in the SDAP. Renner had a post in the library of parliament until he was elected a deputy in 1906; Max Adler became a lawyer in 1902 and shortly thereafter was appointed to an associate professorship of sociology and social philosophy at the University of Vienna; Hilferding was a physician; Bauer became secretary of the SDAP parliamentary delegation; and Friedrich Alder was one of the Socialist party secretaries after 1911. None of them had practical party work experience or more than peripheral contact with workers. One might say that the stereotype of the comfortable Viennese professional who had made politics on an intellectual level his real commitment fit the group very well. Their position as professionals and intellectuals, their marginality as Jews, and their unfamiliarity with the daily life of ordinary Viennese created a natural remoteness from practical questions. What was an intellectual self-exclusion for these men became a bureaucratic exclusion for the activists who commanded the SDAP after 1918.[96]

It is equally difficult to find a common intellectual denominator in the brilliant Marxist studies published by these young men when they were still in their twenties. To be sure, Bauer and Hilferding shared an interest in imperialism as expressed in their respective works of 1907 and 1910.[97] Similarly, Renner and Bauer shared an interest in the state in relation to nationalism.[98] And Max Adler and Friedrich Adler both sought to return the subjective ingredient into the calculation of the course and progress of human events.[99] But in none of these cases can one really find a sense of shared intellectual endeavor or borrowing of ideas and terms. Nor, aside from Marx, do they appear to have shared the same intellectual influences: for Renner it was John Stuart Mill; for Hilferding, Karl Kautsky; for Max Adler, Kant; and for Bauer and Friedrich Adler, Ernst Mach. Though evidence for a school of Marxism is lacking, there are illusive common assumptions, experiences, and tendencies of thought which associated these men and gave their ideas a sense of cohesiveness and orientation actually difficult to pin down in the text of their writing, but still compelling enough to be identified as "Austromarxism" by those who adopted that designation during the First Republic.

Perhaps, this elusive cohesion can best be found not in their ideas but in the locale of their activity—in Vienna itself.[100] The "imperial residence and capital" of the Dual Monarchy was a showcase for what on the surface appeared to be a colorful array of nationalities, but which to the more careful observer represented the gravediggers of a crumbling empire.[101] Vienna was also a bustling metropolis, moving fully into the age of industry, in which an economically established but politically immature bourgeoisie increasingly took confrontational cognizance of a rapidly organizing working class. A further example of the contradictions inherent in the very structure of the city is the dominating baroque quality of the noble quarters and palaces, conjuring up the Counterreformation and reaction, poised against the bourgeois elegance of the Ringstrasse, contrasted in turn with new railway yards, factories, and worker tenements. Vienna was proudly referred to as "the Paris of central Europe," but it was also the capital of a disintegrating empire. This dualism of artistic/intellectual innovation and decay, of "sexuality and death" in the view of Bettelheim, made it the natural birthplace of psychoanalysis.[102]

It is easy to understand why the Austromarxists were radicalized by their surroundings, in which the absence of an Enlightenment tradition, exemplified by rationalism and "liberty, equality, and fraternity" as disseminated by the French Revolution, was still painfully apparent after the turn of the century.[103] The great Anglo-French enlightenment tradition was lacking in Central Europe as a whole. In its absence the Austromarxists clung to German culture as the repository of humanism.[104]

Not only the contradictions and stresses in Viennese society in general but also the specific formative intellectual experience of the Austromarxists in the city imparted a certain common direction to their differing ideas. The Austromarxists first appeared as a group in 1895 as founders of the Independent Association of Socialist Students and Academicians, consisting of socialist university students and instructors. At the University of Vienna the Austromarxists came under the influence of ideas important to their work. From the socialist professor of political economy Carl Grünberg they learned to view Marxism as a social science to be developed by historical and sociological investigations.[105] They also became active members of the Association for Social Science Education organized by Grünberg and the Marxist professor of history Ludo Hartmann.

The university environment further exposed them to a variety of stimulating non-Marxist influences: the ideas of Mach, who was professor of physics and the philosophy of science[106]; neo-Kantian ideas, which enjoyed a great popularity; and Carl Menger and Eugen Böhm-Bawerk of the Austrian or marginal utility school of political economy, whose fierce critique of Marx stimulated responses from the group. They engaged in two collective publication enterprises: in 1904 they began publication of a yearbook called *Marx-Studien,* edited by Max Adler and Hilferding, and in 1907 they started publishing the monthly theoretical journal *Der Kampf* under the editorship of Bauer, Renner, and later Friedrich Adler. All members of the group contributed to both publications, which then and later were widely considered to embody Austromarxism.

Even more important than the university as a gathering place for the Austromarxists was their habitual resort, the Café Central, located in the fashionable Herrengasse and frequented by local and foreign intellectuals.[107] It is here, in the special ambience of the Viennese *Kaffeehaus*, that the Austromarxists are alleged to have developed their remarkable ability to harmonize and reconcile inconsistencies: to be well integrated in society and zealously to repudiate it in word and print.[108]

The outbreak of war ended the cohesion Vienna had given to the group of five. Of these founders of Austromarxism, only Bauer was to make a transition into the postwar world as theoretician and man of destiny. As putative head and theoretical spokesman of the SDAP, he carried the heritage of the prewar progenitors forward. Indeed, as we shall see, Otto Bauer and Austromarxism became synonymous. Two members of the group, Hilferding and Friedrich Adler, left Vienna and the Austrian terrain of Marxist thought and activity. After serving as a physician in the imperial army, Hilferding went to Germany, where he had already been a frequent visitor since 1906. There he joined the Socialist party and became minister of finance in two Weimar administrations. Friedrich Adler emerged as a heroic figure at the end of the war because of his assassination of the imperial chancellor Count von Stürkh. But he refused the offer to become head of both the Austrian Communist party and the Communist International and contented himself with leading the Council Movement into nonrevolutionary waters. He helped found the Vienna Union, popularly called the Two-and-a-half International, of parties seeking a midway course between Russian Bolshevism and social-patriotic social democracy. When the attempt to unify the three internationals failed in 1923, Adler became secretary of the Labor and Socialist International (successor of the prewar Second International) and resided at its headquarters in London and later in Zurich and Brussels.[109]

Although Renner and Max Adler remained on the scene in Vienna, both were vastly overshadowed by Bauer, who on occasion acted as arbiter between them without his own position being placed in jeopardy. Renner served as chancellor during the brief coalition governments of the new republic (1919–20).[110] Afterward he was relegated by Bauer to the margins of the party directorate and contented himself with acting as president of parliament. At crucial times, as we shall see, he reemerged and challenged the whole new conception of Austromarxism in its Bauerian configuration.

Max Adler resumed his quiet academic and intellectual life after a brief period of activity in 1918–19 in support of the workers' councils (though he opposed all soviet experiments) and of the factory councils, which he hoped would become workplace revolutionizing agencies. He spent much time in Germany, where he associated with the left within the SPD and—

together with its leaders, Kurt Rosenfeld, Max Seidewitz, Heinrich Ströbel, and Paul Levi—edited the periodical *Der Klassenkampf—Marxistische Blätter* from 1927 to 1931.[111] He also edited a series on socialist education under the rubric "neue Menschen" for an independent left publisher from 1924 to 1928. He was well received by the circles of the SPD left, where he was generally viewed as the spokesman for Austromarxism. In the SDAP, as we shall see, he made important contributions to Marxist theory, underpinning and justifying the SDAP's cultural experiment, and thereby seconded Bauer's position on the relationship between culture and power on the road to socialism.

What was the actual legacy passed on to the Austromarxists of the republic by this group of prewar theoreticians? Rather than a coherent body of ideas, they seem to have passed on a number of orientations: a neo-Kantian emphasis on subjectivity and human volition in the historical process; a new view of the state; the dissemination of enlightenment through *Bildung*, but with a warped German slant; and Marxism seen as *the* social science. This distillation of the Austromarxist inheritance—though not necessarily my explications which follow—should pass muster before most historians familiar with the subject. I would like to add one more aspect of the Austromarxists' legacy, which is more controversial: making history more human and purposive assumed an individual malleability in the process of socialization and raised the specter of workers as objects, and at the same time embodied the conflict between leaders and masses, authoritarianism and passivity.

I shall try to expand upon the orientations which have been put down in shorthand form above, but hopefully without the painfully confusing jargon and neologisms (such as "ethicality") common to such discourse, and with a reminder to the reader that we are looking only for the legacy of early Austromarxist ideas drawn upon by the Austromarxist activists in the postwar period.[112] Max Adler was certainly captivated by the neo-Kantian model building and Machian attacks on the tyranny of materialism that swept the Viennese intellectual world. Adler attempted to find a Marxist expression for Kant's ethics and categorical imperative, or, as one of the astutest experts on Adler has put it, to adapt Kant for use by social democracy.[113]

Dispersed throughout Max Adler's work is his central attempt to integrate Kantian and Machian ideas with Marxism, in the hope of making the latter both more precise (as the equivalent of the social sciences) and more flexible, so as to allow for the role of human consciousness.[114] He sought to free Marxism from vulgar materialistic determinism without abandoning dialectical economic lawfulness in historical development, while at the same time giving an opportunity to the working class collectively—as the agent of the inevitable socialism—to make its subjective volition part of the objective historical process. The individual—the worker, if you will—attains a higher expressiveness and individual consciousness while being socialized—that is, educated—to the functions of his class in the historical process.

What we have here is Friedrich Engels's sturdy and inflexible "scientific socialist" suit being "knocked off" in a very elastic fabric, one which served well the pragmatic orientation of the Austromarxist republican activists, who set out to create an original proletarian culture. They were particularly receptive to Adler's suggestion that the development of the workers' greater self-consciousness, their intervention in the building of socialism, need not be postponed until the tocsin had sounded the onset of revolution. The workers could act now; they could be prepared to act now; they could be educated to act in their best interest; they could be transformed from being less than human to being "neue Menschen." One can easily see the inspiring quality of such ideals. The problem with—or more precisely, the danger of—this vision lay in the method of its implementation, which is one of the main themes of this book. The danger lay in superimposing the narrow authoritarian structure of the party and its paternalistic and *dirigiste* methods onto a cultural enterprise whose substance and purpose were meant to be liberating.

An important undertone in Adler's marriage of Kant to Marx is the socialization of Kant's ethical universalism expressed through the agency of the "categorical imperative." In Adler's transformation of the latter, the Kantian postulate that man shall never use man as a means emerges as a collective moral commitment to struggle against the reification and alienation of man. There is a lack of clarity in Adler's formulation: if man can become a self-conscious actor in the collective historical process and if, at the same time, dialectical economic lawfulness continues as the motor force of history, then how do linkages between these two processes occur? Is the workers' education only preparatory for the eventual revolutionary event, without any direct means of influencing it—speeding it up, for example?[115]

The Austromarxists' view of the state, stemming largely from Hilferding's *Finanzkapital*, only further confounds the problem.[116] The dialectical struggle is seen as having moved to a higher level in the economic realm, in which finance capital controls production at a distance. The state, accordingly, stands above and apart from the struggle but attains the ability to gain control over the economy without expropriation or a revolutionary seizure of power. Without getting further into the pyramidal contradictions, it must be said that the state theory did not resolve Adler's dilemma; it did pass on to the practitioner socialists the notion that somehow the state could be a neutral force (resting on constitutional assurances, etc.). Looking at the Austromarxist heritage from a considerable distance, it seems apparent that Marx's (or particularly Engel's) dialectical materialism had been left by the wayside.[117]

One of the most powerful orientations inherited by socialists of the republic from their Austromarxist forebears was the commitment to *Bildung*. The postwar SDAP decisively engaged in the enlightenment of the workers to develop their self-consciousness as a socialized class experience. But the "enlightenment" handed down to the socialist reformer was curiously divorced in virtually every way from the great Anglo-French Enlightenment. Particularly absent was the natural law doctrine, guaranteeing defined rights as well as tangible freedoms, and the anticlericalism implicit in theory and widespread in practice in Western Europe. The content of the *Bildung* being offered to the workers, therefore, was largely limited to the venerable ideas of German culture.[118] The reliance on these to the exclusion of the Enlightenment can be explained as stemming from the complicated differing historical development of Central Europe, leading to a certain isolation and provincialism. But there is more to it. The Austromarxist founders had quite clearly concluded that the only culture worthy of the name in Central Europe was German. The earliest writings of Bauer and Renner, which differentiate between "nations" and the state and argue for the maintenance of the empire over culturally entitled nationalities, conclude that the only culture of stature is the German one.[119]

The Austromarxist heritage of *Bildung* was translated by the reformers into what Rabinbach calls the "politics of pedagogy."[120] The characterization is apt in that a general civilizing mission is implied. But pedagogy suggests the primacy of the book very much as the Austromarxist elders had viewed it,[121] whereas the *Bildung* of the cultural experiment in Vienna sought in a comprehensive and even total way not only to give workers knowledge but to directly alter their behavior. Thus the reformer assumed not only the traditional role of pedagogue but that of social engineer as well.

The danger of the neo-Kantian–Machian–Marxist matrix pointed to earlier became clear when the opportunity arose in Vienna to put these orientations into practice. By casting workers as both less than human (because their lives knew only misery) and at the same time malleable, the directors of Vienna's cultural experiment could disregard the existence of worker subcultures and presume a kind of cultural tabula rasa on which they could imprint all that a new proletarian culture would require. The cultural reforms based on the perspective that the workers lived like primitives lacking in standards and norms of common decency could become, in a sense, acts of creation in which the worker as human emerged from the raw clay of exploited and brutish beings.

In recounting the legacy of the Austromarxist founders, it remains to assess their methodology—their view of Marxism as social science. The complicated theoretical process by which the Austromarxists, and Max Adler in particular, attempted to tame science to the task of emancipating the working class did not affect the practitioners of the republic and need not concern us here.[122] The methodological legacy reflected both Kantian and Machian influences in trying to reconcile causality and teleology, determinism and human freedom. From an intellectual point of view, the Austromarxists' effort did not succeed. On the one hand, they posited socialized humanity as a Kantian given, a category of knowledge stemming from reason rather than experience. On the other, they treated Marxism as an empirical science of society compatible with (and derived from) Mach's theory of cognition (sensate positivism).[123] The fact that the dichotomy between deductive givens and inductive epistemology was left unresolved did not seem to bother the Viennese socialist reformers. Indeed, theoretical imprecision in this as well as other aspects of the Austromarxist inheritance seemed to suit their pragmatic inclinations. The could, with some pride of continuity with grand Austromarxist theory, view their own efforts in creating a working-class culture as an application of Marxist social science without having to surrender subjective beliefs.

What the Austromarxists actually passed on to the socialist activists was a distillation of the various orientations examined above. Grand theory had been created in a world far removed from practical politics. Later it was made functional and pragmatic by the socialists, who commanded the SDAP and governed Vienna, by informing and justifying reforms in progress and programs yet to be initiated. For the socialist doers, in other words, prewar Austromarxist theory was a hallowed inheritance more honored than actually understood or made use of save in a talismanic way. It would have been unthinkable for a socialist functionary in republican Austria not to consider or declare himself to be an Austromarxist, for that tradition was a necessary part of his identity. But the actual Austromarxism practiced by the same functionary was a highly political set of strategies and tactics resting upon a defined conceptual framework. The latter, the Austromarxism of practice, was formulated virtually single-handedly by Otto Bauer.

Before we turn to an examination of this "operant theory," it would be useful to have a better idea of who the socialist doers actually were. They included the municipal "city fathers": Mayor Jakob Reumann (and from 1923 on, Karl Seitz); Councillor for Finance Hugo Breitner; Councillor for Health and Social Welfare Julius Tandler; Councillor for Schools and Education Otto Glöckel; and Councillor for Housing Anton Weber. Most significant on the provincial level was Robert Danneberg as president of parliament. Prominent within the SDAP directorate were Seitz as president, Danneberg as secretary, and Julius Deutsch as head of the Schutzbund (paramilitary defense unit). Party educational efforts were directed by Glöckel and Anton Tesarek, founder of the Rote Falken. Party culture was entrusted to Joseph Luitpold Stern, David Joseph Bach, and Otto Felix Kanitz. SDAP publications were guided by Friedrich Austerlitz, Max Winter, and Julius Braunthal. Seemingly in the background, because he held no official position in the party structure, was Otto Bauer. But, as writer of *Die Arbeiter-Zeitung*'s daily leaders and as chairman of the SDAP national parliamentary delegation, Bauer was the undisputed head of the party, with no one to either the left or the right able to challenge his intellectual and moral force. Bauer had the distinction of embodying both the venerable Austromarxist tradition and the Austromarxism of practical reforms.

It would be only a slight exaggeration to say that Austromarxism in the First Republic was synonymous with Bauer's formulations, particularly with his theory of "the balance of class forces."[124] The latter constellation of ideas sought to explain the past, present, and future course of the republic in a way that bridged the most glaring contradiction of the older Austromarxist tradition: to reconcile the subjective volition of the working class, expressed in a heightened self-consciousness, with the lawful, historical

Otto Bauer addressing a worker rally in the early 1920s (VGA)

materialist process. The balance of class forces, formally stated only in 1923–24 but traceable in part to Bauer's prewar writings,[125] was at the same time a justification for the "revolution of the soul" by which the workers would be culturally prepared to undertake their historical mission of establishing socialism.

True to the tradition of the grand theoretical sweep, which he had helped to establish, Bauer put his analysis of socialism's tasks and prospects in a world perspective. Its object was to discover a "third way," a route other than that taken by the Bolsheviks or the passive reformism of traditional social democracy. From the vantage point of the early 1920s, he claimed, a situation had arisen in Europe where neither the bourgeoisie nor the proletariat was able to dominate the state, and they were forced to share power. This condition was the balance of class forces wherein neither class was able to exercise a hegemony without the tacit participation of its opposite. In Italy and Russia this situation had been resolved not by the dictatorship of either the bourgeoisie or proletariat, but by independent state power in the first and by dictatorship of a caste in the second; in both, the resolution had taken place above the classes.

In Austria, Bauer insisted, even though the bourgeoisie had reestablished its political hegemony by 1922, the republic continued to depend on a degree of shared power. As evidence Bauer put the SDAP's control of Vienna and a few rural industrial centers on the scale against the rest of the country and suggested that, although the political powers were not balanced, only a civil war could break the stalemate. The latter, which in addition to causing much bloodshed and suffering would wipe out all the gains and power of the working class, had to be avoided at virtually any price.

It was a daring conception which placed an enormous burden on the SDAP in Vienna to create an institutional network through which the workers could be made culturally mature for socialism. It presumed that neither side would risk a contest for power, and that the workers would not be integrated into the politics of the bourgeoisie or dominated by its culture. Thus Bauer separated the cultural from the political revolution, with the implicit danger (soon to become actual) that the former would be substituted for the latter, that the Viennese cultural experiment would attempt to compensate for socialist powerlessness in the national arena.[126] The dualism of Bauer's conception was no more resolved than the contradictions in prewar Austromarxism had been. The preparatory cultural strategy, which legitimated the whole Viennese experiment, had no real links to the process of coming to power in the future. Nor was the relationship made clear between cultural hegemony through *Bildung* and the laws of capitalist development. In that world of real power, objective and immutable laws of historical materialism were presumed to be grinding on to that happy future when the socialists would "inherit" power.[127]

By some cruel irony Bauer was recapitulating prewar Austromarxism's evasion of the contradiction between subjective and objective power. On the basis of Bauer's analysis, which took little account of the tough-mindedness of the Christian Social party or the Catholic church, *Bildung* became the real politics of the SDAP. The foundation, on which the Viennese cultural experiment was based, seems to have been fragile, and Bauer's optimistic prognostication for the coming of socialism appears to have been based on a very naive conception of struggle. Rabinbach has explained the mystery of such evasions of reality tersely as Bauer's "will to powerlessness."[128]

The course chartered by Bauer for the SDAP exemplified what Dieter Groh has called "negative integration."[129] It consisted of an emphasis on party growth and unity, on electoral success and parliamentary activity, and most of all on *Bildung* to prepare the working class for its historical role and to neutralize worker aggression, diverting it from political action. Bauer's justification of the party's cultural experiment was seconded by Max Adler in a long essay whose title, *Neue Menschen,* was probably more influential than its substance.[130] In this work, dealing with the education of the workers, Adler denounced existing socialist pedagogy as being unscientific and avoided all practical questions. He assumed that the workers were not mature enough for revolution and proposed an educational transformation that would create the new consciousness within the existing state. In that sense, education would act as a vehicle of cultural communication and as an important instrument of the class struggle. The last turn of phrase was probably the essay's main contribution to Bauer's conceptual edifice.

Bauer's formulation of the balance of class forces received critical fire almost immediately after its main exposition, from both outside and inside the party. Hans Kelsen, the liberal legal theoretician and architect of Austria's constitution, argued that there had not been a balance of class power either during the brief coalition period or thereafter, because the capitalist exploitative system and related social order had remained in control.[131] The belief that the equilibrium between classes could be expected to last for some time and provide the necessary basis for the SDAP's cultural strategy was challenged by Otto Leichter, one of the editors of *Die Arbeiter-Zeitung*. The disarray among reactionary forces after the war, he argued, had been produced by the council movement, which had since faded away. In Austria the balance had ceased to function when the Christian Social party regained its power in 1922. Such an epiphenomenon, he concluded, did not merit a whole new conception of the class struggle and the state.[132]

These commentaries neither altered Bauer's position nor deflected municipal officials and SDAP functionaries from continuing to put the cultural program into practice—something they had been doing for some time before the formal discussion of the legitimizing balance-of-forces theory. To end the long saga of Austromarxist theory here, in 1923–24, would offer too idyllic a picture of the fate of Bauer's conceptual structure in its contact with political reality. Strangely enough, the SDAP did experience a few nearly halcyon years in which the party grew and its programs flourished.

The party congress of 1926 took place at the high point of socialist self-confidence and sense of practical accomplishment. This was expressed in the so-called Linz Program, which confirmed the socialists' commitment to social change and cultural improvement.[133] Much of the draft program—as usual, the work of Bauer—was devoted to explicating the party's devotion to political democracy and its institutions. In the process it not only characterized Bolshevism as a failed attempt to elevate socialism to a higher plane and warned about methods of change based on force, but also directly reconsidered the notion of balance of class forces. The latter review was undertaken in response to the growth of the Heimwehr[134] and other antirepublican forces which threatened to overthrow democracy. It projected a future in which the balance would be upset in favor of the socialists, who would come to power by being elected by a clear majority of Austrians.[135] If the socialists then used their democratically gained right to expropriate capitalism, supporters of the latter were expected to defend their property and position of power by resisting. If, the argument continued, the bourgeoisie should initiate a counterrevolution with the object of restoring the monarchy or creating a fascist state, the SDAP would be obliged to use defensive force (civil war) and a defensive dictatorship.

This position on defensive force and defensive dictatorship to safeguard democracy was a compromise hammered out at the congress between Bauer and Max Adler, in opposition to Renner.[136] The latter had argued that the socialists' entry into a new coalition with its opponents would safeguard the balance of forces. Adler had insisted that only the fear of worker self-defense kept the bourgeoisie at bay and the class forces in balance. The final program implied that the SDAP would protect its cultural experiment with force of arms if necessary. Although *Die Reichspost* and other right-wing newspapers characterized the defensive force position as a call for bloody revolution, Bauer's terse slogan—"Democratic as long as we can be; dictatorship only if we are forced to it, and insofar as we are forced"—suggested that perhaps the SDAP's firm stand was only rhetorical after all.[137] A test of how far the party was prepared to go to defend the balance of forces came sooner than the socialists expected.[138]

On July 15, 1927, a spontaneous and massive worker revolt in Vienna directly challenged the SDAP's central doctrine and put the survival of the republic in question.[139] By the end of the day the Palace of Justice as well as the *Reichspost* building had been burned, 89 persons had been killed, and 500 to 1,000 wounded. On the previous day a jury trial in Vienna had found three right-wing activists, accused of murdering a socialist man and boy, innocent of "all wrongdoing." A fiery editorial in *Die Arbeiter-Zeitung* on the morning of July 15 denounced the acquittal as an outrageous example of class justice. Spontaneously workers left their factories, shops, and homes and made their way along the fashionable Ringstrasse to the square facing the Palace of Justice.

In the beginning stages of this developing confrontation between the Viennese working class and the real and symbolic agents of law and order, the socialist leadership avoided taking a stand. Otto Bauer actually hid from a delegation of electrical workers who came to party headquarters to demand orders to shut down power plants.[140] The many thousands who had gathered in front of the Palace of Justice by noon were left to act as a spontaneous mass; no one could blame the SDAP for having ordered or planned anything. The police was completely unprepared and therefore undermanned because their chief, Johannes Schober, had been told by SDAP leaders that no official demonstration was planned.

Mounted police seeking to clear the key streets set off the violence, during the course of which the police were instructed by their superiors to fire hurriedly issued army rifles point-blank into the crowds, while various buildings were set ablaze. In the heat of the struggle there was a widespread demand by workers and members of the Schutzbund (created by the SDAP in 1923 precisely to protect the workers in situations such as this) for the distribution of arms. Socialist and trade union leaders refused to approve a course which no doubt would have led to civil war. Disorder and violence continued in the working-class districts throughout the 16th. A national strike of transportation and information services, called the same day for an indefinite period, was made ineffective outside Vienna, even in industrial towns, by heavily armed Heimwehr units acting as auxiliary police.

In the aftermath the SDAP sought by tough language to force Chancellor Ignaz Seipel to make concessions such as calling new elections, granting a general amnesty, or initiating a parliamentary investigation. But Seipel stood his ground and refused any real compromise.[141] It was apparent that municipal socialism and the socialist party culture in Vienna had made no difference in the confrontation of July 15. Seipel became inclined to depend more on the Heimwehr, which had proved so effective in breaking the general strike in the provinces, and took steps to assure the political conservatism of the municipal police and the national army.

The fortunes of the SDAP seemed to have changed overnight. In both the Viennese and the national elections of April 1927 the party seemed to be well on the way to attaining Bauer's aspired-to, magical 51 percent. In Vienna, in fact, the party gained 123,000 votes and scored 60.3 percent.[142] Nationally, the SDAP reached 42.3 percent of the vote, compared to 39.6 percent in 1923; Seipel's "unity list" of Christian Socials and Pan-Germans received 48.2 percent, down from 57.8 percent in 1923. But Seipel was able to govern with a clear parliamentary majority of 94 to 71 by including the Agrarian League in his cabinet. Furthermore, July 15 called Bauer's optimistic reading of the balance of class forces into question: the state had not been neutral; the opposition had made use of extraparliamentary force; the much-vaunted socialist defense of last resort—the Schutzbund—had revealed itself to be a hollow threat.

A postmortem of the setback at the October party congress turned into a confrontation between Renner and Max Adler.[143] The former charged that the July 15 rising was a direct consequence of the dangerous educational and cultural policy the party had pursued under Bauer's leadership, and that unrealistic talk of revolution had led the workers into an adventure which risked civil war. He also rejected the idea that the workers required a special education to prepare them for socialism, insisting that economic conditions alone would be sufficient to create worker consciousness. Lastly, Renner cut the ground from under both the balance-of-forces theory and the Viennese cultural experiment by demanding that the party enter into a coalition government, a move which would have required a decisive retreat from the SDAP's programs.

Adler responded with an impassioned defense of the Viennese workers, who were generally being maligned in the party as undisciplined.[144] Renner's retreat from a revolutionary educational and cultural policy, he charged, risked the integration of the workers into capitalist society and their transformation into petty bourgeois. "Class struggle," he reminded the party, "is cultural development . . . [and is] by its very nature something thoroughly intellectual . . . the reform of consciousness."[145] In rejecting Renner's call for a new coalition, Adler was joined by Bauer in pointing out that the cost of sacrificing the whole municipal socialist program as well as general advances in creating a proletarian culture in Vienna made such a step unthinkable. The heated controversy shed no new light and was terminated in Bauer's usual fashion by a synthesis of opposites, calling both Renner's and Adler's orientation necessary to Marxism and the party.

Some have called July 15, 1927, the "turning point" in the history of the First Republic.[146] Clearly many weaknesses in the socialists' position as a party and political force had been exposed, and both the balance of classes and the threat to use "defensive force" were put in question. Henceforth the relationship between cultural and political activity shifted markedly toward the former, and red Vienna became more of a beleaguered enclave. Even Otto Bauer, the master of theoretical optimism and tactical compromise, had been challenged and forced to rely on party discipline to safeguard the hallowed unity, the chief catechism in the party's litany.[147]

The nature of such long views tends to reduce poignant choices and complex beliefs in the interest of categorical analysis. If one examined the significance of July 15 from the point of view of a third-level SDAP official, a low-ranking member of the Viennese health or housing service, or a factory worker organized in both his trade union and the party, with all of them looking at Austria from the perspective of Vienna, one might not find the sense of catastrophic descent that later students of the events found to be so inescapable. These participants, with direct experience in the transformation of Vienna since 1919, might well have enumerated a great variety of gains the working class experienced in that short time: the massive growth in size of both the trade unions and the SDAP; the maintenance of rent control and increasing public housing; the extension of social welfare; the expanded possibilities for education; and the various socialist associations which attempted to educate and integrate the working population within the party. Even those workers—surely the great majority—who experienced none or very few of these gains directly, perceived a sea change in the daily social climate of Vienna, where the mayor and most officials who counted were socialists, lending an entirely new dignity to the status of workers.

In view of this new sense of collective importance in the city, it is difficult to imagine that the mood in working-class neighborhoods was simply pessimistic over the events of July 15. If one were to speculate about that popular mentality, one might conclude that a feeling of confidence was mixed with a sense of frustration.[148] There was anger, particularly in the Schutzbund and among other activists, directed in part at the socialist leaders for failing to react to the generally perceived provocation. But in the main it was leveled at the popular Viennese workers' stereotype of the opposition— the "Dorftrottln und Wasserschädln" (village idiots and hydrocephalics) dressed in leather pants or dirndl and decked out with a big cross to mark their true allegiance. There was (mistakenly) nothing in the makeup of the Viennese workers' urban mentality to make them afraid of this comically perceived threat from the provinces. These differing perspectives from within working-class communities kept pessimism at bay in 1927 and allowed the "building of socialism," as then perceived, to continue in much the same spirit as the great leap forward toward municipal socialism undertaken in 1919.

Having gone to some pains to assess the nature and relationship of Austromarxism's theory and practice, I regret having to caution the reader against expecting an orderly causal relationship between the two. The socialists took over the administration of Vienna in 1919 in the midst of postwar uncertainty. They were forced to govern, to face the enormous social problems before them, without being able to wait for party doyens and theorists to hammer out enabling theory. As I have attempted to demonstrate, they were pragmatists in the spirit of Austromarxism; the theoretical justification for their decisions often limped behind the events. As we turn to the unique Viennese experimentation with municipal socialism, it will become clear that during its early years practical improvisation was based on the choices left open by the vagueness of the Austromarxist heritage. Being unfettered by an inflexible theoretical framework was an advantage in dealing with unpredictable and changing realities in daily practice. But as we shall see, that freedom gave immense power to a small group of leaders to fashion municipal reforms using themselves, their personalities and norms of socialization, as the yardstick. The danger loomed large that, instead of the Austromarxist aim of liberating the workers to act with a higher consciousness, a paternalist pragmatism would attempt to impose programs and reforms on workers without including them in the process.

CHAPTER 3

Municipal Socialism

The problems facing the Viennese municipal government during the first postwar years were truly daunting. It took endless months to regulate and reduce the inward and outward flow of populations, to assure the provisioning of the city with victuals from the provinces, and to return transportation and the vital municipal utilities to the requirements of peacetime. The solution of these transitory problems proved more difficult in Vienna than in other capitals because of the newness of the republic and Vienna's place within it, making for the absence of time-worn procedures to be used as models. The socialists, who headed every important department in the municipal council, put the lie to the common shibboleth about socialists being good at thinking and talking but incompetent when it came to practical matters, by competently restoring the normal course of municipal life. But very difficult and seemingly intractable problems remained to challenge their commitment to improving the quality of life for all Vienna's citizens:[1] the condition of public health, undermined by wartime experience and long-term diseases such as tuberculosis; the need of public welfare for the indigent, homeless, and helpless; and the deterioration of the public school system, which had never viewed equality of opportunity as part of its mission.

Overshadowing all these problems and contributing to their gravity was the severe housing shortage. Although wartime dislocations had reduced the population of Vienna by some 165,000, the number of householders increased by from 40,000 to 60,000 at a time when the number of vacant domiciles was never much above zero.[2] This anomaly had several causes. Immigrants to Vienna from the succession states appeared as families, whereas emigrants to the succession states were single persons who sought their fortune there and usually left families behind. A 50–90 percent increase in marriages immediately after the war further increased the demand for housing. Most important, the drastic reduction in rents due to the combined effect of rent control and inflation let many householders rid themselves of subtenants and bed renters in a climate of rising expectations.[3] The problem was fueled on the supply side by the virtual standstill in housing construction from 1915 to 1924. In the decade before the war a yearly average of 9,300 domiciles had been built for the private housing market.[4]

It is to the socialists' credit that they recognized the determining role housing would have to play in their attempt to make Vienna the showcase for municipal socialism. Their plans and expectations went further. Decent housing became the cornerstone of the SDAP's project to create the "ordentliche Arbeiterfamilie," a phrase connoting not only orderliness but also decency and respectability.[5] In other words, the socialists aimed beyond municipal reforms toward an all-encompassing proletarian culture in which the physical context of a certain type of habitation would play a central organizing role. Environmentalism was an important aspect of Austromarxist subjectivism and was the unwritten basis of municipal reform. Theoretically it had a greater affinity to neo-Lamarkianism than to Darwin, whose theory made no allowance for human intervention in evolution and precluded the socialists' belief that they could be the midwives in the creation of "neue Menschen." As we shall see in chapter 6, on the question of birth control the city fathers, led by Julius Tandler, the councilor for health and social welfare, adopted a eugenic view akin to social Darwinism.[6]

The idea of creating a total cultural environment grew gradually and haphazardly out of the socialist city fathers' attempts to bring some relief to the housing crisis. Whereas there might have been a theoretical affinity between these two aims, in practice they were frequently at odds, as the challenge to make available the largest number of livable domiciles (with at least some of the basic amenities of the promised decent life) conflicted with a growing socialist commitment to creating a special kind of living environment for the controlled socialization of the working-class family. Without a doubt the nature of public housing in Vienna, not only because of its obvious visibility but also because of the underlying reasons for its particular characteristics, became the touchstone of the attempt to create a socialist party culture.

Public Housing: Environment for "neue Menschen"

Between 1919 and 1934 the municipality built 63,924 new domiciles in Vienna, 58,667 of which were in apartment dwellings and 5,257 in one-family houses.[7] In practical terms this meant that every tenth dwelling was a new creation of the public authorities and that almost 200,000 Viennese were fortunate enough to reside in them.[8] One cannot but be impressed by the genuine accomplishment embodied in these statistics. The socialists' housing program was truly remarkable, but it hardly made Vienna the "Mekka to which socialists the world over were drawn," as a leading socialist publicist claimed.[9] Almost from the beginning, the genuine attainments in housing and other municipal endeavors were embedded in a mythology that exaggerated their importance and refused to countenance either criticism or a confrontation with the significant social problems which the well-intentioned municipal reforms had been unable to alter. To marvel at the number of Viennese workers accomodated by the new housing, for instance, without taking note of the fact that the number of homeless accommodated in shelters had tripled between 1924 and 1934 to 77,419 per month, or that only 18 percent of worker households had gas, electricity, and running water, while an equal number had none of these, was a common characteristic of all socialist publications.[10] Such obfuscation has continued down to the present day in the work of otherwise responsible historians who tend to make heroic precisely what they have set out to analyze and evaluate.[11]

The socialist municipal government was largely reactive to the housing crisis until 1923. In addition to the reasons already given for the rapid increase in housing seekers, the rent-control law made it possible for crowded worker families to dispense with the subtenants and bed renters they had formerly needed to pay the rent. These persons were thrown onto a housing market in which the rents of subtenancies in larger apartments were not controlled. The dire housing needs were further aggravated by completely substandard accomodations such as cellar flats, barracks, huts, and wagons, which together housed nearly 10,000 Viennese families.[12] Moreover, domiciles in the heavily populated working-class districts consisted overwhelmingly of two rooms (living room/kitchen and bedroom or half bedroom) or less.[13] With tenants protected from eviction and thereby liberated from the previous virtual gypsy life, rising expectations created a

Courtyard in one of the poorest tenements (VGA)

hitherto unknown house-proudness and a desire for privacy and for more space.[14] The cumulative effect of nearly 100 percent occupancy, old and new demands for housing, and popular expectations for a more settled life was a constant increase in domicile seekers from 42,642 in 1922 to 68,175 in 1924.[15] The ready means of dealing with this situation available to the municipal authorities allowed them only to respond to the gravest cases.

In an attempt to literally put a roof over their heads immediately after the war, thousands of workers became squatters on public land on the periphery of the city and proceeded to erect rude shelters of various kinds. Some of these settlements took on permanence with the assistance of loans from cooperative societies and trade unions in the spirit of the garden-city movement emerging in other European cities.[16] The SDAP greeted the settlers' movement with a certain amount of suspicion as a degeneration into petty bourgeois aspirations. At stake no doubt was the party's role as planner and mover, which spontaneous innovation outside the party structure seemed to threaten. After 1924 the municipality took over the building of one-family housing communities, but soon abandoned such efforts on the grounds that they were too costly in comparison to superblock apartment buildings. That the latter were cheaper and therefore served larger numbers cannot be controverted. But one senses that the reasons were more complicated and that the self-management prevailing in the garden-city enclaves was a contributing cause: it was seen almost as a kind of anarchism disrupting the customary channels of party activity. From 1919 onward the municipality attempted to undertake some building with the aid of mortgages, but discovered that the available funds were very limited and the carrying costs much too high to allow for low rents.[17]

The most important means of dealing with the housing crisis available to the city government was a series of emergency decrees of 1918–19 allowing the municipality to interject itself between landlords and tenants. Based on the assumed right to requisition empty or unused habitations, a housing requisitioning law *(Wohnungsanforderungsgesetz)* was passed in 1921 at the national level, giving municipalities power over the housing market. But even before that, the Viennese housing bureau had assumed control over all housing relations, including the registration and assignment of all vacant dwellings.[18] The socialist administration was able to use this law to great advantage in making apartments available on the basis of a point system. By the end of 1923 some 30,000 domiciles had been assigned to tenants; by the end of the program in 1925, the number had risen to 44,838. Through the ingenious use of this requisitioning law, the Viennese housing office was able to alleviate some of the pressure. But many of the domiciles assigned to the needy were already inhabited by them as subtenants, and little additional housing was created by this means. The Christian Social party, however, treated it as a brutal law of expropriation and exacerbated its attacks on rent control to the point of forcing the socialists to give way in parliament in 1925 on the renewal of the requisitioning law.[19]

Despite these valiant efforts to reduce the number of those seeking domiciles, it was already clear at the end of 1922 (with over 42,000 registered) that drastic steps would have to be taken by the municipality to create new dwellings and to improve substandard ones. Two concurrent developments made the times propitious for initiating a housing program: Vienna had attained the status of capital and province with the power to raise certain taxes, and Austrian currency was stabilized through foreign loans, reducing the financial instability of the inflationary period. In the autumn of 1923 the city council passed a unique method of taxation to support the building of 25,000 domiciles over a five-year period; actual building commenced in the following spring.[20]

The creative financial plan of City Councillor Hugo Breitner rejected the traditional method of mortgage financing, with its burdensome carrying charges, and relied on an annual housing tax borne by all householders. Although this tax was steeply progressive, so that the wealthiest few paid an amount equivalent to all working-class contributions, it raised only about 20 percent of a rent tax collected from landlords and passed on to tenants before the war.[21] The resulting building fund was supplemented by luxury taxes levied against all objects and means of entertainment associated with middle-class conspicuous consumption.[22] Despite the ingeniousness of Breitner's scheme, which earmarked the entire income of these two taxes for the building fund, almost half of it had to be furnished from the municipal budget.[23] In their enthusiasm over providing basic needs for workers at the expense of the bourgeoisie, the socialists overlooked the fact that the building program was at least in part dependent on the tax apportionment coming to Vienna as province from the federal government. Their jubilation about this Viennese coup and its constant refrain in the socialist press increased pressures by their opponents on the national level for the abolition of rent control and its associated benefits.[24]

The municipal housing program came into being in 1924 because the housing market had collapsed and makeshift solutions had run their course, and not because a long-standing socialist strategy had been moved to the top of the SDAP agenda.[25] In fact, the party was largely unprepared for the role which the municipal government had to play. Before the war the Austromarxists had paid little attention to piecemeal reforms in the expectation that housing and other social needs would be satisfied by a general social transformation.[26]

Between 1870 and the turn of the century, the housing question had been widely discussed by liberal reformers, who had put forward various schemes for improving the living conditions of the lower classes. One proposal of 1874 called for the construction of "worker barracks" including communal facilities such as laundries, bathhouses, clinics, central heating, and common dining rooms.[27] The Stiftungshof and Lobmeyerhof, two housing complexes for the lower classes created by a private foundation in 1896, were even more exemplary for the socialists as they devised their housing plans in 1923. These two projects occupied only 45 percent of their allotted land (instead of the usual 85 percent), and the remainder was used as an interior courtyard with greenery and gardens. In this typical Viennese interior courtyard various common facilities were grouped, including communal kitchens, central baths, a medical dispensary, a library, a lecture hall, free-enterprise shops, and a male and female old-age home.[28] Two-thirds of the 383 apartments consisted of a living room/kitchen and bedroom. Rents were about 10 percent below the market price, and subtenants were prohibited.

It seems clear that the municipal housing erected by the socialist municipal administration between the wars was not particularly original. It rested on liberal reform ideas and experiments at the turn of the century and, as we shall see, even from the point of view of layout and design continued the courtyard tradition of Viennese domestic and public architecture.[29] To point to such traditionalism is not to diminish the contribution of humanitarian concerns to the socialists' conception of decent housing for the working class.

As in so much else, Otto Bauer set the tone in his projection of the road to socialism published as the SDAP came to power in Vienna.[30] He proposed the means by which the municipality might obtain land, avoid mortgage costs by raising new taxes, and thus assure that the rents for workers would be based on operating costs alone—techniques which the socialist city fathers were to use so effectively to make a substantial building program possible. Claiming decent housing to be a fundamental right of all citizens, Bauer demanded the building of municipal housing projects which would include (in every building block) "central kitchens and laundries, central heating, play and classrooms for children, common dining rooms, reading and game rooms for adults, and the cooks, laundresses, and child-care specialists required for the functioning of these communal facilities." The actual operation as well as the institution, supervision, and control of communal facilities and activities was to be carried out by the tenants themselves, organized in committees constituting a decentralized form of administration. Bauer saw immediate benefits arising from what he called "partial socialization": working women would no longer be victims of the double burden of job and household; children would be better provided for; and at long last men would be able to enjoy a comfortable home.

Once the first ground was broken in 1924, the speed of erecting the housing projects exceeded the estimated annual 5,000 apartments. Before long the first inaugurations of new superblocks began to appear in the socialist press, with photographs of jubilant lucky tenants surrounded by masses of the less fortunate but proud, all being addressed by Mayor Seitz or some other municipal socialist leader in an atmosphere of popular celebration.[31] Indeed, a substantial number of former residents of municipal housing, on record in extensive oral history interviews, attest to the tremendous impact of the growth of what amounted to new worker enclaves dispersed throughout Vienna.[32] Whereas various strands of evidence now inform us about public reaction to the building program, we know very little

Inauguration ceremony of Reumannhof (VGA)

about the planning within the SDAP and the municipal council which must have preceded such a large and costly public enterprise.

A host of questions arise, once one looks at the photographs and plans of the housing developments revealing the architectural styles, size, and facilities of the apartments. How were decisions arrived at? Was there open discussion in the SDAP, and at what levels? Was the subject aired in the press? Were there lengthy debates or controversies in the municipal council? Was any attempt made to survey the potential worker population of these new housing projects to determine their needs and preferences? In short, did the decision-making process reflect some special socialist orientation, or was it left to the good intentions of an oligarchical structure?

Strange as it may seem, these questions have not been addressed in the existing literature. The reason is not hard to find: there is a dearth of sources for a detailed analysis of the decision-making process. One looks in vain through the calendar of discussions of the municipal council for evidence of extensive treatment both of the building program and its objectives, and of the details of its contents.[33] The actual discussions which took place seem to have been desultory, because the socialists knew they had the necessary votes, or because they were interrupted by the Christian Social opposition with outrageous objections raised out of sheer frustration.[34] A case in point is the proposal before the municipal council in 1923 and again in 1925 to help finance a cooperative building with a centralized kitchen and other communal facilities—the famous *Einküchenhaus Heimhof*.[35] The original structure contained thirty-five one- and two-room apartments with a central kitchen and dining room, light cooking facilities in each apartment, a central laundry, and a staff of housekeepers and cooks who professionally performed the normal housework of each tenant.

In other words, the Heimhof conformed remarkably to the partial socialization Bauer had proposed four years earlier. But the socialist rapporteur of the demand for credits to the cooperative never considered the advantages of this model or the feasibility of extending the experiment. He merely responded to an earlier objection by the Christian Social councilwoman, Gabriele Walter, that the building's collective arrangements undermined the housewifely function of women, by arguing that only a small number of people were involved in the venture and that the population at large would not be affected by it one way or the other.[36] Heimhof was again on the council agenda in 1925, when municipal financing for the extension of the cooperative to 246 apartments was proposed. Again, there was no real debate about the merit of this type of housing, only obstructionist arguments from the minority and a demand for the acceptance of an atypical housing venture by the socialist majority.[37]

The apartments in Heimhof turned out to be too expensive for worker budgets, because the construction techniques and maintenance of a single small complex were too costly. But the high quality of life for its fortunate tenants was never in doubt. The expansion of 1925 included a roof terrace with showers, and between mealtimes converted the attractive dining room into a café amply supplied with current reading material.[38] The idea of professionalization of housework in the new building projects of the municipality died with this experiment. But the SDAP had never really presented the positive aspects of this housing model to the workers. One searches in vain through the pages of *Die Arbeiter-Zeitung,* for instance, for a discussion about adapting the Heimhof partial socialization for mass housing. What one finds is the negative assessment of such possibilities by the socialist luminary Otto Neurath.[39] The workers, he claimed, did not want such centralization of personal needs on a communal basis; such innovations could only be realized in the future. But how did Neurath or any socialist party functionary or municipal councillor know "what the workers wanted"?

Whether working-class women understood the possible advantages of professionalized housework (especially communal kitchens) remains doubtful. Leichter's study of industrial workers reveals a great deal of confusion about what such socialization would involve.[40] Some women expressed the fear that it would rob them of the individuality and feeling of control experienced at home, replacing it with the monotony and complusion they experienced in the workplace; others thought the cost would be too high. Younger, single women were more favorably disposed to the idea. But none seemed to be well informed, to have read about the possibility of combining family individuality and collective facilities, or to know about the existence of Heimhof.

Throughout the two building periods from 1924 to 1933, when the 377 housing projects were planned and built, the SDAP failed to conduct a single survey either at workplaces or in working-class neighborhoods about worker expectations and needs concerning the housing being created principally for them. It was not the lack of trained social scientists in sympathy with the SDAP which prevented such investigations, for they were carried out on other subjects at the direct or indirect request of the party leadership.[41] It stemmed from the transformational heritage of Austromarxism according to which the workers had to be educated by the party to reach a higher state of being than their present precivilized or at best unformed state (see chapter 2).

That explained why the workers did not need to be consulted about the domiciles being prepared on their behalf, about other aspects of the municipal reform program, or about the institutional party network fashioned to bring workers into a proletarian culture that would both transform and incorporate them as "neue Menschen" and "ordentliche Familien." It remains one of the tragic ironies of this period of great expectations that, despite the genuine commitment to democracy by Bauer and other principal leaders of the SDAP, they failed to translate this belief into action by allowing the workers to behave as subjects in what was, after all, a common enterprise. Like virtually all interwar socialist parties, where the olgiarchy "knew best" and always "acted in the common interest" of party and rank and file, the SDAP operated as a parternalist machine.[42]

Lest this judgment be considered excessively harsh, let us take a quick look at the organization and decision-making bodies of the SDAP. The party structure remained remarkably stable from its creation by Victor Adler before the war until the fall of the republic. It followed a common form in which a powerful executive made all decisions, operated through a secretariat for their implementation, and based its authority on annual party congresses.[43] The power of the executive was transmitted down the hierarchical ladder of the party to the smallest organizational units: the district, street, and house. Lower levels had no direct access to party policy; their only connection lay in the delegates elected to party congresses from their sectors. But only slightly more than half of such delegates were actually elected; the remainder were assigned along oligarchical principles to members of the executive and secretariat, the parliamentary and provincial socialist delegations, heads of the trade unions, and others from within the party machine deemed to be "essential."

That the career socialists clearly ran these party congresses between 1919 and 1933 can be seen from the following telling examples[44]: not a single candidate for the executive was ever put forward who had not been proposed by the nominating committee; not a single candidate on the official list was ever not elected; nor was a member of the executive ever recalled from office. The SDAP executive was an exceptionally stable oligarchy. Of some twenty members and alternates active between 1919 and 1934, four retired because of old age and five died in office at an advanced age.[45]

In the early 1930s the SDAP had close to 1,500 paid functionaries commanding the party's political, publication, and cultural infrastructure. To these must be added some 21,500 unsalaried cadres *(Vertrauensmänner)* in Vienna alone, without whose dedicated performance of daily routines the far-flung activities of the party would have been unthinkable. Through its control of the municipal government the party leadership disposed of an unknown but certainly sizable number of civil service jobs in the administration of the various city bureaus, the most important and lucrative of which were assigned to major party functionaries.[46] In addition to such old-fashioned patronage, the party leaders also designated candidates for national, provincial, and municipal elective office. The sociopsychological effect of this economic structure of the party—well-paid functionaries, unpaid volunteers, and the working-class rank and file—was to increase the distance between the decision makers and those for whom they spoke and acted. That the salaries of paid party functionaries were two and a half to four times the wages of skilled metalworkers at a time (1932) when virtually every third party member was unemployed, helps to explain the subject/object attitudes among party leaders.[47]

The reasons for this brief detour into the structure of the SDAP will soon become apparent. In initiating a high-quality municipal socialism as well as the much more ambitious experiment in working-class culture emanating from it, the socialists were faced with difficult choices. Those controlling the city government had to develop their housing and other reform programs with the slender tax resources of an impoverished country where wages were about 35 percent below those in Germany.

Little could be done by the municipal council to alter this harsh economic reality. Yet at the same time the socialists had tremendous power in Vienna to use these resources on projects of their choice and in ways they found suitable. We have established that the SDAP's subjective choices were made by a small oligarchy. Unfortunately, we can go no further in designating the real wielders of power, except to remark that Otto Bauer was certainly central to all important decisions. A further refinement is made impossible by the dearth of sources about such important leaders as Bauer, Danneberg, and Seitz.[48] About the hopes and desires of the mass of Viennese workers our knowledge depends on oral histories of aged survivors. The SDAP simply made no provision for communication among the party base through local newspapers, forums, or initiatives from within the new housing projects. The weekly meeting to approve the programs of the party leadership discouraged the asking of questions or grass-roots initiatives.[49]

By 1923 it had become clear that something substantial had to be done by the municipality to deal with the housing crisis. But this problem, which demanded a creative and rapid solution, was far from simple. A host of important questions had to be confronted and answered: Should the worst slums be cleared and replaced? Should some of the decaying private housing stock be acquired by the city and renovated? On the same basis of acquisition, should basic utilities such as electricity, gas, and running water be introduced in a substantial number of apartments that lacked some or all of these? What proportion of the newly created building fund, based in large part on the creative new taxes, should be devoted to the building of new houses and the other options outlined? Unfortunately, we have no way of knowing what discussions took place at the highest party levels, since only the final decisions have come down to us.

The socialists in city hall decided to build 25,000 new apartments in five years (1924–28) and thereby to oppose a number of "peoples' palaces" to the *Zinskasernen* (tenements) which were the characteristic dwellings of the Viennese working class.[50] To justify this course of action, the socialists exaggerated the deficiencies of worker housing so as to make it all seem like an undifferentiated slum.[51] Whereas there were a few real slums in Vienna, such as the infamous "Kreta" in the 10th district,[52] as well as cellar flats and barracks in other working-class neighborhoods, the large majority of tenements were quite durable structures (surviving down to the present) in quite well-kept if not very attractive streets, and with many a green lawn or garden in the vicinity. The main problem with living in most but not all of these tenements was the overcrowding of flats and the absence of the aforementioned basic utilities.[53]

In the deprecations of all tenements as unfit habitations, one finds no reference to the fact that a large part of the Viennese petty bourgeoisie (skilled artisans, small shopkeepers, lesser white-collar employees) lived in the very same places as many of the workers, and together with them constituted neighborhood communities (with all the usual complexity of friction and mutuality). The socialists' choice to build only new structures, therefore, was based on painting the existing housing conditions so black that only enclaves of new projects scattered throughout the city could provide both healthy housing and spiritual uplift.[54] It would not be an exaggeration to say that renovations were never on their agenda,[55] simply because housing for the socialists represented the framework of a far larger undertaking: the organization and molding of the working class into a new cultural form, opposed and superior to the dominant bourgeois one. If Vienna was to be a larger cultural laboratory, the municipal housing project was to serve as the crucible for the experiment. In making this choice, the socialists were forced to abandon other possiblities so far as housing was concerned.

The matter of making choices did not end with having decided to build 25,000 new domiciles in project-like complexes. The process of construction itself became an issue, because the private building industry had collapsed and the municipality was in a position to do virtually what it liked— a truly unique circumstance. Since 1919 the municipality had been buying land in the city at ridiculously low prices, because inflation combined with rent control made it seem unlikely that any money could be made in real estate. By the beginning of 1924, therefore, the municipality was already the largest landowner in the city, guaranteeing a basic very low land cost for its building program and allowing it to choose housing sites within the city close to the existing infrastructure.[56]

As the only builder in the city between 1919 and 1933, the municipality was in a position to take over the entire industry and restructure it away from a market to a communalized form. In the absence of any serious economic competitor, such a "municipal socialization" did not face any more serious obstacles than other parts of the socialists' reform program. As sole customer, the municipality succeeded in influencing the pricing policy of some firms supplying building materials.[57] But it made no attempt to replace the host of small construction companies by encouraging the formation of production cooperatives, or to consolidate and control them. By failing to do so, it kept these less-than-efficient private enterprises alive. It seems clear that the SDAP's housing policy was expedient and temporary, forced on the municipality by rent control which brought private building to a standstill. The SDAP appears not to have considered the possibility of a long-range strategy to partially alter the economy of the city through the quiet socialization of an industry.[58] Given the small size and large number of companies, this would not have been easy. Even so, the feasibility was not considered.

Perhaps this failure to capitalize on socialist possibilities created by the city's housing program can be explained by the antimodern architecture chosen, as well as by the building materials and construction methods used. Functionalist architecture, which made its appearance at this time in a considerable number of European cities in radical forms or moderated adaptations, was not chosen by the socialist city fathers as appropriate to symbolizing the worker dwellings they were constructing.[59] A "Viennese mélange" of art nouveau, art deco, and pseudomodernism with feudal stateliness and baroque decorations, all adding up to a remarkable monumentalism for the public housing of a rather poor city, seems to have been preferred by the city fathers.[60] The socialists' taste in architecture as well as other aspects of high culture is something we will turn to later in this chapter as well as in a digression of chapter 4.

This mixture of architectural forms, devotion to outmoded decorations, and preference for the monumental reflected a taste for traditional Viennese styles on the part of members of the municipal housing bureau and leading socialists. In all likelihood this taste was shared (or aspired to) by the workers, for whom functionalist styles probably would have seemed stark and strange. Two further considerations merit attention. First, the leaders made their choice without consulting the workers, who were far less concerned with the outward appearance of buildings than with the size and quality of apartments and their amenities. Second, by and large the functionalist architects showed little concern for the users of their structures, which were ends in themselves; they also viewed tenants as objects.[61]

The socialists' decision to realize their housing program largely through the existing building industry, and their rejection of radical departures from existing architectural styles, in large measure determined the construction materials and methods to be used. The city fathers opted for brick and mortar on the grounds that this labor-intensive technique would provide employment to an additional 11,500 workers a year.[62] Such an employment bonus to the housing program tended to cut off discussion, but it should not have.

Monumental central structure, Reumannhof (VGA)

There is no evidence that the municipal council investigated the experience of Frankfurt or Berlin, not to mention Hamburg, Lyons, Amsterdam, or London, in using newer building materials such as pressed or preformed reinforced concrete ceilings, floors, and other structural as well as secondary building parts.[63] In Frankfurt, where all the public housing was modern/functionalist, all these were in use as well as standardized windows, doors, ovens, bathtubs, kitchens, and lighting fixtures. The newly prefabricated structural parts were assembled at the building sites by specially designed cranes on a year-round basis.[64] Brick construction in Vienna was limited to eight months of seasonal temperature and was extremely slow.[65] It was apparently also more costly per square meter and unit than was the case in Hamburg and other German cities.[66] The fact that the Frankfurt building trade unions, usually very suspicious of new techniques as being "labor-saving," supported the new construction materials and methods suggests that employment was probably also increased by them.[67] With these comparative situations in view, one begins to doubt the socialist city fathers' arguments that their noninnovative housing program was cheaper and faster, and that it created more jobs.

Whatever reservations one might have about how the municipal housing

Interior courtyard, Sandleitenhof with 1,587 apartments (VGA)

was conceived or carried out, the most significant of the 370 structures continue to make their imposing presence felt. At the time these were called "peoples' palaces," reflecting both their monumental appearance and their popular use.[68] Despite a great variety of architectural styles, the basic courtyard orientation was used, giving buildings and complexes an inward-turned, both protective and excluding aspect.[69] One could quite easily and without exaggeration view them, as the city fathers did, as proletarian oases in which sun and light, space and color set the tone of a new form of decent and dignified living.

But whereas there was a striving for monumentalism in the exterior of the projects, the interior of the apartments suffered from minimalism. In the first program of 25,000 units, 75 percent had 38 square meters (410 square feet) of space, and 25 percent had 48 square meters (518 square feet), typically with a living room/kitchen and additional bedroom or half bedroom. In the second program, after 1928, the majority of apartments had 40 square meters (432 square feet), while a smaller number had 49 or 57 square meters (529 or 615 square feet). The typical layout in this later group reduced the kitchen to a functional small room separated from a living room. Standard in all apartments were electricity, running cold water, gas for cooking, a toilet with a foyer separating it from the other rooms, tiled kitchen and toilet floors, and hardwood parquet flooring in the rooms.

No doubt these municipal apartments represented a considerable physical improvement over the typical tenement habitation. But they also fell

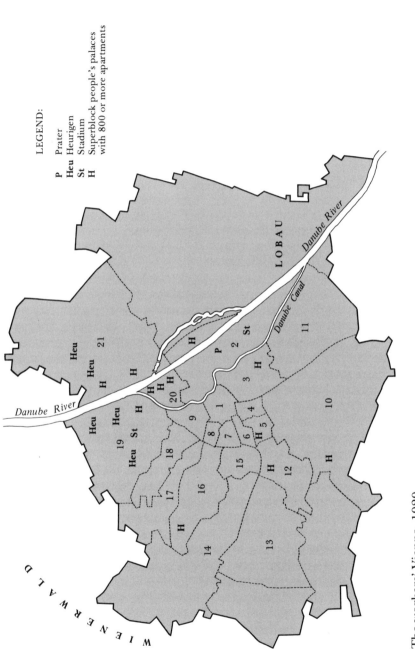

The workers' Vienna, 1929

short in important ways of being the revolutionary departure in worker living space extolled by SDAP spokesmen at the time.[70] The quality of municipal apartments was greatly diminished by the absence of private bathrooms, hot water, and central heating, which greatly increased the housework of women, who were obliged to haul coke and ashes to and from the basement (as they had done in the tenements), and to keep vats of hot water on the boil for long hours to meet the needs of various forms of washing (dishes, clothes, floors, and so on) and bathing.[71] Private bathrooms were not feasible, given the small size of apartments and probable high unit cost. But the failure to provide central heating and hot water could not be explained away as being too expensive. Could the price of installing these facilities have significantly altered the rent structure? Would the tenants have been prepared for a small surcharge for such conveniences? One wonders, considering the installation of expensive parquet flooring, whether the rationalizations offered did not simply cover up the faulty understanding of domestic facilities and tenant needs by some (male) planners.

The fact remains that in Frankfurt, Berlin, and Hamburg, to mention but a few contrasting examples, the municipal flats built were larger (fifty to sixty square meters) and included hot water, bathrooms, and central heating.[72] The shortcomings of the Viennese apartments, which the socialist reformers attempted to explain away, most probably stemmed from the construction methods used, as well as from what appears to be resistance to technical innovation. The so-called Frankfurter Küche, a modular built-in kitchen, was designed by the Viennese architect Margarete Schütte-Lihotzky but manufactured in Germany for 238 Marks and widely used in German public housing.[73] No doubt this "kitchen as rational factory" went counter to the large living room/kitchen to which Viennese tenement dwellers were accustomed. But the municipal building program of 1928 created apartments with a tiny kitchen space (lacking the advantages of the Frankfurter Küche) and provided the worker tenants with an unaccustomed living room, neither of which they had desired.

In part, of course, the shortcomings in individual domiciles were compensated for by a wide array of communal living facilities. These included mechanized laundries, bathhouses, kindergartens, playgrounds and wading pools, meeting rooms, medical and dental clinics, libraries and lecture halls, shops of the consumer society, and youth and mothers' consultation clinics.[74] A full array of such amenities was available only in the largest projects. Small and medium-sized ones shared a limited number of collective facilities between them. Central laundries and bathhouses tended to be overtaxed with use.[75] Many municipal housing tenants were forced to seek showers in other projects at quite some distance from their homes, or to rely on the old municipal bathhouses *(Tröpferlbäder)* which continued to serve tenement dwellers.

It must be borne in mind that where these communal facilities existed, they required the payment of small fees (laundry, bathhouse, kindergarten, and clinic) and strict adherence to rules established by the central housing

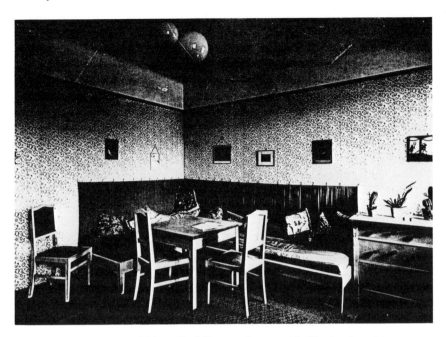

Model living room in Karl-Marx-Hof. Few workers could afford to buy this functional furniture. (VGA)

office.[76] The city fathers made no attempt to experiment with providing even partial professional domestic help in any of the projects. That the idea of a central kitchen and dining facility (or similar "partial socializations") did not appeal to the leading socialists can be seen from their personal choices in habitation. In a building on the Albertgasse, constructed for senior socialist party and municipal employees, the typical apartment consisted of four rooms plus a kitchen, bathroom, toilet, and maid's room.[77] The SDAP elders were not inclined to experiment either in architectural styles or in new living arrangements. Their preferences, values, and tastes were far closer to an inherited "bürgerliche Tradition" than their Austromarxist amour propre would allow them to admit.

Who was actually lucky enough to become the proud tenant of a new municipal apartment? The consensus seems to be that younger working-class couples with one or two children made up the majority.[78] Since the initiation of the housing-requisition law in 1920, the municipality made use of a point system to determine the allocation of domiciles on the basis of need. The same point system was used in selecting tenants for the new municipal projects, and the socialists were emphatic in insisting that the selection was fair and untinged by political considerations. Opponents claimed the opposite, charging that the housing office, which made the final choice among applicants, operated in secret and by its bureaucratic demands discouraged those who did not have the correct political connections.[79]

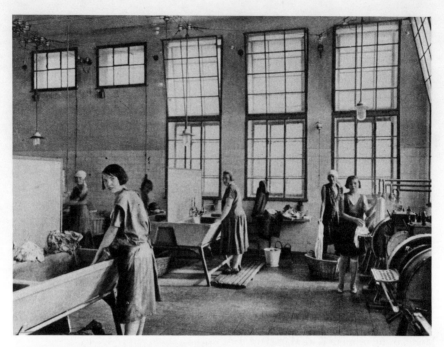

Communal laundry, Sandleitenhof (VGA)

A closer look at the actual schedule of points seems to suggest that the socialists faced a dilemma. Their hopes for creating a working-class culture based on municipal institutions rested on young worker families less encumbered by the past and more receptive to the multifaceted education being prepared for them. But the old point schema (in which ten points put an applicant in the highest category) allotted five points to those living in uninhabitable quarters and five points to invalids as well.[80] If the scheme had been strictly enforced, it would have populated the municipal projects with many of the poorest, least skilled, socially most unsettled and needy, and physically most frail members of the Viennese working class.[81] Such a strict adherence to the rules also would have completely discounted the important voluntary work done by cadres of the party. It would seem that a certain amount of political patronage and, above all, selection of the "more promising" worker families was the actual and quite reasonable practice followed.

The very small size of apartments and the prohibition of subtenancies revolutionized the family structure in the new municipal projects. The open family of the tenements had consisted of diverse relatives, subtenants, and bed renters, in cheek-by-jowl proximity. They had formed an intricate network of social relationships characterized by both friction and mutual aid. Oral histories of life in the tenements abound with recollections of socializing with neighbors on the landings and in the stairwells, on the sidewalk in front of the building and on nearby lots and lawns.[82] These reminiscences go beyond the daily contact points at the hallway water faucet and coal cellar to the exchange of services such as baby sitting, and to large festivities involving collective dining, singing, and dancing among relatives, friends, and neighbors.[83]

The nuclear family of the municipal houses—parents and one or two children—experienced an unaccustomed privacy which alternated with necessary participation in highly controlled public facilities.[84] Municipal housing thus created two forms of the worker family: one was isolated as a small family within its four walls and basic utilities, and shut off from spontaneous peer contact through landings with only two to four apartments and narrow stairwells; the other was part of the large building family of shared facilities representing the community, their class, and their party.[85] Some aspects of daily family life remained impervious to the new surroundings: parents and children continued to share the same bedroom; the living room/kitchen in the smaller apartments remained the center of family life; and the large pieces of furniture from tenement days dwarfed the rooms and could not be replaced with the costly modern, modular units praised and recommended in the party publications.[86]

It would surely be an error to view the municipal housing program of the SDAP merely as a means of improving the life-styles of the workers—of providing additional conveniences, space, air, light, and so on. Judged in this respect alone or primarily, this part of municipal socialism is open to serious criticism on the basis of what alternatives were possible in Vienna and accomplished in other cities. The "peoples' palaces" were from the first intended to be more than better housing. They were to provide the all-important environment in which the worker family would be socialized so as to become *ordentlich* and be educated by an emerging party culture in the direction of "neue Menschen."[87]

Life outside the cell-like apartments was strictly regimented by the housing management. The paternal managerial structure, which drew its authority directly from the housing bureau of the municipal council, included a concierge charged with prescribing and enforcing building rules (the time and place to beat rugs and deposit refuse; how and where children should play in the courtyard; the appearance of hallways, cellars, and balconies; etc.).[88] There was also a laundry supervisor who scheduled the monthly wash days of each family, kept all but the women out of the washing facility (on the prudish grounds of protecting female modesty), and supervised the use of machinery; an apartment inspector who made monthly visits to all domiciles to ascertain their state of maintenance and to receive reports of infractions of the rules from the concierge (children playing on the grass in the courtyard were duly marked down in a book of infractions); and an array of "experts" in the clinics, consultation centers, kindergartens, and libraries whose function was above all tutelary.[89]

The tenants of the new municipal housing were confronted with structures, spaces, rooms, facilities, and rules of operation devised for them, all in place and impervious to influences or demands from below. It was not so much that private initiatives by tenants to regulate and control their collective living spaces were discouraged, but that no mechanism for their expression had even been conceived.[90] To be sure, tenants' committees were elected in each of the housing projects, but their function was vaguely advisory. The will of the housing office was transmitted to the housing project level by stalwart party cadres. There was no mechanism for the grumbling among tenants to be made public, save the tenants' meetings, which were choreographed by the SDAP.[91] A considerable number of communist newsletters, hastily conceived and hectographed, made the lack of democracy in the management of the municipal houses their leitmotiv.[92] But it is very doubtful whether their exaggerated broadside attack on the SDAP had much effect on the conditions complained about.

The socialist-dominated municipality's rather simplistic approach to complex social structures and problems, and its reliance on so-called "experts" to bring about a "better" working-class family, were based on a disregard for the subjects' own framework of experience.[93] Small wonder, then, that municipal socialism was often viewed as regimentation from the top by its beneficiaries themselves, a condition to which workers were already subjected in full measure at the workplace. The socialists' propensity to act in loco parentis reinforced the paternalistic/authoritarian tendencies present in the socialization of working-class families in the prerepublican era and in the contemporary dominant culture. Ironically, here in the cradle of the "neue Menschen," initiatives of the smallest variety, such as planning and carrying out a demonstration to mark the historic opening of the Karl-Marx-Hof, were throttled by the party in the name of discipline and control.[94]

To say that the socialists were immodest about their housing accomplishments would be an understatement, for they crowed their uniqueness from the rooftops at home and abroad.[95] In their Olympian stance they failed to recognize or credit socialist housing programs in other places. They did not take note that Hamburg also used a building tax to partly finance a larger number of worker flats; that the brilliant socialist architect and planner Ernst May developed a general housing plan for Frankfurt based on the harmony of form and function in a new design of worker urban living; that the socialist city government of Villeurbanne-Lyons had created spectacular worker skyscrapers, and the suburban Parisian "red belt" of worker housing had been constructed against great odds; or that worker housing in England was increased by one-third between 1919 and 1939, with massive slum clearing in London.[96]

Such a comparison might have found the Viennese socialists' municipal housing program lacking in various ways and certainly less exceptional. It would not, however, have detracted from its accomplishment of providing the environment for a potentially unique political culture. Education of the workers was to be carried out not only among the limited number fortunate enough to have won a place in the "peoples' palaces," but among all of Vienna's workers, who would be drawn to them as symbols of a working-class presence and strength. This strength was exemplified by such architectural fortresses as the Karl-Marx-Hof and Karl-Seitz-Hof, in which massive walls and huge archways protected the inhabited interior from the outside world.[97] The visual force of these enclaves throughout the city contributed to a sense of political power among the workers, encouraged by the socialist leaders, that was more apparent than real.[98] It is this vastness of vision, pretension even, to fashion a new proletarian culture in its context that made the Viennese municipal housing a far more impressive totality than the sum of its parts.[99]

Public Health and Social Welfare: Shaping the "Orderly" Worker Family

The same environmentalism which underlay the socialists' housing program—the concept of creating an enclosed and protected living framework in which the worker family could be assisted to a higher standard of civilization and a new humanity—was the guiding thought behind the socialist city fathers' approach to health and welfare. Their mission became not only to reform these by extending the very limited programs initiated during the monarchy, but to change their focus and ultimate purpose.

When the socialists assumed control over the city administration in the summer of 1919, the ravages of war were everywhere apparent: the virtual breakdown of public sanitation; a population weakened by four years of malnutrition; the danger of epidemics; a sharp increase in the traditional killer disease, tuberculosis, and in venereal diseases; overcrowding of less than adequate hospital facilities; a sharp growth in the number of indigent and homeless; and a general shortage of fuel and foodstuffs needed for a return to normal public health. The socialist municipal government moved quickly to arrest and reverse these adverse conditions mainly by investing more public resources in the expansion of clinics, family assistance programs, and aid to children. Closely associated with measures to arrest the deterioration of public health and welfare was a drive for cleanliness and hygiene in public places, made possible by the introduction of sprinkler trucks and a new method of mechanized garbage collection.[100]

Socialist approaches to these problems remained piecemeal and lacked a focus until the summer of 1920, when Dr. Julius Tandler became city councillor for welfare. He came to his office with the experience gained in public service as undersecretary for public health in the short-lived coalition national governments. Tandler was a distinguished anatomist, one of the few Jewish chaired professors on the medical faculty of the university, and a man with strong socialist and scientific beliefs.[101] With an enlarged budget at his disposal, made possible by the new Breitner taxes, Tandler proceeded to alter the perspective and practice of public health and social welfare.[102] In place of the notion that health and welfare were matters for Christian *caritas* or other private charitable organizations, the socialists in the municipal council adopted Tandler's view that health and welfare were the right of every citizen.[103]

Although this view seems in many ways exemplary and humanitarian at first sight, its explication as both theory and practice aroused considerable resistance not simply from the church or the Christian Social party, where one might expect it, but also from the workers in whose interest it was developed. In subsequent publications Tandler went to great lengths to explain and justify his approach, in order to allay what he believed (quite correctly) was suspicion among the workers.[104] What emerged from these explications were a number of advertised principles underlying the Viennese welfare system: that society is committed to assist all those in need; that individual welfare assistance can be administered rationally only within the context of family welfare; that constructive welfare aid is preventive welfare care; and that the organization of welfare must remain a closed system.[105]

What were the practical accomplishments to which the municipal council pointed with great pride as being remarkable? The decline of the death rate by 25 percent and of infant mortality by 50 percent from prewar levels stood high on the list.[106] The incidence of tuberculosis, which had been rampant particularly in the working class, was only somewhat reduced and continued to be the major threat to health among schoolchildren.[107] A comprehensive system of aid to children was put in force.[108] It included school lunches, school medical and dental examinations, provisions for publicly sponsored vacations and summer camps, and newly created after-school centers. The number of kindergartens increased significantly from 20 in 1913 to 113, with almost 10,000 children, in 1931.[109] Municipal bathing facilities including swimming pools, with some 9 million patrons in 1927, were also high on the list of attainments in public hygiene. Prophylactic medical examinations for adults and children in municipal clinics reached 123,000 in 1932, and welfare workers carried out 91,000 home visits in the same year. No doubt this was a commendable record of accomplishment for the municipal administration, but it was by no means as unique as the socialists claimed. If we compare it to a number of contemporary German cities, Düsseldorf for instance, we find an almost identical roster of health and welfare measures and achievements.[110]

Like other aspects of Viennese municipal socialism, it was the instrumental role of the health and welfare programs in the lives of the workers that gave them a special character. Under the forceful direction of Tandler, the welfare department pursued an overall policy of population politics. It assumed responsibility for improving the quantity and quality of the population at large. This mission was predicated on the duty and power of the public authority to intervene in the life of the family. Population politics, particularly concern about the steady decline in population and the need to add not only to the number of workers but also to improve them biologically, was a major concern among leading socialists in the party and municipal government.[111]

Socializing schoolchildren for dental care in clinics (VGA)

Tandler's own version combined elements of neo-Lamarkian and social Darwinist ideas, proposing that changes in the human environment could be transmitted through the germ plasm and that a "natural selection" carried out by responsible officials would enhance and improve the genetic pool of future generations.[112] At times Tandler slipped into eugenic fantasies of sterilization and other means of reproductive denial by society, quite frightening in their implications.[113] Finding no contradiction between these ideas and his commitment to socialism, he set out to fashion a powerful organization of social intervention to put population politics into practice.

Under the auspices of the Public Welfare Office, a number of institutions were created to assist the family—considered the germ cell of a healthy population—in its task of rearing the next generation. Where the family failed to provide optimal conditions, the Public Welfare Office was to provide temporary or alternative care. The municipality thus empowered itself to remove children from their parents, if it judged them deficient in their nurturing capability and responsibility.[114] Tandler was attacked repeatedly in the municipal council by Christian Social members who accused him of alienating children from their parents in order to indoctrinate them with socialist ideas. His stock reply was that he considered the family sacred, but only if it was capable of performing its vital function.[115] Under Tandler's direction the Public Welfare Office put the population under surveillance with the argument that preventive welfare, aimed at raising the moral climate of families, necessitated that it act in a supervisory capacity.[116] The methods employed combined persuasion with compulsion, voluntary cooperation with juridical force.

The realm of municipal family supervision was organized in lockstep virtually from conception to adulthood, when the cycle continued as the former child became a parent. The most original and controversial agency was a marriage consultation clinic created in 1922. Its function was to advise couples intent on marriage about their sexual health, genetic deficits, hereditary weaknesses, and prospects for producing normal and healthy children.[117] The clinic offered to issue certificates to prospective conjugal sexual partners that they were free of disabilities such as syphilis and tuberculosis and hoped thereby to improve the quality of the population. But very few individuals were prepared for such intrusions into their private lives, and after ten years the venture was admitted to be a failure. Throughout its existence the clinic met with violent opposition from the church and the Christian Social party ("the Jews are touching the holy state of Christian matrimony").[118] Tandler's true intention in founding the clinic—that of using marriage consultation for weeding out those "eugenically unfit for reproduction"—and the possible legal misuse of the case records no doubt did not escape the general public.[119] The clinic made a point of refusing to have anything to do with sex counseling or birth control advice, subjects which might have made it attractive and useful. The SDAP approach to these sensitive subjects is taken up in chapter 6.

The marriage consultation clinic was the only failure; all other agencies of the Public Welfare Office concerned with the life cycle of the Viennese family were successful in promoting its aims. At the beginning of family control were the municipal hospitals, in which 83 percent of all births took place.[120] Social workers in the maternity wards registered the newborn infants, arranged for a subsequent home visit, and recommended that mother and child regularly attend a mothers' consultation clinic for further assistance in infant care. By 1927 there were thirty-four of these clinics, concerned with infant and child care to the age of six, distributed throughout the city.[121] Doctors advised mothers on breast feeding and infant and child care and hygiene, and resident social workers followed up these instructions with home visits to see that consulting mothers had carried them out.[122] This was but one way in which the Public Welfare Office found entry into the home in order to observe and judge the adequacy or insufficiency of family nurture for the children. The right to regular inspections of families receiving any kind of municipal assistance was statutory and, as we shall see, gave the welfare machinery tremendous power not only in dealing with individual cases but in setting the norms of family health and behavior.

In 1927 Tandler proposed to the Municipal Council that the Public Welfare Office be granted the right to distribute, regardless of need, infant layettes to all newborns as a "birthday present" from the municipality. After an extended and heated debate in which the socialists were accused once more of using city hall to make political propaganda, the measure was forced through by the socialist majority.[123] These gifts were packed in attractive red cartons with a reproduction of a famous mother-and-child image by the sculptor Anton Hanak on the cover and a listing of the thirty-four mothers' consultation clinics with addresses on the inside. Before long, some 13,000 of these parcels were being distributed annually.[124] Without diminishing the virture of such "need-blind" distributions, it is necessary also to consider the "Trojan horse" aspects of these gifts. Their distribution by social workers made it possible for the Public Welfare Office to look into homes which were otherwise outside its purview.[125]

By and large, however, the welfare authorities did not depend on invitations to pass judgment on family life. The city council as early as 1921 claimed guardianship over children born out of wedlock, foster children, and those in institutional care.[126] After this group, subject to the most intense form of control, came all those who received public assistance in any form. Such families were subject to regular visits by social workers who kept close watch on the standard of housekeeping, especially cleanliness, the condition of beds and clothing, food preparation, and family relations. Adolescents with problems were referred to or expected to seek assistance at youth consultation clinics throughout the city. Families were also subject to home visits because the school doctor had reported some health problem of their children, or because court proceedings—for eviction, for instance—drew attention to the family as being "troubled."[127] Kindergartens and after-school youth centers worked hand in hand with welfare efforts to produce the orderly family. Special attention was paid to the professional training of

A grand total of 53,000 municipal layettes had been distributed by 1931. *(Der Kuckuck)*

social workers.[128] They were exclusively women, because Tandler and most leading socialists believed that a special "female empathy" was necessary for the often emotional demands on welfare workers.[129]

Finally, we come to the question implicit in all the welfare activities: what happened when the social worker making a home visit concluded that the

Municipal Socialism 71

family did not meet the municipalities' norms of the orderly family? A report was made to the childrens' diagnostic service *(Kinderübernahmestelle)*, a modern observation center under the direction of the child psychologist Charlotte Bühler, charged with deciding the fate of children from problematic environments. A court order was issued requiring the parents to surrender the child to the diagnostic service, which in the course of some four weeks would make a recommendation that had the force of law.[130] The child might be put in the care of foster parents, be sent to a children's home or correctional institution, be admitted to a hospital, or be returned to the parents. Of the various reasons given for remanding the child to the diagnostic center, relatives admitted to a hospital, poverty, and homelessness ranked highest, and neglect and delinquency were frequent, whereas endangered morals and parental conflict were rarely mentioned.[131]

How working-class communities reacted to the municipality's attempts to transform them into the orderly families of its definition is not easy to discern. Contemporary means of grass-roots communication among workers were virtually nonexistent. Once again we have only a scattering of local communist newsletters which, despite their expected political line, throw some light on how the welfare system worked and was perceived from below. In these, complaints were all centered on the behavior of social workers and other municipal officials: their treatment of the needy, their arrogance in dealing with sick workers needing medical and pharmaceutical referrals, their tendency to treat workers like children, and the poor functioning of kindergartens.[132] There were also claims that welfare payments to families were being cut on spurious grounds, as for instance a family's having "gained property" in the form of a bag of apples, so that it needed less support. More directly aimed at the subjective judgments of the welfare department was the charge that a social worker took a child to the police rather than dealing with the mother, who was estranged from her husband.[133]

Oral histories of working-class families and social workers of the First Republic depict public welfare as a coercive system. The judgment of what was respectable, orderly, and decent on the part of social workers was arbitrary.[134] Very often the fact of being poor—a condition ever more widespread after the impact of the depression and mounting unemployment—was sufficient cause to put a family on notice for further investigation, or to threaten to send children to the diagnostic center if a mother did not improve the general appearance of her domicile, the condition of the beds and childrens' clothes, or the family meals.[135] But, given their increasing poverty, Viennese workers had to pawn household objects and bedding, cut corners on food by serving children bread spread with lard *(Schmalzbrot)*,[136] and repair all clothing items to the point where they were ragged.

The tendency of the welfare system to look upon the poor, the unemployed, those evicted, or those who consumed a glass too many, as deviants or at best as marginal families needing the full measure of social control to bring them up to normal standards, made welfare workers appear as hostile government agents in working-class communities. Indeed, the removal of children to the diagnostic center, as described by one social worker, resembled a police raid, with social worker and bailiff arriving unexpectedly to reduce the amount of parental resistance to what the parents considered a violation of their rights.[137]

The socialists had put their health and welfare program into effect against the vociferous objections and condemnation of the Christian Social party in the municipal council. It is interesting to observe that the whole *Kulturkampf* being waged did not arise from differences about the right or necessity of higher authorities to intervene in the life of the family. It was the nature of the intervener which lay at the heart of the struggle: whether it should be the state acting through the agency of a socialist municipal government, or the church acting through its apostolic spokesmen. Neither side paid much attention to the similarity of positions; only the differences were fought over. Invariably, as in most conflicts between the two political camps, anti-Semitism was a weapon of choice for the Christian Socials. The position of the SDAP in view of these attacks was to attempt to deflect them without meeting them head-on.

It is to Tandler's credit that he provided a rare example of unabashed resistance to anti-Semitic slander used by the clerical forces. The incident involved the construction of a municipal crematorium in 1923, at the request of various citizens' groups including the workers' crematorium society, Die Flamme.[138] The archbishop's pastoral letter responded to Tandler's proposal by threatening excommunication to Catholics who took a hand in such an un-Christian enterprise. The fierce debates in the municipal council included a large measure of anti-Semitism directed against Tandler and the SDAP leadership. After the first cremation had taken place, the Seipel government directed Mayor Seitz to close the crematorium. He refused on grounds that he was bound by the decisions of both the municipal council and Viennese provincial senate. The dispute, having turned constitutional, was handed over to the highest court, which ruled in favor of the province of Vienna.[139] The conflict demonstrated that the church was not invincible and that a determined stand against anti-Semitic attacks could prevail (at least in 1923). But this remained an isolated case, as did Tandler's courage in not giving way to hate mongering. The SDAP remained leery of challenging the church or of fighting anti-Semitism with a raised visor.[140]

The socialists' population politics was an explicit aspect of the Viennese system of welfare. Environment and biology were to be combined to produce "neue Menschen" and orderly worker families.[141] The municipality's system of social control included welfare and health care, the police and judiciary, as well as such pedagogical institutions as kindergartens and after-school centers. The most intense application of the combined force of these institutions took place in the largest municipal housing projects, where branches of many of these agencies (mothers' and youth consultation clinics, kindergartens and health clinics, etc.) were part of the communal facilities. In short, the socialists' conception of welfare as well as of public housing was predicated on the creation of a superior environment (domicile or family) through intervention by leaders and experts in the daily lives of workers. The notion that workers were malleable objects ready for transformation from above harked back to a presumption of Austromarxism that workers suffered from a cultural amnesia, permitting the imprinting of ideas and practices of a higher socialist order. In using a powerful municipal welfare policy to further their goal of creating "neue Menschen," the SDAP sought to open the way for a fundamental change in the behavior of workers. Municipal socialism prepared the working-class communities for the introduction and diffusion of a socialist party culture that would complete the making of a new proletarian form of existence.

Public Education: Equality for Workers and Rising Expectations

If the reform of housing, health, and welfare formed a unity in the SDAP's quest to implant municipal socialism, its educational goals were both part of that framework and went beyond it. The socialists' view of the complex institution of education harked back to one of the fundamental tenets of Austromarxism, that *Bildung* for the workers was the foundation upon which a transformational structure for the creation of "neue Menschen" would have to be built. But *Bildung* was much more than educational formalism based in public institutions. In included a wide range of party organizations and activities to "educate" workers and workers alone. In education more than any other socialist municipal reform effort, that distinction between the central and select public of workers and the public at large was decisive. For that reason, perhaps, the socialist effort to make fundamental changes in public education appears to have been less than a complete commitment. As we shall see in chapter 4, party *Bildung* served not only as a supplement to public education; it also assumed the role of an alternative.

In attempting to reform public education along egalitarian lines of access to higher learning—so as to give working-class children an opportunity to go beyond existing class-based limits—the SDAP challenged both class and religious interests in the Christian Social party. By holding out the promise of equality of opportunity, the SDAP attempted to create a climate of rising expectations, laying the ground for the party's transformational cultural program. But equality of access to higher education based on merit was an ambiguous quest. If successful, it would produce a substantial number of professionals from among the workers (welfare officers, school and kindergarten teachers, physicians, lawyers, engineers, etc.). If they could not all be absorbed into the municipal service—which was highly unlikely then or in the near future—these sons and daughters of workers, advantaged by ability, would be in danger of abandoning their class for the positions and emoluments only the bourgeois world could offer.

The three cardinal goals of education under the monarchy—to create submissive subjects, to accept the hierarchical upper and lower orders, and to support the military and the church—were quite typical for conservative states.[142] "The state," Stefan Zweig recalls, "exploited the school as an instrument of its authority. Above all we were to be educated to respect that which existed everywhere as perfect: the opinion of the teachers as unerring; the words of one's father as incontrovertible; and the organization of the state as absolute and eternally valid."[143] Liberal legislation in 1867–69 had established the important principle of separation of school and church, whereby the ending of religious practices in and by the schools was implicit.[144] By the time of the founding of the republic, however, the meaning or spirit of this law had been blurred; religious observances such as communion, mass, and assembling for processions were carried out in and by the schools by priests stationed there to give religious instruction. The other significant liberal reform of the 1860s was the institution of mandatory public elementary education which was to last for five years and be carried out in public schools open to all.

Around the turn of the century liberals, Freemasons, and socialists in separate as well as common organizations attacked the monopoly of education by the propertied classes and demanded the creation of a merit-oriented system of higher education that would provide an equal opportunity for everyone. A reform group called Die Jungen, consisting of socialist elementary schoolteachers, was founded in Vienna in 1898 and joined forces with liberals and Freemasons to create the reform organization Freie Schule in 1905.[145] In both these organizations two teachers, Karl Seitz, later mayor of Vienna, and Otto Glöckel, later municipal councillor for education, played a central role in preparing a program for reform. It included complete separation of school and church; replacement of religious education by instruction in morality and law; free books and materials to all, as well as financial support for needy students; the creation of kindergartens and after-school centers; a maximum class size of thirty, and freedom of teaching methods; and above all, equality of opportunity in higher education.[146] Many of these proposals had been anticipated in the SDAP congress program in 1900.[147] Shortly before and during the war, interest in and support for educational reform became more widespread, and reprints of the 1905 program appeared in the socialist pamphlet literature. But only war, defeat, the collapse of the monarchy, and the creation of the republic put education on the agenda for action.[148]

The appointment of Otto Glöckel as undersecretary for education during the socialist-led coalition government (1919–20) put the right man in the right place at the right time. He moved quickly to restore Austrian education to the principles laid down by the law of 1869 by issuing a directive terminating the obligation of students to participate in religious practices in the schools (mass, confession, processions, etc.). He left intact, however, mandatory religious instruction in all public schools and, as we shall see, allowed the church to remain a significant presence in public education.[149] In the short time available to him—he left with the fall of the coalition in October 1920—he put in draft form the essential features of the reforms which had been developed but dormant before the war. Central to smaller practical measures was a plan to restructure compulsory education from six to fourteen years of age in such a way as to provide the highest level of education for everyone in common elementary and high schools, thus postponing work and career tracking to the terminal date of compulsory schooling.[150] In the federal government Glöckel was replaced by Christian Social ministers who did their best to shelve his proposals for improving the national educational system. They were unable, however, to revoke his decree barring religious practices in the schools, since it was upheld by a constitutional law that could be altered only by a two-thirds majority in parliament.

Like other socialist reformers, Glöckel found his arena of activity reduced to Vienna, where he became chairman of the city's educational council in 1922. There, school reform became an integral part of the municipality's social policy, especially its concern for child welfare.[151] Thus health care automatically became part of the school program, with 50 physicians, 210 social workers, and 11 dental clinics participating directly.[152] Practical improvements initiated by the municipal council's education department included free schoolbooks and educational materials for all pupils, regardless of need[153]; abolition of physical punishment; creation of school libraries for pupils and a pedagogical central library for teachers; admission of women to the study of law, engineering, and agronomy; abolition of political criteria in the selection of teachers; ending the required celibacy of female teachers; and the creation of parents' associations in each local school district.[154]

Whereas these reforms exemplified sound municipal government, they were hardly earth-shaking. Far more ambitious was Glöckel's attempt to restructure compulsory education so as to provide equality of access to higher education for working-class children, who up to then had been virtually excluded from such opportunities. He proposed that middle schools be created for all children between the ages of ten and fourteen, with a curriculum that provided common, enriched core subjects (mathematics and German) for all, yet at the same time was flexible enough to allow the study of foreign languages and other specialized subjects required by the Gymnasium and the Realschule, stepping stones to university education and the higher professions.[155] The novelty of this project lay in the postponement of tracking until the end of compulsory education at age fourteen. That goal was to be guaranteed by the core curriculum. But beyond that, student-initiated streaming would lead to various postsecondary choices: the higher education already mentioned; vocational workday training along with apprenticeship; and technical studies in special institutes for subprofessional positions.

Such radical structural changes also required the development of new curricula and teaching methods. As Glöckel later put it, the "drill school" of the grandparents' generation had served to instill the authority of church and crown plus a little of the three Rs; the "learning school" until 1919 had prepared the general population for work and a select few for skilled technical jobs and the professions; the "work school" being created would base its curriculum on the life experience of students, replace rote learning with independent study and self-discovery, supplement intellectual with manual learning, and make the city as well as the classroom the school environment.[156] The work schools were to become the incubators of democracy growing out of the cooperation of teachers, students, and parents.[157]

Experimental restructuring was carried out first in three, later in twelve middle schools in Vienna, which by 1926 contained about 9 percent of all secondary students.[158] To prepare teachers for the desired new motivational methods, a pedagogical institute was created by the municipality. Its lecturers included Alfred Adler, the developmental psychologists Karl and Charlotte Bühler, and Anna Freud. How influential was this institute? The tendency has been, then and now, to suggest that the influence of modern psychology and pedagogy available to a very small elite of teachers in training was somehow generalized throughout the Viennese school system, where it became instrumental in reducing the authoritarian features of instruction.[159] As important as the methodological innovations offered at the pedagogical institute may have been, it is very doubtful that they had any effect on what went on in most classrooms, where teachers trained at other times and with other orientations prevailed. It is also very unlikely that more than a few of the new "progressive" teachers could have been placed, considering the radical decline of students by nearly 45 percent between 1915 and 1923, which produced a surplus of teachers until 1929.[160] By then the experiment in restructuring had collapsed.

The new middle-school concept was first tried out in six former military academies in and outside Vienna, which Glöckel converted into federal educational institutes *(Bundeserziehungsanstalten)* with the directive to select the most gifted among working-class children and to implement the full range of curricula and methodological innovations being devised.[161] There is little evidence, except for the strong opinions of former "old boys" or latter-day enthusiasts, to determine whether the teachers in these institutes were sufficiently different from the normal run to make the expected democratization of the whole school experience possible. One alumnus recalls that the guiding principles were "democratization of the school, animation of instruction, and socialization of education."[162] A graduate of the new Glöckel institute for girls in Vienna reports that the teachers were simply taken over from the institute for the education of the daughters of army officers it had replaced, a situation which produced pedagogical tensions.[163] Another alumnus recounts in greater detail that all reform efforts were sabotaged by the teachers—either the former instructors of cadets, who as clericalists and monarchists detested the republic, or relocated instructors from German schools in Czechoslovakian cities who remained loyal to their Pan-Germanism down to supporting the Austrian Nazis in the early 1930s. This eyewitness found little evidence of democracy: workers were denounced as riffraff in the classroom, and Marxists were designated traitors.[164] Whatever character their program in practice might have had, these institutes were experiments that could not be replicated or generalized.

Throughout the period the socialists, and especially Glöckel himself, lost no opportunity to proclaim the excellence, inventiveness, and incomparableness of their educational reform program. It is small wonder, said Glöckel, in view of the organization and attainments of its working class, "that today Austria stands in the front rank of civilized countries in educational innovation and that Vienna has become the destination of pedagogical experts the world over. Justly, Vienna can claim the proud title of 'city of school reform.'"[165] Recent Austrian histories of the working class have made their contribution to repeating what, in the face of only a cursory glance at reforms attempted and accomplished elsewhere, is a patent exaggeration.[166]

A quick survey of educational reform efforts in the Weimar Republic (Berlin, Frankfurt, Hamburg) reveals important parallels with the Viennese experiments, both in postponing tracking to increase working-class access to higher schooling, and in attempts to enrich curricula and teaching methods.[167] The British Labour party's very strong educational reform program of 1923 demanded access to secondary school at age eleven or over for all children, and the Hadow Commission established by Labour succeeded in implementing compulsory school attendance through age fifteen as of 1926. This meant that by the early 1930s a majority of English and Welsh children, who began at age five, had ten years of schooling. As for experimentation with new methods under the influence of psychology, both A. S. Neill's Summerhill, founded in 1924, and Bertrand Russell's Frensham Heights, founded in 1926, in conception and scope put the Viennese attempts in the shade.[168] The most obvious challenge to the Viennese socialists' claims to leadership in educational reform comes from the United States. A long list of its egalitarian structures and modern methods might be devised. Suffice it to say that university education was available and sought by a substantial number of working-class youth (who had finished the compulsory common secondary school) in state and municipal colleges and universities where tuition was free.[169]

With a certain confidence that belied political realities, the SDAP made the Viennese school reform a key element of its national program at the Linz party congress in 1926.[170] It was a time of relative calm between the two political camps, which also marked a standoff between Glöckel's operation in Vienna and the ever hostile Ministry of Education. The appointment of Richard Schmitz, a militant Catholic and close associate of Seipel, as minister quickly brought the conflict between the federal government and the Viennese municipality to a head.[171] Schmitz presented a list of guidelines to be used in drafting a national education law to propose to parliament. Its salient feature was an unqualified rejection of Glöckel's attempt to reduce the class monopoly of higher education by postponing tracking until the end of a common secondary school.

Negotiations by the SDAP and Christian Social party about the guidelines were interrupted first by the election of 1927, in which the socialists scored a marked increase to 42.3 percent, and by the violent events of July 15. What occurred in the negotiations thereafter, when the socialists were no doubt still stunned by the implications of an apparent but unadmitted setback, remains unknown. We learn about an alleged compromise from the socialists in late autumn of 1927, when a national education act was passed with the mandatory two-thirds majority, indicating that the SDAP had approved the measure.[172] It was a case of the socialists claiming victory in the face of defeat when they labeled a "compromise" what was a patent revocation of the Viennese municipality's reform effort. In place of the common, obligatory, enriched middle school allowing for tracking at age fourteen, dual tracks at age ten were adopted with the proviso that a select number of the best students in the Hauptschule would be permitted to take examinations to enter college preparatory schools. This federal law ended the Viennese school reform; no further attempts were made to deepen or extend the experiment.

The socialists' virtual capitulation remains a mystery. The collapse of the Viennese school reform has been universally interpreted as a consequence of the machinations of the Christian Socials at the national level (though without elucidating why the socialists capitulated).[173] I would like to offer an alternative or at least supplementary explanation for the socialists' failure in democratizing the access to formal education and thereby to make available to the working class this traditional path to *Bildung*. The Glöckel experiment failed for a number of reasons: the reformers' reluctance to deal with the powerful opposition in the schools exercised by the church; their failure to reach the corps of teachers, who were unprepared for radical changes in their status and professional activity; their assumption that education could be value-free; and their blindness to the educational expectations and related economic limits of working-class families, children, and youth.

From Glöckel's first efforts to the socialists' capitulation in 1927, the most powerful opponent of any secular reform of the schools was the church. In April 1919 Glöckel had struck an important blow for secular education by decreeing the return to the principles of the 1869 reform law in abolishing religious practices in all schools.[174] This merely reduced, but did not eliminate, the influence of the church on education. Priests remained in the schools by virtue of the continuation of compulsory religious instruction and served as a "fifth column" engaged in guerrilla warfare against the reforms in Vienna. Religious instruction remained strictly enforced, consisting of one afternoon a week when the Catholic priest would come to the classroom, dispatching children of minority religions to their lessons elsewhere. Attendance at religious instruction was mandated by law; without a grade in the subject, one could not be promoted to the next class. Only those could be exempted whose parents had formally renounced their confession before their children had reached their seventh birthday. Thus, considering the large number of resignations from the Catholic church after 1927, many parents who had become freethinkers were forced to send their offspring for instruction, until the age of fourteen, in one of the religions recognized by the state.

Since the church laid claim to controlling education in place of the municipality and state,[175] its continued presence as a hostile force undermined the socialists' reform efforts from the beginning. Could the socialists have banished the church from the schools? There was at least one important historical precedent for such action in a Catholic country. In France the famous Jules Ferry Laws of the 1880s, in prohibiting all religious instruction by teachers in state schools, created a public school system that was free, lay, and compulsory at the elementary level.[176] The corps of secular teachers created at that time became the stanchest defenders of republican France and acted to diminish the influence of the Catholic church in public life in general.

I am inclined to answer the question about the feasibility of banishing the church from the schools in the affirmative, but only at one crucial point in time. In April 1919, when Glöckel issued his decree, Soviet republics held sway in neighboring Bavaria and Hungary, Bolshevik ideas had currency among demobilized soldiers and in the radicalized workers' councils, and the old power structure was discredited by military defeat. Fear that this contagion of revolution might infect Austria as well was widespread among the Christian Social leadership and in the Vatican. The newly created republic was then under a coalition government dominated by the socialists, who by playing on the fears of the opposition could have forced through a two-thirds vote abolishing the presence of religion in the schools. Such an opportunity did not present itself again; what amounted to a "clerical republic" was accepted by the socialists in place of the democratic one to which they aspired.[177]

Although Glöckel and other socialist educational reformers realized the importance of converting the corps of public schoolteachers to their views, they found it nearly impossible to change ingrained conceptions of professional prerogative among them or to bring in a significant number of young teachers converted to the new objectives and trained in the new methods. It is to Glöckel's credit that he kept on all the teachers who normally would have been rendered superfluous by the decline in students of some 45 percent in the immediate postwar years. This humanitarian decision left the teaching staffs heavily balanced against democratization and other reforms. By 1930, among teachers organized in associations, the Pan-German Österreichische Lehrerbund had 10,000 to 11,000 members, the Christian Social Katholischer Lehrerbund für Österreich had 10,000, and the SDAP's own Sozialistischer Lehrerverband had 5,000.[178] Glöckel's attempt to neutralize a potentially hostile political climate among teachers, by demanding that all politics be kept from the schools, was totally unrealistic in a municipal and national environment supersaturated with confrontational politics.[179]

Moreover, it exposed him to quite justified attacks within his own ranks for having adopted the non-Marxist view that pedagogy was somehow a neutral science.[180]

That the socialist city fathers suffered from a lack of realism was especially apparent in their belief that the educational reform program they struggled to put in place was desired by the Viennese working class or was high on its agenda of needs. Given the low wages and the need for young people residing with their parents to contribute to the meager household budget, higher education, even if the fees were paid by someone else, still meant that the family had to undertake the impossible task of supporting one of its members for years to come.[181] As a consequence of this blocked path, working-class families had little use for talk of the professions for their offspring. Vocational training leading to better skills, a more qualified job, and higher pay was another matter. But that lay within existing working-class norms of expectation. There were more incentives during the period for skills created by the increase in white-collar jobs, and the one-child family was in a better position to make sacrifices for the economic advancement of their offspring. Judging from some absenteeism records in a typical working-class district, schooling for its own sake, including enrichments, was not highly regarded.[182]

Who among the workers in the socialist camp were able to benefit from educational reforms? The traditional answer would be the single children of better-off skilled workers able to shoulder the financial burden of maintaining a noncontributing youth past the age of fourteen. But such families also suffered from insecurity caused by continuous high levels of unemployment (among skilled workers in metal trades, for instance). More likely, it was the children of paid party functionaries and socialist municipal employees with secure and better-paid positions who could take advantage of enrichment and more equal access to higher education.[183] This brings us to a central question regarding municipal socialism and ultimately the Socialist party culture as well: which workers were the actual audience, who and how many participated or were influenced indirectly? These questions appear not to have troubled socialist leaders at the time. If their municipal reform attempts fell short of setting the stage for creating "neue Menschen," the party's own program of *Bildung* was being devised and extended daily to do just that and in a context that was securely socialist.

CHAPTER 4

Socialist Party Culture

By 1931–32 the SDAP's loose network of more than forty cultural organizations registered an aggregate of some 400,000 members in Vienna.[1] Even if one adjusts this figure because of the duplicate listing of subordinate organizations and their parent bodies and the typically multiple memberships of individual workers, an impressive number of participants in the party's cultural activities remains.[2] The size of cultural undertakings is not surprising when one considers that SDAP membership in Vienna was 425,000, that of the trade unions 375,000, and that of the tenants' organizations 150,000.[3] Such a high level of organization alone made the Viennese cultural experiment unique among socialist parties outside the Soviet Union in the interwar years. Although there were similar cultural efforts by the German Socialist and Communist parties, nowhere else did the attempt to create a proletarian counterculture rest on a socialist-controlled metropolis that had created a solid foundation of municipal socialism as the basis for a party culture. Despite all the shortcomings of the SDAP's experiment (which will become apparent in the course of this chapter), the collective effect of memberships and municipal power structure on the mentality of the Viennese working class, though by no means measurable, should be kept in mind.

Despite the SDAP's commitment to creating a specific worker culture, it at no time exercised complete control over or was able to coordinate the far-flung collection of organizations that laid claim to furthering socialist culture. Indeed, the relationship between a good number of these and the political aims of the party, or even a vague socialism, was neither apparent nor real. Some of them dated from the formative years of the party or even preceded its formation (singing, cycling, nature); others were created by small groups of workers sharing common interests (chess, Esperanto, animals). Many of these served the interest of party cadres, who gained satisfaction and recognition from making the organization work and its programs run.[4] Too often such organizations shielded themselves from criticism within the party in their single-minded devotion to the numerical growth of their enterprises and proliferation of time-serving meetings, which sapped their members' energies without providing much cultural content. The sheer number and diversity of organizations led to a flitting from place to place and, because they all demanded modest membership fees, constituted a financial burden on interested workers.[5]

Complaints about the dissipation of cultural efforts in useless duplication and competition among organizations, and demands for a rationalization of cultural work, fell on deaf ears.[6] Middle functionaries and party leaders engaged in the bureaucratic apparatus were too far removed from the daily life of numerous esoteric organizations to establish real contacts and make their criticism felt. And, as we shall see, criticism of the cultural program as a whole was fairly freely expressed in party publications without having the slightest impact on the content or direction of the work being carried on.

Not all the SDAP's cultural efforts were as dispersed and nondirectional as the interests of socialist teetotalers, folklorists, or freethinkers might suggest. Two organizations, created not at the grass roots but by the party's executive committee itself, attempted to give a focus to the theoretical and strategic place of a directed and carefully guided cultural program in the SDAP's attempt to create "neue Menschen." The Sozialistische Bildungszentrale (socialist cultural center), rudiments of which existed as early as 1908, coordinated and controlled a variety of activities: press and publication, the lecture department, worker libraries, schools for party functionaries, festival culture, and excursions and vacations. It operated on a national level, although most of its efforts were pursued in Vienna.[7] The Sozialdemokratische Kunststelle (socialist art center) was charged by the Viennese municipal council in 1919 with bringing music, theater, and the arts to workers, employees, and students. It too had roots in earlier worker music and theater organizations (Arbeitersymphoniekonzerte and Freie Volksbühne) and eventually included radio and the film among the artistic offerings under its direction.[8] From time to time these dominant bodies reached out to admonish or reorient one of the myriad primary organizations such as the Radio Club or the Association for Sports and Body Culture.

Throughout the First Republic the Socialist party's cultural program was guided, however indirectly, by several important perspectives which contained insurmountable contradictions: (1) to appropriate for the workers the best of elite/bourgeois culture, and at the same time to create a closed proletarian counterculture committed to the class struggle for a socialist society; (2) to provide for collective rather than individual cultural development of workers; and (3) to safeguard and work within the democratic institutions of the republican state and to combat the class state at the same time. Although the intention of the SDAP leaders was to blend culture and politics in their transformation of the working class, increasingly after 1927, as the political opposition became more combative and Bauer's theory of the balance of class forces became more illusory, the cultural experiment became a compensation for political powerlessness.[9] In what follows

I shall examine Socialist party culture from several perspectives: elite culture and culture theory; the power of the written and spoken work; and the attempts to enrich and ennoble the workers' artistic taste. The impact of these was restricted to a minority of the rank and file. Sports and worker festivals, my last subject for discussion, were the most important cultural forms attempting to engage the mass of workers in activities that were both actual and symbolic.

Elite Culture Rejected and Desired

From the earliest days of prewar social democracy to the end of the First Republic, socialist leaders wrestled with the role of culture in the party's efforts and aims. A subtext of almost all pronouncements on the subject was the difficulty of striking a balance between the workers' right to the cultural inheritance which had been denied them and the need to fashion a culture expressive of their own class. Reflections on the subject by the doyens of the SDAP reveal a continuous contradiction in the relationship between a desired socialist culture and the existing bourgeois one. Victor Adler, father of the party, saw socialist culture arising in a struggle against lifeless and superficial bourgeois forms, as a transformation of these in the process of creating a socialist society in which a "true" culture could be developed.[10] Engelbert Pernerstorfer, another founder of the party, declared that the goal of social democracy was to bring the worker to high culture and especially to the classical (German) inheritance.[11] The ambiguity of these early socialist positions reflects the deep roots of the party in the liberal bourgeois Austrian tradition.[12] It was a central element in the formation and socialization not only of the older generation but also of the cultural party spokesmen during the republic. The old liberal slogan "knowledge is power, education makes you free" was taken over by the socialists and given a dual meaning: workers had a compensatory right to the cultural products of society; workers must raise their consciousness in order to wage the class struggle.[13]

Despite an obvious and steadily increasing commitment to creating a party culture on the part of leading socialists, no clear theoretical formulation emerged that might have been translated into an overall party policy.[14] The enshrined writings of Austromarxism's founders were too amorphous to serve as a guide. In 1924 Max Adler attempted to clarify the role of party cultural work in the shaping of "neue Menschen."[15] This treatise had the strange fortune of being cited by virtually every socialist cultural director as a basis for the work being carried on, but without being understood by any of them, and for good reason. The cultural development and education of the workers, Adler postulated, cannot be neutral but must be a part of the proletarian class struggle. As such, working-class culture must make a determined break with the old bourgeois world. It must be achieved through special organizations of the proletariat and serve as instrument of the class struggle aimed not at the preservation of present conditions but at the destruction of class society. The goal of such a revolutionary culture and education was "neue Menschen"—"the future-oriented preparation of a new society in the soul of man."

Although the hortatory tone was clear, Adler said next to nothing about translating these vague aphorisms into practical cultural work, save that the proletariat could not learn from its own experience and had to forswear material and professional concerns in striving for the socialist goal. The surest guide to that, he concluded, could be found in the books of the great socialist teachers as well as the classics of German philosophy, together with natural science, history, and the economic laws of societal processes (Marxism).[16]

I have dwelt on this work at some length in order to demonstrate two important points: that one can clearly detect the important contradiction between rejecting bourgeois culture and upholding its classics as a necessary part of the socialist cultural diet; that such verbosity left the means by which the SDAP's cultural program was to be carried out up to those directing the various activities. Elsewhere Adler gave vent to a bombastic self-congratulation of the SDAP (an all-too-common weakness, as we shall see) for directing the greatest mass movement and cultural movement in history.[17] The most prominent review of Adler's tract, by Otto Neurath, subtly exposed the empty phrase mongering of the work.[18] His central complaint was that Adler had failed to characterize the proletarian/socialist essence and that he had said too little about the "neue Menschen" which was the subject of his work.

Otto Neurath completely rejected Adler's view that only socialist man can create the new society by positing the opposite: that only socialist society can create socialist man. Neurath was the founder of Vienna's Social and Economic Museum, in which his own creation, visual statistics (pictograms), was featured.[19] As a member of the famous Vienna Circle (Moritz Schlick, Otto Hahn, and Rudolf Carnap), he was a rationalist; as an Epicurean he rejected German philosophy as being too theological. For him, societal development was based on the quality of life, and from that perspective he believed that the socioeconomic development of capitalism was preparing the ground for socialism.[20] Although he demanded that the outward forms of proletarian culture, architecture for instance, should not cling to bourgeois forms but use modern ones that suggested future proletarian power,[21] he rejected contemporary SDAP attempts to foster special life-styles among the young (antialcohol and antitobacco), including bobbed hair and short dresses, as superficial enthusiasms unrelated to the class struggle and life experience of the working class as a whole.[22] In order to change proletarian life-style, the entire existing power relationship and social order had to be changed. Only after victory, he concluded, could the proletariat command art and science, which were currently in the hands of the bourgeoisie.[23]

Neurath's often trenchant critique of the SDAP's cultural efforts and of Adler, Stern, and other culture experts failed to have any impact. It was part of the strange fate of criticism in the party that it appeared quite frequently in its publications but in no way disturbed or altered the cultural machine at work. Not only on cultural questions but also on political strategy and tactics, the SDAP was quite open to criticism, so long as it did not threaten the command structure of the party or create more than temporary divisions.[24] Neurath's critique was special because it came from outside the party's cultural establishment. What made it even more unacceptable to the party's cultural practitioners was the fact that it really left no room for the SDAP's heavy commitment to a proletarian counterculture and justification for the work in progress.

The SDAP's cultural directors—Joseph Luitpold Stern, David Joseph Bach, and Richard Wagner—refused to enter the thickets of high theory. They were satisfied to operate with the liberal heritage that knowledge equals power, which led them to give full expression to an ambiguity about elite or bourgeois culture. Wagner, the leading figure in developing cultural work in the trade unions,[25] insisted that the countless temptations of bourgeois culture were to be combated by working-class organizations.[26] What should a worker culture do beyond strengthening democracy? he asked. In the long run, he responded, it should build class consciousness and the power of socialist construction; in the short run it should gain a share of the cultural goods of the capitalist world.[27] Though Stern and others gave lip service to the class struggle, politics and ideology virtually disappeared in the central place they accorded to culture and pedagogy in preparing for a socialist future.[28] They saw their mission in elevating the worker from his present immaturity, with its preference for cultural products of low quality (*Schund* and *Kitsch*), that he might consume his rightful share of society's best cultural products.[29]

The question was: which products? The answers, by David Joseph Bach and others, tried to steer a course between bourgeois values to be condemned and the best products of that culture, which in their artistic excellence contained progressive elements and even revolutionary overtones that a proper class analysis would make apparent.[30] Elite culture was both a right and necessary part of workers' political education, Stern insisted, but only after social interpretation had filtered out its specific bourgeois orientation.[31] On closer examination, what constituted excellence was generally more narrowly defined. The older generation of SDAP leaders, who set the tone of the party's cultural work, were committed to passing on to the workers the bourgeois-humanistic classical heritage in which they had been socialized.[32] In effect that meant a narrowly defined Germanism. It is interesting to note the total absence of reflection about the subjective nature of cultural values, about the force of inherited traditions which had shaped the leaders themselves, and about the dialectical process by which mature cultural judgments were personally arrived at. Youthful predilections for trash and kitsch seemed to be forgotten; the "classics" sweated through in the Gymnasium were elevated into universal standards of excellence.

Stern's background and attitudes on that subject are particularly revealing. His intellectually formative years had been spent in Dresden, where he worked on the *Kunstwart* in close association with Ferdinand Avenarius, a passionate follower of the German nationalist ideology of Julius Langbehn.[33] Stern appears to have brought this orientation with him when he assumed direction of the Bildungszentrale in 1918. From then on in the SDAP's program, uplifting the worker to a higher level of culture was equated with initiating him into German national culture. But Stern was no exception in this narrow definition of cultural excellence. Max Adler, as we have seen, as well as Bauer, Renner, Deutsch, Danneberg, and the whole team of SDAP leaders, subscribed to that view with varying degrees of intensity. It tended to make the party's cultural orientation backward-looking. Bach's jubilation about having made a new production of *Faust* available to working-class audiences, for instance, was embarrassingly so.[34] He justified his enthusiasm by claiming that socialism always selects that art which points to the future, even if it comes from the distant past during the time of its appreciation.

The SDAP's heavy tilt toward elite culture, despite disclaimers about resisting bourgeois values, did not go unchallenged. The party's cultural efforts were attacked, with a veiled reference to Bach and the work of the Kunststelle, as providing a photographic replication of bourgeois culture and often of the past generation.[35] The principal task in developing a proletarian culture, the party was reminded, was to liberate it from elite culture so as to make the celebration of March 1848 more important than the reading of Goethe's poetry, and the "Lied der Arbeit" more important than a Mozart sonata. In what became a constant critical refrain, Richard Wagner bemoaned the absence of a cultural theory and policy comparable to the party's political one.[36] In worker festivals, in the overevaluation of bourgeois theater, in worker song societies, and in worker sports, he charged, class orientation had been neglected in favor of aesthetics.

If considerable vagueness persisted in the SDAP's theoretical orientation regarding proletarian culture—especially the relationship to the dominant bourgeois culture, and the connection of the party's cultural effort to the ongoing class struggle—it did not mean that the party was uncertain about enveloping the workers in a total culture of its own creation. Indeed, with the onset of a more ominous and confrontational politics after 1930 (seen in the aggressive actions not only of the domestic Heimwehr but also of local Nazis in the arena of street politics), the socialists became more adamant about immuring the workers in a closed world from the cradle to the grave. In 1932 the SDAP executive decreed, for instance, that party rank and file were not to assume any function in bourgeois sports organizations; furthermore, party functionaries (including cadres) were forbidden to become members.[37] Perhaps it was the lack of clarity about its cultural program and especially about its natural limits, defined by existing traditions and social structures, combined with the senior leaders' reluctance to accept the possibility of defending the party by force,[38] which produced the frenzied belief that a greater cultural effort could somehow safeguard the republic and the party as well. The party's slogan "against the idea of force, the force of ideas" turned out to be a costly illusion.[39]

Magical Powers of the Word

In the SDAP's attempt to raise the workers to a higher cultural level, a strategy in the creation of a proletarian counterculture, the word and particularly the printed word played a central role. For the party's chief educational reformer, Otto Glöckel, "the book is the strongest weapon in the class struggle. . . . It raises the question of why . . . and the why is the means to intellectual development and knowledge. . . . Once people have the courage to gain knowledge, they must become socialists."[40] Such innocent idealism echoes the special importance accorded to the printed word by German liberalism. Another strong influence on the socialists' overevaluation of the power of the book, it has been suggested, was the book's high valuation in Jewish tradition, given the predominance of Jews among original Austromarxists and their practitioner epigones in the republic.[41] We shall look more closely at the socialists' intoxication with the word and their expectations about its magical powers of transformation in the context of the party press and publications, lectures and party education, and worker libraries.

By 1930 the SDAP, trade unions, and cooperative societies published 127 newspapers and journals with a total print run of 3,161,000 copies. This included 7 dailies, 68 specialized periodicals (addressed to tenants, consumers, teetotalers, cadres, mothers, women, and those interested in culture, to mention but a few), and 52 trade union weeklies.[42] As Langewiesche has pointed out, this avalanche of socialist and associated publications betokened not reader interest but a lack of coordination.[43] A critic of this publication mania hypothesized how many books of 250 pages each would be available to every member of a socialist organization based on the print run of 3.16 million just cited, and concluded that everyone would receive forty books a year. To illustrate how workers could not possibly deal with this flood of publications, he conjured up a typical periodical diet a Viennese party cadre might be exposed to. Besides a subscription to *Die Arbeiter-Zeitung,* he would get two trade union publications, *Der Vertrauensmann,* and the Viennese party organ, *Der Sozialdemokrat,* as well as other publications of the municipal party organization. He might also receive automatically publications such as the tenants' association organ, the Schutzbund and cremation society newsletters, as well as those of any of the forty cultural organizations he belonged to. If he was married, his wife received an equal pack of material.[44]

The point is not to lampoon the SDAP's publishing efforts but to express some doubt about the relationship of the huge publication figures given and the actual number of worker readers. A closer look at who read what leads one to doubt the statistical information on this as well as other subjects offered in party yearbooks and other official periodic reviews about gains on the cultural front. Numerical growth of party organizations of all sizes and kinds seems to have been a foregone conclusion used to demonstrate the increasing popularity and strength of the party. The circulation and readership in Vienna of *Die Arbeiter-Zeitung,* the party's principal daily, is extremely revealing. If one assumes that the daily circulation was consumed by SDAP members alone, then in 1906 every second member bought the paper, in 1926 every fifth, and in 1931 every eighth.[45] A study of Viennese newspaper readers conducted in 1933 concluded that only one-third of *Die Arbeiter-Zeitung* readers were workers.[46] The number of Viennese party members purchasing the main theoretical journal, *Der Kampf,* in the late 1920s and early 1930s was roughly one in eighty.[47]

In 1927 the SDAP created *Das kleine Blatt,* written in a livelier and more popular style, with more "human interest" content and a small tabloid format, in the hope of weaning workers away from the reactionary *Illustrierte Kronen-Zeitung* and *Die kleine Volkszeitung.* By 1931 it had attained a circulation of 173,000, which rose to 200,000 by 1934.[48] At the same time two illustrated weeklies, *Der Kuckuck* and *Bunte Woche,* were added to the party's roster of publications. All three combined the creative use of photographs with features that were both entertaining and educational. These concessions to popular taste included a heavy dose of revelations about the demimonde in a seductive tone, while pointing out that prostitution was a social problem. In the *Kuckuck* particularly, travelogues about strange and exotic places were amply illustrated with bare-bosomed women.

The most successful of the newer publications was *Die Unzufriedene,* started in 1926 to attract women to the party. Set in an attractive format, it offered a mixture of political subjects from a socialist perspective; tips on health, beauty, clothing, and cooking; a column to let "women speak from the heart"; serialized novels; and a large number of advertisements offering everything from contraceptive devices to the newest housekeeping implements and modern furniture. By 1931 its circulation was 154,600.[49] Most of the party's major publications featured serialized fiction as a means of building a loyal readership. Among the works published, social novels predominated, though a significant number of romantic potboilers, of the type the SDAP's cultural establishment considered trash and kitsch, were included.[50]

Several publications addressed to the already committed party or trade union members raise further questions about actual readership. *Die Frau* (*Die Arbeiterinnen-Zeitung* until 1923), originally a weekly but demoted to a monthly for lack of funds, claimed a circulation of 218,660 by 1931. But in a large-scale survey of mainly organized female industrial workers, only 2.4 percent indicated that they read this monthly.[51] Only 3 percent of the same sample read the trade union papers, yet the total circulation claimed for these was 860,000. The discrepancy between the huge editions of many party and trade union periodicals and apparent readers is explained by the fact that these were distributed free of charge as a matter of course to members of various organizations. It thus becomes clear that the readership, particularly of the big-run publications, was mainly fictitious. That the party was investing considerable sums without reaching its intended audience must have been recognized at the time. The creation of *Das kleine Blatt* and the illustrated weeklies, and attempts to give other publications a more popular tone, attest to that. How large the worker readership of particular publications was is not easy to determine. Even where a clearly targeted audience existed, the relationship between circulation and readers is confusing. *Der Vertrauensmann,* addressed to the party cadres, is a case in point. In 1931 it had an edition of 12,600.[52] But the cadres in Vienna alone numbered 20,636.[53]

Contemporary surveys of worker reading habits offer some insight into the preferences and frequencies of readership, as well as into the particular features in the periodical literature preferred or shunned by workers. Of the large sample of female factory workers surveyed, 48.2% read *Das kleine Blatt* and 28.8% read *Die Arbeiter-Zeitung* among the dailies; 21.4% read the weekly *Die Unzufriedene,* and 9.3% read papers of other political persuasions.[54] Another survey of leatherworker apprentices aged fourteen to nineteen revealed that 80 percent read a daily paper. Their specific choices were as follows: *Das kleine Blatt,* 41%; *Die Arbeiter-Zeitung,* 30.7%; the reactionary *Kronen-Zeitung* and *Volks-Zeitung,* 27.5%; and papers of other (non-working-class) orientation, 26.2%.[55] Although the editors of *Bildungsarbeit* applauded this effort to gain an "objective" understanding of worker youth in shaping the future direction of cultural work, no attempt was made to expand such surveys or to do more than give lip service to their importance.

The 1933 study of Viennese newspaper readers found that the majority of workers read the four popular tabloids, of which only one *(Das kleine Blatt)* was published by the SDAP, and the others by its political opponents.[56] Lead articles were read regularly by only 35.5 percent because the language was too difficult and the text too abstract.[57] The workers preferred parliamentary reports, because they were colorful and dramatic; they showed little interest in economic subjects. Feature articles, stories, serialized fiction, and travel accounts were most popular; theater, film, and art criticism were near the bottom of worker preferences.

The SDAP publishing house was modest in its editions of accounts of important party events and writings of the leaders. Protocols of the party congresses, which presented the annual program, had the following print runs (sales are not known): 1920—600; 1921—500; 1925-30—1,000; 1931—2,000.[58] Important political texts also had small printings: Julius Braunthal's *Kommunisten und Sozialdemokraten* (1919)—5,000; Otto Bauer's *Bolschewismus oder Sozialdemokratie?* (1920)—10,000, (1921)—5,000; and *Austerlitz Spricht* (1931)—2,000.[59] The party's publication of *Wiener Dreigroschenbücher* presenting good literature in an inexpensive edition, and its attempts to provide such belles lettres through book clubs, also reached out to only a very small audience.[60]

It remains to gain some perspective on the foregoing numerical no-man's-land. The SDAP sought to bring the working class into its fold with the sheer numbers of its varied publications—to overwhelm workers with a flood of printed material easily available. Only after 1927 did the party consider other means based on the interests of their audience. But even with the introduction of more popular organs such as *Das kleine Blatt*, its publications with huge editions, such as *Die Frau*, remained virtually unchanged. The party leaders evidently assumed that captive audiences were available, since as members they received a good number of publications free of charge. *Die Frau* was distributed to all female party members, but Leichter's study reveals that it was hardly read. The same may be said about trade union papers, the format and painful dullness of which must have put off all but the most motivated activists and functionaries.[61] Even among the elite students of the Arbeiterhochschule (workers' college), surely among the most committed socialists, only 57 percent read the party newspapers and periodicals heaped upon them.[62] The proffered pamphlet literature met the same kind of reader resistance. A plaintive voice from the rank and file revealed that Otto Bauer's important work *Der Kampf um die Macht* was hardly read by anyone at his workplace, not even the cadres. Even if bought, it was claimed, it was stuck into the toolbox unread.[63]

The SDAP's reliance on the quantity of its publications failed to make them the transmitters of the aspired worker culture for several reasons. Foremost was the paternalistic assumption that the readers' interest, current habits, and mentality need not be taken into consideration. Implicitly, the party had its own ideas about how to better inform the workers, raise their cultural level, and promote their consciousness. Failure to understand the worker reader was reflected in the difficult style and linguistic usage, as well as in the format, of most publications. To the average Viennese worker, who spoke a broad dialect and was accustomed to the simple formulations of the boulevard press, the party's sober publications may have suggested hard work (like school) rather than recreation.[64] Since the masses of workers could not cope with the flood of party publications, vast numbers of those publications must have remained unread.

Who then were the loyal readers prepared to deal with a good part of the printed matter made available to them each week? It was probably a central core of party cadres, paid functionaries, officials of the trade unions and cooperative societies, socialist civil servants, and others whose secure employment indirectly stemmed from the socialist municipality. All in all this group probably at best comprised about 30,000 to 35,000 people. This is not to argue that a core of dedicated party members were the only readers of SDAP publications (even cadres, as we have seen, at least sometimes simulated the reading of the party's principal writings). No doubt many workers consumed some of the available party literature, but they seem to have been highly selective and far from constant in their choices. As important instruments of the SDAP's cultural program, its press and publications appear not to have made much headway in reaching the party rank and file. The core group engaged most successfully represented only some 7–10 percent of party membership. It had been significantly larger in the prewar party. The rapid growth of the SDAP after 1923 created a formal membership that in reality still had to be won over to the socialist cultural perspective of SDAP leaders.

Next to the printed word, the spoken word in the form of lectures was a well-established form of worker education and cultural enlightenment dating from the prewar period. It was a major accomplishment of the Bildungszentrale to expand these into a vast network of speakers available to worker groups at the local level. In Vienna, single-lecture evenings grew from 3,000 in 1924 to 6,500 in 1932.[65] These were conducted at neighborhood party branches, at *Volksheime* (people's clubhouses) in Ottakring, Leopoldstadt, Landstrasse, and Brigittenau, and in the meeting rooms of the new municipal housing. Topics offered included contemporary politics, the socialist movement, history, religion, womens' subjects, education, culture, and sexuality, to mention but a few. Attempts to organize series of lectures on single subjects failed to attract a sufficient audience.

Despite the numerical success of these lectures, which at least allowed the Bildungszentrale to put what the party considered important subjects before worker audiences, they suffered from some fundamental shortcomings which reduced their value as vehicles of a socialist party culture. Frequently lecturers faced the insurmountable task of having to present complicated subjects without sufficient time to offer the necessary detail. This meant that whatever time had been set aside for this cultural event was used up by the speaker, so that the much-desired discussion (one of the central aims of such projects) could not take place. As one sharp critic of the lecture program complained, the format led to a superficial experience about which one might expect reports such as: "The speaker had a powerful voice, therefore, he was well liked."[66] A far more fundamental critique of the lecture program and other aspects of the SDAP's cultural efforts came from Max Adler very late in the day. A vast gulf had opened between the masses of socialist workers and the party and trade union bureaucracy, he charged, as was apparent in cultural endeavors where listeners gave their respectful attention but remained unengaged.[67]

Without minimizing the importance of the SDAP's lecture program in helping to give worker participants a sense of belonging to the socialist camp, one must express some doubt about the degree to which the content of programs could be called educational or cultural in the sense the party intended. By far the most popular subject on the Viennese circuit were readings of a light and entertaining nature. They comprised about 25 percent of all lecture topics and were clearly popular entertainment rather than the sober fare considered to be culturally uplifting by the SDAP's culture establishment.[68] One other aspect of this activity remains puzzling. Granted that 6,500 single lectures in 1932 was an impressive number, no one has been able to ascertain how large the audience of worker listeners actually was. If one assumes an average group size of twenty-five (averaging small local audiences with larger ones), the total yearly number would be about 160,000. Leaving aside the likelihood that a large number of these were repeaters attending several events, it is imperative to look at attendance figures for popular leisure-time activities. In the same target year of 1932 soccer matches *on a weekly basis* drew from 150,000 to 200,000 spectators, and cinemas 500,000 *weekly* viewers (see chapter 5). No doubt the lecture program played an important part in integrating a limited number of workers into an aspect of party life, but the masses of workers were not reached by it, nor could the high socialist goals of the SDAP's cultural program be really advanced by it.

The SDAP's commitment to the power of the word was also evident in the various special schools created to prepare its leadership for the task of implementing the party's programs.[69] These schools were given the task of transmitting the essentials of Austromarxist thought, of strengthening the loyalty of lower-level leaders to the higher echelons, and of inculcating a revolutionary élan to be transmitted through the cultural and other programs of the party. They represented a specialized and very limited aspect of the Bildungszentrale's cultural activities and were only peripherally related to its attempts to reach the masses of workers. The three types of schools, in ascending order, were Arbeiterschulen for party cadres, Parteischulen for functionaries, and an Arbeiterhochschule for future high-level leaders. The curriculum in all of them stressed socialist theory, party organization, and current politics, with a progressive increase in theory and difficulty of subjects. The cadre schools offered courses running about ten evenings; the functionaries received evening instruction for a three-month period; and the workers' college offered six months of instruction in a resident campus setting.

The numbers involved at all three levels were small, accounting for about 12 percent of both cadres and functionaries for the first two, and a total of 114 persons in the four years the Arbeiterhochschule was in operation. The social distribution in the party schools says much about the composition of the party leadership in general. Workers were greatly underrepresented among the students, while employees and civil servants were greatly overrepresented.[70] These varied attempts to prepare the party leadership ideologically for its varied tasks were certainly commendable, but one notices a decided absence in the curriculum of any instruction about the workers themselves, their traditions, life-styles, communities, and general sense of milieu which would have been crucial in translating the SDAP's heavy investment in culture into ways that might have found popular response and acceptance.

For the socialist cultural leaders, neither periodicals nor lectures embodied the magical powers of the word as well as books. The development of a large network of worker libraries became the central aim and crowning achievement of the Bildungszentrale. Today, when we have become somewhat skeptical about the transforming power of books, the socialists' long list of expectations seems refreshingly idealistic but also naive.[71] It was believed that through books the worker could be weaned away from cheap amusements such as *Gasthäuser*, that his basic ignorance about the world could be reduced, that he could gradually be guided along an incremental path of literary quality to appreciate the serious works of social science and Marxism, and that through them he would be made ready to transform the chaos of capitalism into a rational socialist order. The prescribed path led from knowledge attained through hard work and diligence to the class struggle at a higher level. It demanded much—too much, as we shall see—from the worker who was not the ideal type imagined by the SDAP's experts.

The worker libraries, like many other socialist cultural institutions, were already well developed before the war. In 1910 various dispersed libraries were centralized on a districtwide basis. Two years later Joseph Luitpold Stern took over the network under the aegis of the Bildungszentrale. In 1914 he published a handbook for librarians which became the bible for the organization of worker libraries and the tasks of librarians.[72] In it he proposed a model structure for all centers and branches, and for centralized purchasing; a uniform system of cataloguing; and a mandatory gathering of statistics by each unit on membership, titles received, and books published. Stern's main concern in promoting the worker libraries was to combat the trash and kitsch to which he believed the workers, more than others, were exposed.[73]

This quest for the ennoblement of cultural products to be consumed by the workers became a main component of the SDAP's cultural program.[74] If books were to play a central role in reorienting the workers, Stern insisted, libraries would have to undergo the same rationalization as that being introduced in industry.[75] The purpose of such centralization, aside from savings to be gained, was control by the Bildungszentrale over the books made available as well as over the readers themselves. "The worker-librarian," he maintained, "must act as the moral confessor of his worker comrades."[76] His central task, accordingly, was to guide readers and to encourage them to advance from simple belles lettres to more difficult and valuable social and analytical texts of socialism and science. Above all, detailed records were to be kept, and branches were to be judged on the basis of "the more vigilance, the more success."[77] Stern's moralizing attitude in his campaign against trash and kitsch and on behalf of cultural ennoblement set the tone for the SDAP's ideological approach to the function of reading. In place of an evaluation of the needs of readers and their choices carried out by a Marxist analysis (which one might have expected, given the equation of Marxism with social science by the founding Austromarxists), we find sentimental preaching.

Despite the prevalence of such narrow views among the cultural directorate, the worker libraries appeared to flourish.[78] By 1927 more than a million books were loaned out annually. The combination of attractively designed and furnished branches (many of them located in the new municipal housing), 53 well-stocked central libraries, and 1,064 dedicated volunteer librarians[79] brought book borrowing to 2,395,547 in 1931. The actual number of worker library members was far more modest, for only by the payment of small fees were these and other libraries accessible to the public. In Vienna this membership amounted to 40,000 in 1931, or about 10 percent of SDAP membership. This was significantly higher than the library membership of the Viennese per se, but only after 1930. Before then it was not much different from the record of worker central libraries prior to 1914.[80] As a corrective for these bald statistics, one must assume that many worker library memberships served families rather than individuals, and that the circle of readers was substantially greater than subscribers.

But even if one makes such an adjustment, the number of workers who actually drew on the worker libraries was markedly smaller. A breakdown by percentage of the social composition of readers for 1929 establishes that 55.8 were workers, 23.3 were employees, 4.1 were in business, 3.4 were independent professionals, and 13.4 were housewives.[81] This distribution very closely paralleled the social composition of the SDAP, which electorally had begun to resemble a peoples' party extending beyond its working-class base.[82] But it would be a mistake simply to reduce the 40,000 worker-library members of 1931 by some 80 percent who might be considered workers, to arrive at a figure of 32,000 and assume that to be the total of workers inscribed as library members in Vienna.[83] A fact which the extensive publicity of the SDAP for its own libraries quite naturally never mentioned was that an alternative library system existed of which workers constituted 40

Worker library, Sandleitenhof (*Wohnhausanlage Sandleitenhof*, Vienna, 1929)

percent of the borrowers.⁸⁴ In 1897 middle-class liberals founded the Verein Zentralbibliothek, closely associated with the University of Vienna, to provide scientific enlightenment free of all religious and political influences.⁸⁵ By 1910 it had twenty-four branches in Vienna and Lower Austria which loaned out 3.5 million volumes. In 1918 that number exploded to 10 million. Owing to the decline of liberalism after the war and the lower fees charged by the worker libraries, the books borrowed declined to 3 million in 1936.

What books were borrowed from the worker libraries? Did the librarians succeed in carefully guiding workers from "light" reading to the sciences and social sciences which were considered the true cultural enrichments? Belles lettres (novels, short stories, poetry, drama) consistently accounted for 83–87.3 percent of the books borrowed. Works on science, which included largely escapist travel literature, constituted 5–9 percent, and social science, the most important category from the point of view of the SDAP, amounted to 5.8–8.4 percent.⁸⁶ An internal critic charged that the statistics about social science works were heavily inflated, that librarians admitted that such works, even when borrowed, were frequently returned unread. He urged that the worker libraries abandon the fetishistic attachment to this category and pay more attention to the realities and fantasies in workers' lives.⁸⁷

The readers of the worker libraries may have failed to live up to the high intellectual standards expected of them by Stern and others, but they did not fail themselves in providing for their own entertainment, relaxation, and enlightenment. A listing of the most popular authors in Viennese worker libraries for 1933 included the German-language writers Stefan Zweig, Jakob Wassermann, Erich Maria Remarque, and Klara Viebig; the internationally famous Jack London, Upton Sinclair, Emile Zola, Theodore Dreiser, and B. Traven; as well as the popular but (for the Bildungszentrale) undesirable writers Ludwig Ganghofer and Hugo Bettauer.⁸⁸ The choice revealed a very strong interest in social novels and a cosmopolitan taste, even though the adventure fiction of James Fenimore Cooper, Alexander Dumas, and Jules Verne continued to be popular. The SDAP's culture experts should have been delighted with their readers' choices. But they seemed fixated on converting worker readers into collective-minded and future-directed champions readied by their mastery of socialist classics to take up the class struggle at a higher level. From their perspective, the ability to choose could not be left to the workers.

The party's struggle against trash and kitsch was all-consuming and reached ridiculous but also dangerous heights. An example of this mania is the attack on Karl May, the author of numerous American Wild West adventures featuring the white hunter Old Schatterhand and the noble Indian chieftain Winnetou, and a series about Arabia as well. As early as 1910 Stern attacked these novels as unrealistic and romantic, claiming that they led youth into dangerous adventures.⁸⁹ The attack was continued after the war, and May's books were removed from the worker libraries and, as part of

Glöckel's upgrading of books, from the Viennese public-school libraries as well. Even those who claimed some understanding of the workers' need for relaxation and enjoyment, and for a change from a grim everyday reality, insisted that worker libraries should not stock kitsch but help the workers to appreciate "good" books.[90] A lone voice went full tilt against the party and charged that it alienated the cultural establishment from the individual needs and experience of the workers and turned them into mere consumers.[91]

At the SDAP congress in 1932 Stern claimed that the worker libraries had succeeded in bringing about the intellectual transformation of a metropolis and had set a standard for the Socialist International in the use of cultural methods in the class struggle.[92] Such bombast was all too common among the party's cultural directors. It was a self-deception that brought the magical powers of the word no closer to the workers it was intended to transform.

Enrichments of Taste: Music, Theater, and the Fine Arts

The task of bringing artistic culture to the workers was entrusted to the Sozialdemokratische Kunststelle. Under the virtually dictatorial direction of David Joseph Bach, who also was artistic advisor to the Viennese municipal council, artistic editor of *Die Arbeiter-Zeitung*, and editor of *Kunst und Volk*, it acted as a membership organization providing reduced-price tickets to theater and music performances and initiated and supervised the musical, film, radio, artistic, and festival undertakings of the SDAP. Between 1919 and 1923 it was subsidized by the municipality with 6–10 percent of the local luxury taxes.[93] Its publication *Kunst und Volk*, begun in 1926, reported on the Kunststelle's activities and from time to time included critical commentaries on the direction of various aspects of the SDAP's cultural program.

The Kunststelle's most demonstrable success was in transmitting inexpensive tickets to Viennese theater, concert, opera, and operetta performances. By 1922 it had 40,000 subscribers, which declined to a stable 20,000 by 1926.[94] In responding to a sharp criticism (see later discussion) that the Kunststelle was nothing more than a ticket bureau, Bach proudly cited the more than 2 million tickets distributed by his organization in five and a half years, including 200,000 for concerts, 377,000 for opera and operetta, and 1.4 million for theater.[95] As in other aspects of the SDAP's cultural efforts, numbers were used to obscure the precise reality and to silence critics. A closer look at the 193,639 tickets brokered in 1929, for instance, reveals that only 3 percent were to the state opera and 6 percent to the Burgtheater, the two pillars of Viennese cultural life.[96] Bach's intention of providing workers with their just share of the best in art was clearly reflected in the preponderance of German classic and naturalist plays performed by the Viennese theaters on which his ticket program depended.

Such a heavy exposure to elite culture was consonant with Bach's often-stated view that "all true art is revolutionary and acts in a revolutionary way," and "that the conquest of true art [by the workers] is a part of the cultural ideal of socialism."[97]

What exactly was meant by "true art" remained an open question that neither Bach nor any other socialist cultural director answered with any precision. It included dramas of a clearly social, socialist, proletarian, and revolutionary content and modern form. Claims have been made, then and more recently, that Bach succeeded in promoting the performance of modern expressionist plays which would otherwise have been avoided by the theaters.[98] Karl Mark, the former party secretary of the district of Döbling, rejects the assertion that the Kunststelle attempted to sponsor more modern, socialist-oriented plays.[99] He recounts that Bach was prepared to let the production of Brecht and Weill's *Dreigroschenoper* close because of bad reviews, and that its continuation and success was forced by the Döbling party section, which had bought out half the house. It was only the initiative of individual districts, he insisted, which made the production of plays with a decided socialist content by Ernst Toller, Friedrich Wolff, Karl Credé, and others possible. Bach, he concluded, was simply not interested in bringing theater to working-class neighborhoods distant from the city center.

Whatever Bach's inclinations, as long as he purchased blocks of tickets in existing theaters, he was forced to abide by their scheduled productions and to include a good number of trivial plays for the few original ones he could promote. It was clear that, without at least one theater of its own, the SDAP's aim of providing workers with plays that fit its notion of cultural enlightenment could go no further. In 1928 the Kunststelle acquired the Carltheater, which closed after two productions because of financial difficulties. The long-standing dissatisfaction with Bach's management both within the Kunststelle and outside it was aired in public. The opening salvo had been fired by Karl Kraus when he attacked the diet of trivial plays and operettas offered to the workers as being (in a play on words) "an Stelle der Kunst." Kraus also demanded to know why the Kunststelle was content to be a ticket bureau when, in view of the large audience it made available to the theaters, it should have influenced and determined the programs themselves.[100]

A more fundamental critique of the entire orientation and program of the Kunststelle came from Otto Leichter, one of the editors of *Die Arbeiter-Zeitung*.[101] Despite the fact that the Kunststelle had more possibilities than any other organization save those in Soviet Russia to create proletarian cultural forms, he charged, in the ten years of its existence it had failed to realize its potential. This failure could be attributed in part to the unwillingness of party functionaries to engage in cultural work that was open to criticism, and in part to the municipality's inability to provide adequate financial support, given its heavy investment in public housing. But most of all, he insisted, cultural demands had aimed too high, and too much had been expected of the workers in a tutelary manner. The charge, thinly veiled, was directed at Bach, who had consistently demanded that the highest art be aspired to because it was also the most revolutionary.[102] Cultural programming, Leichter continued, had paid too much attention to the taste of coffee house aesthetes and not enough to the proletarian in his neighborhood *Gasthaus*. That proletarian had a right to popular, light, and distracting entertainment without being made to feel guilty if he chose less than the highest quality. It was up to the Kunststelle to wean the workers from these and to lead them to a higher cultural level.

No matter how telling, Leichter's critique offered no suggestions about how to solve the fundamental problems that plagued the SDAP's cultural enterprise: What were these higher, more proletarian, more socialist-directed, specific cultural forms? How should workers be "led" from their traditional involvement in commercial culture and growing consumption of mass cultural offerings to the higher goals? If Leichter was simply rhetorical on the subject of alternatives, he was disingenuous on another important point. He seems to blame the party for not giving the cultural programs sufficient support, but exonerates it by citing the municipality's heavy financial commitment to housing. But it surely was no secret that the municipality also gave substantial subsidies to the major institutions of Viennese elite culture—the state opera and Burgtheater, for instance. Was it unthinkable for the socialist-dominated municipal council to reapportion its cultural subsidies to include a more substantial support for proletarian cultural activities? Or were these pillars of elite culture too much of a "holy cow"—too admired by the SDAP leaders themselves as symbols of Vienna's cultural excellence—to be tampered with?

Bach's response was to reject all criticism and to insist that the Kunststelle's cultural program was a great success.[103] He did this in the by-now-customary way of listing the large number of tickets sold. At the same time he rejected the charge that the Kunststelle was nothing more than a ticket bureau. It had a cultural-educational function and therefore sought to provide the best artistic quality. The workers, he suggested, were free to seek "mere entertainment" in the majority of theaters, which offered such fare in large measure. He neglected to say that only the block buying of tickets by the Kunststelle made reduced prices possible, and that the workers were far from free to exercise their preferences.

It was in the realm of music that the Kunststelle came closest to Bach's aspiration of providing the "finest and therefore most revolutionary art" to potential worker audiences. The vehicle for this attempt was the worker symphony concerts, begun in 1905, which had become a centerpiece of the SDAP's cultural program. From 1926 on the concerts were under the baton of the Schönberg student Anton von Webern. During the 1920s an average of eight programs were offered per year and generally repeated two or three times, declining to three programs without repetition in the early 1930s.[104] The musical orientation of the concerts was plagued by an important ambiguity. As in all cultural endeavors of the SDAP, a new form of expression and content was to be developed to aid in the transformation and class iden-

Socialist Party Culture 99

tification of the workers. This meant developing an alternative to both classical and modern music—something akin to the Soviet Russian experiment of combining the technology of machines, factory sirens, cannonades, and fog horns with mass choruses and instrumental music.[105] Bach remained wedded to the classical heritage up to contemporaries like Schönberg and von Webern, clearly the domain of elite culture. But Bach refused to see this heritage as elite. Art, he held, conforms to laws of its own, unrelated to societal reality; great artists retain their value because they rise above their class.[106] Thus for Bach, as well as for other socialist culture experts, bourgeois cultural forms and practices attained a revolutionizing value simply because the workers apprehended them.[107]

If we look at the programs of the worker symphony concerts, it comes as no surprise that they hewed to the German classical tradition, with Beethoven being worshiped as a musical Prometheus.[108] Forays into contemporary forms—the rhythmic, harmonic, and melodic combination of classical music with jazz, or experiments with twelve-tone music—did not diverge from practices in the dominant culture.[109] Who was the audience of these concerts? For once we are not inundated with a proud display of numbers, for no record seems to have been kept. It does appear, however, that workers were only a small minority, compared to employees, managers, and intellectuals.[110] Private listening was possible through recorded music, but phonographs and records were beyond the reach of most workers. From time to time worker symphony concerts were broadcast on the radio, but they ranked near the bottom of programs preferred by radio listeners (see chapter 5).

Not only were worker symphony concerts a misnomer with respect to the audience they served, but workers who did attend were expected to conform to the norms of bourgeois dress and comportment. A glowing description of the first concert in 1905 reports with pleasure that the teamsters who attended did not disgrace themselves or the party by dressing differently.[111] It concludes by observing that workers have learned to behave properly: they understand that classical music has to be appreciated in festive clothing.

Bach apparently believed that the worker symphony concerts had made an important beginning in rectifying the musical deprivation of the workers and were providing them with their just share of the elite cultural tradition. But he, as well as the party and trade unions, was far less certain about how to deal with traditional musical practices in worker communities, or whether these should even be considered as potential contributors to a working-class culture.[112] But the very rich musical expression in everyday worker life, going back to the late nineteenth century, could not simply be neglected. It included singing ranging from the family setting to two- and three-part harmonies at local feasts, to organized choruses singing sentimental popular ballads, folk songs, hymns, operetta hits, tendentious lyrics, and declarations of the workers' movement.[113] Nor was there a shortage of musical instruments in the worker tenements, where the zither, mandolin, guitar, and accordion predominated, but violins, wind instruments, and the piccolo piano were played as well.[114]

Most of this grass-roots music making remained untouched by the Kunststelle. Only the formally organized efforts, such as singing societies and mandolin orchestras, were made part of the SDAP's cultural program. Transforming the singing societies into worthy exponents of the party's future-oriented vision proved extremely difficult. The eighty-odd separate Viennese organizations with a total of 5,000 members, of which 90 percent actually were workers, had individual characteristics and traditions.[115] Of these, their conviviality and love for banners, their feasts with alcohol, their frivolity, and especially their preference for light entertainment were condemned by the Kunststelle as petty bourgeois habits which had to be overcome.[116]

As a corrective, attempts were made to upgrade the musical content by emphasizing the classical heritage, in the hope that professionalization of the singing societies would also reduce forms of undesirable sociability.[117] But only a tiny number of the societies could master the complicated choral compositions; most remained wedded to folk and light popular music.[118] From the late 1920s on, when most cultural organizations were affected by increased political struggles, the societies were joined into mass choruses to perform newly created socialist cantatas, oratorios, and militant songs. But these were hardly new cultural forms of the working class. "Pompous mass choruses, hymnlike compositions . . . the musical style often resembled that of bourgeois late romanticism. The preferred form of the whole, however, was the religious oratorio."[119] The singing societies may have participated in these spectacles, but they retained their original sociability and musical preferences.

Fine arts seem to have had a low priority in the SDAP's attempt at cultural reform. Aside from the placement of contemporary proletarian-oriented sculpture, such as that of Anton Hanak, in the new municipal housing, and a limited production of inexpensive art reproductions, few practical measures were undertaken to improve worker taste or to turn it in a socially positive direction. But from time to time articles by associates of the Bildungszentrale denounced the bad taste of workers, claiming that it reflected the degree to which the bourgeoisie still held the working class in thrall. In a particularly vicious attack, no doubt read only by party functionaries, the culture expert Richard Wagner lambasted the average proletarian for the pictures, knick-knacks, framed proverbs, postcards, antimacassars, and other purportedly artistic objects he used to decorate his home.[120] Public museums were not visited, he charged, because the worker's home was a museum in itself of the means by which the bourgeoisie waged the class struggle against him and at his own cost.

Wagner hardly went beyond a diatribe; others tried to be more helpful. In response to a survey about pictures in the proletarian home, one commentator suggested that the typical "cheap reproductions of landscapes, still lifes, and genre paintings in bad taste" be sold or burned by the worker housewife and that she should paint her walls in plain colors to attain a calming effect.[121] Only then should pictures be chosen carefully—a portrait of one of the socialist leaders or a familiar and pleasant landscape. "In short, a picture that suits every mood and does not disturb any guest."

If this approach to art in the worker's private sphere was to substitute one form of kitsch for another, others were concerned about enhancing the worker's taste in art by exposure to the very best. Can the contemporary proletariat achieve the true enjoyment of art? one commentator asks.[122] The bourgeois answer would be no, because workers lack the tradition and sensibility. But the worker can be brought to full enjoyment if he visits museums. Once there, he should avoid painters like Rubens and Raphael, whose subjects are far removed from his own experience. He must seek out "social, accusing, and rebellious art that portrays his sorrows and needs," such as the work of Rodin, Millet, Meunier, Kollwitz, and Zille.

This attitude, that the worker will feel most comfortable with and most appreciate reflections of his own everyday hardships and struggles—a kind of reinforcement of misery—rather than the real and imagined world outside his ken (invariably called escapist), permeates the SDAP's orientation toward books, music, theater, and art. The emphasis on social content in such realistic renditions as those of Kollwitz or Zille closely resembles the turn toward socialist realism accompanying the rise of Stalin in the Soviet Union. However well intentioned, asking the worker to sit down to such a meager and prescribed cultural meal, seasoned with admonitions in paternalistic tones, did not succeed in enlarging the realm of socialist party culture.

During the late 1920s the Kunststelle added the management of worker feasts and festivals to its activities (to be discussed in the next section). With the increasing politicizing of these as well as most other cultural activities that put propagandistic values in the forefront, Bach's Kunststelle with its "elitist art chatter" became redundant.[123] Latent conflicts between Bach and youthful critics in the organization came to a head and led to the departure of the Socialist Performance Group, committed to agitprop and cabaret.[124] In 1931 *Kunst und Volk* ceased publication, allegedly for financial reasons. The annual report of the national Bildungszentrale for that year no longer listed the Kunststelle as a member organization.

One is forced to agee with Otto Leichter that the Kunststelle failed in its mission—not, as he insists, because it did not live up to its potential, but because it pursued false and hopeless goals. Most of its artistic program was based on turning the worker into a passive consumer. Whatever active artistic resources existed in working-class communities were rejected or ignored as petty bourgeois and therefore in bad taste and subversive of class interests. Ultimately, the Kunststelle's main goal was to ennoble the art appreciated by the workers by bringing them in contact with elite culture.[125] The presumption underlying that ennoblement was that a particular socialist orientation to elite art could be devised, so that the worker consumer would be set on the path of socialist values in his transformation.

The difficulty with this assumption was that none of the SDAP's culture experts could conceptualize what such a socialist orientation might be. In practice, vague admonitions and rhetoric obscured the emptiness of the party's pretension of converting elite culture for its own aims. This led to ridiculous claims like the assertion that a work of Beethoven played by the workers' symphony was of a higher order than the interpretation of bourgeois orchestras, because the true revolutionary essence was being extracted.[126] The Kunststelle failed not because it was unable to formulate and set into practice a socialist conception of art. It failed because it undertook an impossible task: to devise an abstract model unrelated to the social realities of working-class life. The reformist SDAP refused to accept a reformist cultural goal: that the workers' partaking of bourgeois culture is part of a dialectical historical process. The party's high-toned Austromarxist rhetoric did not allow it to subscribe to such an interim goal.

In retrospect, and despite the glorifying exhibits with catalogs mounted in Vienna in the past decade, the efforts of the Kunststelle can be seen as marginal to the majority of Viennese workers. It cannot be denied that the elite of workers were reached and did benefit from its programs. But it remains doubtful whether this was enough, whether 5–10 percent of the workers would have been a leaven sufficient to permeate the whole working class in later, less troubled decades, and whether a party that was democratic rather than "vanguard" should have entertained such hopes. It is an important question about the whole cultural experiment, to be discussed more fully in the conclusion.

To Culture Through Action:
Sports and Festivals as Symbols of Power

The Socialist party culture discussed so far was based on the intellectual transformation of workers through education. Newspapers and journals, lectures and libraries, theater and concert performances, vocal and instrumental practices, and artistic initiations, though aimed at all the workers, reached only a varied minority. Two closely related aspects of that party culture, organized sports and mass festivals, were able to attract a larger number of Viennese workers. They succeeded mainly because they offered an easy form of association with mass experience, catered to psychological needs of workers for relaxation and expressions of self-worth in compensation for the increasing tempo and depersonalization at the workplace, and provided symbolic assurances of collective strength.

A recent assessment of the SDAP sports program exaggerates in calling it "a form of cultural revolution . . . that contributed more to creating a socialist consciousness and conduct than the best-intentioned experiments of socialist education."[127] Yet there is some truth to this assertion. Historians of worker sports have for some time attributed mass participation to the shortened workweek, allowing for more leisure time, and to the intensity, sterility, and tension-producing quality of modern production, which created anxiety and failed to offer personal satisfaction. Sports provided the means for achieving a sense of physical accomplishment in a setting of community and even solidarity.[128] By 1931, socialist organized sports in Vienna included 110,000 active members distributed over a wide variety of clubs practicing virtually every kind of sport.[129] That membership, based on the payment of small fees, had remained stable for half a decade despite the fact that by 1933 nearly every third worker was unemployed and the trade unions had lost one-quarter of their members.[130] It remains an open question, however, whether this stability reflected the loyalty of members or a stagnation of the organization. For an answer we will have to turn to the development of the SDAP sports program, its suppositions and practices.

The worker sports movement began before the turn of the century largely as a reaction against alcohol and its debilitating effect on the workers. But early on this utilitarian purpose was superseded by the hope that sports, through physical competence, could lead to the attainment of a measure of socialism here and now.[131] Various independent organizations such as the gymnasts, cyclists, and Friends of Nature were merged into an umbrella organization in 1919. Even then organizations such as the worker chess club, whose relation to sports was doubtful, were included. The final organizational form was attained in 1924 with the creation of the Arbeiterbund für Sport und Körperkultur in Österreich (ASKÖ), responsible to the SDAP through the Bildungszentrale. It coopted the clubs for swimming, handball, fishing, tennis, and heavy athletics (wrestling, weight lifting, hammer throwing, and later boxing), as well as organizations whose primary purpose was not sports, such as the paramilitary Schutzbund, the Socialist Worker Youth, the Trade Union Youth, and the Worker Radio Club.[132]

The growth and centralization of worker sports did not proceed without conflicts. In 1923 the gymnasts demanded that all worker sports organizations be forbidden from contact with bourgeois sports activities. The worker soccer association, which played in the middle-class soccer league, refused to conform. This led to a temporary split between the soccer players, for whom sport was sport and politics was in the party and the trade unions, and the gymnasts, who insisted that sports were part of the class struggle.[133] In view of the unity attained by ASKÖ in the following year, this might seem to be a minor episode, but the role of sports in the SDAP's cultural program, and particularly the relationship of sports to politics, remained a constant problem and source of dispute.

From its earliest days worker sports had been regarded with suspicion if not hostility by leaders of the Socialist party and the trade unions. The common view was that sports were merely pleasurable diversions from the real struggles of the workers' movement and, lacking any political content, they did not deserve recognition or support.[134] A proposal that worker sports organizations be represented at the party congress in 1923 was rejected on the grounds that these members were already represented through their party and trade unions.[135] Only when a political content for worker sports was devised by the SDAP, were they grudgingly accepted as part of the cultural program. Even then, sports were never considered on a par with other activities which committed party members or cadres with political maturity were expected to engage in.

Politicizing worker sports meant first of all to differentiate them from their bourgeois counterparts, and much ink was spilled on this point. Bourgeois sports, it was charged, laid claim to a neutrality which deflected workers from their class interests, oriented them toward bourgeois morals, and turned them into gladiators.[136] Bourgeois sports, morevoer, were based on an individual star system and record performance reflecting the capitalist order, where the strong triumphed over the weak.[137] Their greatest sin was the conversion of the playing field into an immoral stage on which performers entertained passive spectators.[138]

Supporters of worker sports maintained that their fundamental difference from bourgeois sports, which appealed to the lowest instincts and indeed to Mammon, consisted of their progressive education in socialism. The body and its physical development, it was argued, was the natural starting place for other cultural and educational advancement, and the unity of body and mind was more productive than purely mental activity. But such physical development was not to be considered an end in itself; it had to be associated with education aimed at the socialist goal. Every sports or gymnastics leader was directed to become an "apostle" of that idea.[139]

Above all, the argument ran, worker sports were a collective experience. In distinction to art, which was the work of experts and demanded spectators, sports could be participated in by everyone and were the domain of amateurs. Spectator sports were at best a private pleasure,[140] but even they could be given a socialist content. If a worker soccer team played and workers watched passively, Anton Tesarek, head of the Rote Falken, claimed, an educational process was taking place.[141] Competition which tested the workers' strength and increased their performance was considered natural, so long as it did not become excessive and hinder the complete development of the individual.[142] Hans Gastgeb, secretary of ASKÖ, summarized the SDAP's expectations: a combination of mass sports with political enlightenment. This meant sports not simply for distraction or recreation but for the creation of a proletariat mentally and physically prepared for struggles to overcome the reactionry capitalist system.[143]

If one compares the SDAP's antibourgeois admonitions with the proposed socialist distinctions of worker sports, the vagueness of the formulations becomes apparent. In practice these alleged differences must have largely disappeared. How could competition in team sports really be kept within bounds, and victors not be acclaimed?[144] Even in individual sports such as gymnastics and cycling, were there not excellent and inferior athletes, and could the losers of a race be really satisfied with "having done their best"? One looks in vain for the insights of Adlerian individual psychology allegedly absorbed by party functionaries (but more of that later). The distinction between good proletarian participant and decadent bourgeois spectator is also ambiguous. Could not a worker active in ASKÖ sports be both, especially where team sports were involved?

The most problematic aspect of socialist formulations was the process by which politics became implanted in worker sports (leaving aside the notion that participation in worker sports alone was a political signifier). No doubt SDAP sports functionaries were correct in claiming that the sense of community and solidarity provided by organized worker sports indirectly compensated, at least in part, for the feelings of inferiority and resignation instilled in the proletarian at the workplace and in capitalist society in general.[145] But such an autonomic effect took place without the party's intervention or control, which were considered essential in giving sports a political content. A more direct method was introducing a political lecture or discussion before each practice session preceding a sports activity or event.[146] This requirement led to complaints by leaders of team sports that practice sessions were simply being avoided by the players.

The technique most widely employed to instill a sense of relatedness to the ongoing class struggle was the enforcement of strict discipline in all sports activities. Drill, athletic uniforms, demonstrations of precision, and obedience to team leaders and instructors were accepted early on by the gymnasts, and by the mid-1920s were generalized throughout ASKÖ. Public demonstrations of the discipline of worker athletes, uniformed and well groomed, marching in step and moving like a mass ornament, were expected to counter the public image of the disorderly worker.[147] They went hand in hand with the SDAP's attempt to discipline the worker family to make it more orderly (see chapter 6). They did not contribute to making worker sports more attractive, particularly to youth, and it is doubtful whether they enhanced political education in the sports organizations, save to make them usable for political purposes such as demonstrations and festivals extraneous to their primary interest. It was probably the increasing emphasis on discipline to the point of militarization in the early 1930s, more than the effects of massive unemployment, which prevented ASKÖ from reaching larger numbers of workers who chose other alternatives.

The paramilitary orientation of worker sports began in 1925 with the creation of combat gymnastics as a separate organization in ASKÖ. It was limited to eighteen-year-old males and concentrated on such military skills as hand-to-hand combat, hand grenade throwing, obstacle course running, and maneuvers.[148] In 1926 the Schutzbund joined ASKÖ, Julius Deutsch became the head of both, and the movement toward militarization accelerated. After the bloody events of July 15, 1927, virtually all member organizations were required to take up combat training, which took the place of all other attempts at political activity and came to be viewed as the natural contribution of worker sports to the class struggle. Rigid discipline, a sworn oath, a leadership cult, and the banning of all political discussion marked the end of the SDAP's original goals[149]; a socialist cultural movement was turned into a reserve army in the class struggle. Widespread resistance to this development among essentially pacifistic workers and especially youth

Mass calisthenics in the Prater during the International Worker Olympics, 1931 (VGA)

had little effect on Deutsch, for whom the increase of Nazi votes in the elections of 1932 meant only that the tiger was at the gates.[150] One of the little-discussed side effects of the increasing masculinization of worker sports through combat training was to affirm the second-class status of women athletes, already largely segregated into the kind of sports "appropriate for females."

Even though the SDAP failed to turn worker sports into an institution for the political education of the masses or to transform a basically recreational activity into an instrument of class struggle, worker sports produced symbols of political importance. In addition to the obvious signs of discipline and orderliness, solidarity and community, the worker athlete's body itself projected an image of power and control.[151] It counteracted the picture of workers weakened by toil, in dirty work clothes, disorderly, and victimized.[152] The body of the worker athlete, muscular and subtle, individually an example of aesthetic purity but collectively a symbol of orderliness, power, and control, projected an image of the "party soldier" ready to defend his class, antithetical to the worker slave of bygone days. That powerful image was projected to workers who remained outside ASKÖ's formal structure and to wider audiences outside the working class. It transmitted a political message of strength independent of the SDAP's confused attempts to link sports and politics. It was recognized and fostered by SDAP and socialist municipal leaders in the final defensive years as an emblem to dispel the reality of political indecision.[153]

The high point of the emblematic quality of worker sports was reached during the second International Worker Olympics held in Vienna on July 19–26, 1931.[154] Preparations for this event included the completion of a modern stadium with 60,000 seats and the reduction of transportation fees for the expected participants.[155] The summer games in Vienna attracted 77,000 participants from nineteen members of the Socialist Sports International, including 37,000 Austrian athletes. For four days, 200,000 spectators filled the stadium to watch competitors in every sport, including combat gymnastics. On the last day 100,000 participants marched in disciplined ranks across the inner city, traversing the Ringstrasse on the way to the stadium, and were watched by countless spectators.[156]

For nearly a week Vienna became a workers' city as never before or after. The symbol of Austrian worker strength in the context of the international working class was projected in two directions: as a summons to the unorganized and nonparticipating workers to join these powerful ranks, and as a message to political opponents that the workers stood ready to protect themselves and the republic. That the military aspects of this spectacle were also obvious is reflected in the comment by Friedrich Adler, secretary of the Labor and Socialist International, present in Vienna for the organization's fourth world congress. He characterized the worker olympics as "an international military parade more powerful than anything else the working class has succeeded in so far."[157] One of the most spectacular events of that week was the mass festival performed in the stadium by the whole panoply of SDAP cultural organizations. It clearly revealed the important links between worker sports and socialist festival culture in representing symbolic power.

The desire to emotionally attract and bind the workers to the party dates back to the earliest days of Austrian social democracy. Victor Adler's adage

Banner parade at the International Worker Olympics, 1931 (VGA)

that "the brain is an inhibiting organ" stood in stark contrast to the fundamental intellectual orientation of Austromarxism.[158] With the phenomenal growth of SDAP membership in the early 1920s, the idea of integrating the majority of members, who remained outside the party's programs and activities, through emotional appeals again attained currency. The proletarian feast, with its roots in a preurban past and shaped by everyday practices in worker communities, became the vehicle for this orientation.

Integrating such feasts into the SDAP's cultural program meant not only to invest them with a particular socialist political content but also to cleanse them of undesirable characteristics ranging from their setting in *Gasthäuser* to their association with the consumption of alcohol and light entertainment (folk music and marches).[159] In the earliest phase of SDAP attempts to transform worker feasts into proletarian festivals, a calendar of socialist holidays was devised to compete with those of the Catholic church.[160] To these were added a variety of working-class holidays and celebrations of the birthdays of persons significant to the movement. But there were frequent complaints registered with the Bildungszentrale that at the local level such serious commemorations either took place in an unworthy setting or retained an unsocialist, folksy character.[161]

The SDAP sought to overcome these local deficiencies by centralizing worker celebrations. At the behest of Joseph Luitpold Stern the Kunststelle took over the organization and supervision of festivals in 1930. In that year it was already able to organize seventy-three of these on behalf of party locals, trade unions, and other party organizations. Even though it devised the programs sent out, it could not control their execution, so that seriousness and educational values did not predominate over entertainment. Local worker feasts remained a grass-roots affair throughout the period, resisting SDAP attempts to make them conform to its concept of party culture. In retaining their original integrity, worker feasts reflected the ongoing struggle between indigenous forms of worker culture and the party's attempt to impose its own blueprint.

But local worker celebrations were hardly the means of emotionally drawing the masses of workers into the party. That aim was to be realized in the development of mass festivals, in which the varied forms of party culture were to play a role. An impetus for such a marshaling of cultural resources, it has been suggested, was July 15, 1927, when the workers en masse stormed the Palace of Justice and set it on fire. This act sent shock waves through the party leadership, who judged the workers to have behaved like illegitimate masses—like "pitchfork revolutionaries," in the words of Karl Renner.[162] It resulted in an aestheticizing of politics in which worker festivals played an important role, partly to expunge that defeat from memory and partly to tame and control the masses.[163]

Though the timing of the SDAP's effort to make mass festivals a central part of its cultural program was no doubt determined by the trauma of July 15, 1927, other motives helped to produce this shift in focus. The most persistent of these, time and again stated as a primary aim by the culture experts, was to integrate the party masses previously barely or totally unaffected by the party's organizing efforts.[164] It was understood that attracting the passive workers also meant to compete with radio, film, and spectator sports, which had a growing popular appeal. Another important motivation was the continuing success of the Catholic passion and mystery plays and oratorios, performed on high holidays, which in updated versions celebrated God's mystery, and the church as the all-inclusive community. A similar source of concern for the socialists was the tremendous success of neo-Catholic spectacles mounted in Salzburg by Max Reinhardt, Hugo von Hofmannsthal, Hermann Bahr, and Richard Strauss. There the attempt was made to fuse players and spectators in order to leave the chaotic present behind for a medieval unity and baroque protectedness.[165]

The idea of an all-encompassing mass festival akin to Richard Wagner's conception of a *Gesammtkunstwerk* had fascinated SDAP leaders for a long time.[166] But the actual form of the mass festival was derived from Russian *proletkult* experiments in the early 1920s, as transmitted by German communist festivals of that time.[167] It was adapted by the SDAP in Vienna during the late 1920s, when it had already been abandoned elsewhere. The masses, Sergei Eisenstein said in an interview in 1929, had been used up as a theme.[168] In its final form the SDAP's mass festival made use of the *Sprechchor*, the group recitation of prepared texts, which also had been transmitted from the Russian *proletkult* via the German communists.[169] The SDAP's cultural leaders celebrated it as the true art form of the proletariat, involving commitment, collective effort, and unity and communality.[170] It was proclaimed as the voice of the masses expressing the collective will in a disciplined manner.[171] But its formal presentation in mass festivals was not left to workers at the grass roots. The demand for perfect spoken German (rather than the Viennese dialect ubiquitous among workers), for precision attained only after extensive practice, and for discipline, required professional performers.[172] A special *Sprechchor* of trained speakers was established by the Kunststelle for recitations in the party's important festivals.

The most successful mass festival was mounted in July 1931 during the Worker Olympics. The playing field of the newly created stadium was transformed into an immense stage on which musicians, actors, speakers, gymnasts, singers, Rote Falken, and Socialist Worker Youth in blue shirts—4,000 in all—performed a historical spectacle against a background of artistic props and stage effects.[173] In the course of an hour the history and future of struggling, laboring humanity, shown in tableaux vivants, symbols, and mass movements, led the audience from the Middle Ages to industrial capitalism.[174] Toward the end, a huge gilt idol representing capitalism was overthrown by collective strength, and thousands of red flags carried by youth in white ushered in the socialist future. From a high red tower a splendid voice intoned an oath in the manner of statement and response typical for the *Sprechchor*, only this time the audience was prompted to give the response: to swear that collectively it would fight for the liberation of working people and to remain faithful to the high ideals of socialism. The singing

Mass festival in the new stadium, with participants shown toppling the symbolic gilt idol of capitalism, 1931 (VGA)

of the "International" ended the festivities. This spectacle was repeated four times before a total audience of 260,000. The success of this festival—resulting in part from the concurrent Olympics—in putting the SDAP on display before the larger Viennese public, led to euphoric accounts in the party's publications.[175]

Looking back on this most important of the socialist's mass festivals, it is difficult to share the party's sense of triumph. This and other celebrations held between 1928 and 1931 had a compensatory character for the actual political reversals of the SDAP. They pointed to a future of success while leaving current events out of the picture. During that time the festivals may indeed have served as a transmission belt to the mass of unintegrated workers, but at best this effect was not truly political but more in the realm of *mentalité*—part of a general climate of working-class presence and importance—which municipal socialism and, to a lesser degree, the party's cultural program were able to enhance. But the mass spectacle also may have been a vast theater of illusions in which the power to topple the capitalist idol was nothing but a magical deception within the protective concrete walls of the Vienna stadium.[176]

Outside those walls two months earlier, the Styrian Heimwehr leader Walter Pfrimer had attempted a putsch. It failed, but signaled the intensification of politics by force.[177] The 1931 festival, which made the masses themselves the hero with the power to direct the course of history, did not confront this and other real threats to the republic on whose survival the SDAP depended. Performers and audience intoning an oath in refrain had an exorcistic effect in which "the power of individual cognition was surrendered and the power of decision making reduced in favor of a romantic and emotional attachment to a value system."[178] That such exorcistic rites, in which the individual lost himself in the all, could be even more dangerous is revealed in the ritual of Socialist Worker Youth around their campfires at the festival marking the summer solstice. In the prescribed litany of what amounts to a socialist auto-da-fé, all enemies were to be consigned to the flames: "trashy films, bad books, beer, liquor, wine, swastika, pipe, cigarettes, and all foolish fashions."[179]

In the increasingly threatening political environment after 1931, the SDAP abandoned the chiliastic, pseudoreligious message of mass festivals and attempted to transform them into mass demonstrations with a more radical, ritualized content.[180] The move toward cultist expressions was in keeping with the aims of a newly created propaganda center within the Bildungszentrale under the direction of Otto Felix Kanitz. Its demand for more, better, and still more propaganda, without discussions of tactics by the rank and file, brought underlying differences between the party youth and leadership into the open. Kanitz and Deutsch, representing the leadership, argued that, in view of the increasing Heimwehr and Nazi threat to the republic, the disciplined use of propaganda was the order of the day and that democratic decisions on tactics would have to wait. Ernst Fischer, speaking for the socialist youth, countered that it was impossible to mobilize the masses without prior political discussion and that sheer obedience to party directives would lead to a vague romanticism.[181] Although the points made seemed academic, they had very serious undertones. Turning the participants of festival demonstrations into party soldiers was to depoliticize them at a time when the party leadership was more and more following a policy of retreat.

During this final party crisis the festivals were stripped of their cultural aims and replaced by attempts to confront reaction and fascism in the streets by symbolic means.[182] The SDAP became a participant in a kind of street theater in which it demonstrated its public presence with flags, emblems, blue shirts, party badges, slogans, greetings, and demonstrations in response to those of its opponents.[183] But the raised right arm with clenched fist to counter the Nazi salute, the three-arrows emblem to negate the swastika, and the greeting "Freiheit" to drown out "Heil Hitler" were empty symbols, lacking in socialist substance and goals.[184] They neither encouraged confidence as expressions of defense, because the real struggle was not a street drama, nor could they mobilize the masses of workers for socialism's higher goals, which events had pushed to the margins of daily reality.

As we have seen, socialist party culture was not some well-thought-out scheme directed by a powerful central agency, but a gathering of diffuse organizations including many whose relationship to socialism (burial society, chess club) or to workers (vacation society, aviation club) was tenuous at best. The SDAP tried to infuse these with long-range socialist perspectives—to strip them of their plebeian and petty bourgeois content and ennoble them with infusions of elite culture. Such enrichments frequently contradicted the party's attempt to differentiate its cultural program from bourgeois forms and practices. A case in point is its attack on spectator sports at the same time that the Worker Olympics and mass festivals marshaled thousands of spectators.

Considerable resistance was experienced in the efforts by organizations and publications to transform the workers into "neue Menschen" ready to enter the socialist promised land. Singing societies and worker feasts resisted centralization and concentration and clung to their subcultural forms of entertainment. Rank-and-file workers disregarded the avalanche of printed matter heaped upon them, especially the large number of free publications. Members of worker libraries could be brought to appreciate the social novel, but not the difficult social and scientific literature the party considered the goal of cultured reading.

The SDAP's cultural experts deceived themselves with the alleged numerical success of their efforts. Workers subscribing to nonparty libraries or reading the popular tabloids were not taken into account in the self-congratulations. Nor was the fact that for every worker in ASKÖ there were twice as many in nonparty sports organizations. These miscalculations stemmed from a failure to realize or accept the fact that workers disposed of their newly gained leisure time in various old and new ways. Older forms of recreation such as *Gasthäuser,* the *Heurigen, Varieté,* the Prater, and hiking were still enjoyed. The newer forms included radio, cinema, and spectator sports, which had an increasing appeal. No allowance was made for unorganized and spontaneous leisure activities such as sports, which afforded large numbers of young people a self-expression that the discipline-oriented ASKÖ denied them.[185] In short, the idea that the SDAP's culture could fill the workers' whole private sphere was unrealistic and ignored existing life-styles in worker subcultures.[186]

Whereas Socialist party culture was in theory a democratic outreach to the masses of members, it affected only a minority of them. This was hardly admitted or possibly not even recognized by the functionaries in charge, partly because, for many, keeping their particular organization or activity going became an end in itself, and partly because they were victims of their own claims of success. Perhaps their unfamiliarity with a mass party so different from the prewar committed vanguard helps to explain their difficulty in reaching unfamiliar masses,[187] and in understanding that the desired cultural transformation of workers depended more on their receptivity than on the theories and programs themselves. A number of leaders of the party's left wing were well aware of these shortcomings and during the last years of the republic protested against the "elitism" and "social conservatism" of functionaries satisfied with reaching only an "aristocracy" of workers.[188] But by then, time had run out.

Much has been written about the influence of Adlerian individual psychology on the SDAP's cultural program.[189] Although it certainly influenced a small group of university students, educators, and social scientists, most of these did not set the tone of the party's cultural program. Its directors' conception of Adlerian theory consisted mainly of the notion that the individual's psychological makeup was malleable and that developmental feelings of inferiority and inadequacy could be rectified. The SDAP's program was predicated on the ability to ameliorate the presumed deficiency of selfhood (a consequence of capitalist alienation) in the worker's personality.[190] But there was a negative side to the idea that the worker was clay to be shaped and a blank page to be filled by the party. This notion was based on the view that the worker existed at a brutish level subject to alcohol, sexual excesses, and the manipulations of capitalist society, a dark image which only the party could turn into one of light.[191] It would seem that the Adlerian influence was an impediment to the SDAP's cultural aims, because its inherent environmentalism allowed functionaries to believe that the worker, being clay, had as yet no existence worthy of the name. Therefore the worker did not need to be understood or reached but only shaped. This misconception might explain why so many of the cultural programs aimed wide of the mark.

The socialist cultural leaders had the illusion that their aims would be achieved in the short run. "We in Vienna," Marie Jahoda remembered, "lived with the great illusion that we would be the generation of fulfillment, that our generation would bring democratic socialism to Austria."[192] Such idealism and blind optimism revealed the tendency to view history as an act of volition rather than as a process—a contradiction of the hallowed tenets of Austromarxism. Socialist Party culture provided new opportunities to a part of the working class, but its lofty theories about actualizing socialism kept most workers at a distance. Coupled with the reforms of municipal socialism, it created a climate of opinion congenial to the workers of Vienna. This achievement was greater than the sum of all the members and participants in the activities themselves.

CHAPTER 5

Worker Leisure: Commercial and Mass Culture

In the SDAP's quest for a total worker culture, party leaders used their power base in the Vienna municipal administration to transform the workers' public sphere. As we have seen, reform efforts in housing, public health, education, and welfare were aimed at producing a new, orderly, and class-conscious working class.[1] These efforts, and especially the party's large cultural program, went beyond attempting to establish a more favorable worker environment; they were clearly aimed at changing worker behavior as well. In seeking to alter the public sphere, party reformers encountered a traditional political opposition from the Christian Social party, the Catholic church, and economic pressure groups. In attempting to dominate the workers' private sphere as well, far more elusive obstacles and opponents were arrayed against the party: worker subcultures; the dominant bourgeois culture, about which the socialists' position was ambiguous; and an emerging mass culture. Mass culture, particularly in its ability to commercially permeate everyday life, was a powerful adversary to the party's attempt to shape the workers' leisure time and private space. Its most unique quality, perhaps, was its ability to transfer the mass production of the working world to the arena of leisure time.[2] As we shall see, party leaders wavered between rejecting and wanting to ennoble what they considered vulgar influences in entertainment and consumption, and struggled against this newest manifestation of pleasure for its own sake.

In the period under consideration it would be a mistake to draw a sharp distinction between commercial and mass culture. The latter may be seen to emerge in Austria and elsewhere in Europe after the turn of the century and to begin to replace older forms of commercial culture. We cannot lose sight of the fact that mass culture is commercial. But it involves commerce of a more advanced type, geared to the production and consumption of large quantities, reaching various targeted markets, but aiming at a comprehensive one. It clearly parallels the most advanced techniques and aims of the most advanced industry at the time.[3]

In the years following the war, Viennese workers were exposed to and consumed products of the older commercial culture—the *Gasthaus,* circus, amusement park, boxing and wrestling ring, dance hall, variety theater, home decorations, patent medicines, and so forth—as well as those of the emerging mass culture such as cinema, radio, dime novels, spectator sports (especially soccer), and cosmetics. A mass clientele became increasingly important to both older and newer forms of commercial culture, and the working class was perceived by that culture's purveyors as providing a mass market.

Working-class subcultures were exposed to the advertised allures of the new culture products, and these became part of worker experience insofar as constrained worker budgets allowed. Within the worker subculture there is little evidence of resistance to the blandishments of commercial or mass culture. To acquire or participate in these products and activities was a means of extending worker life-style at least in small measure in the direction of those who were better off and better able to consume.[4]

Socialist leaders and city fathers attempting to create a total worker counterculture within the dominant bourgeois culture were essentially unprepared for the challenge and competition offered to their efforts by commercial/mass culture in the marketplace. Much as had been the case among middle-class reformers, the socialists denigrated this competitor as crass and vulgar, and its products as superficial pleasures unworthy of the emerging new style of worker. Among those who made socialist culture policy there were few throughout the period who were able to go beyond this very narrow rejection of mass culture as just another evil of the capitalist system. But even among the lonely voices who realized that commercial/mass culture could not simply be wished away by condemnation, hardly anyone could see further than the need to ennoble mass culture by uplifting it to the quality (and content, often enough) of elite culture.[5]

It is important to keep in mind that in the realm of worker leisure commercial, noncommercial, and mass cultural forms coexisted and competed with the Socialist party's attempts to develop an all-encompassing cultural program of its own. The commercial culture included such older entertainment forms as the *Gasthaus/Wirtshaus,* Prater, circus, *Varieté,* dancing, balls, boxing matches, and parades. The noncommercial included children's street play, swimming in the Danube (at the Lobau beaches), unorganized sports (especially soccer), and hiking in the Wienerwald. The newer mass-commercial forms—radio, film, and professional spectator sports (soccer)—partly because of their industrial form of organization and distribution—offered sharp competition to older forms of entertainment.[6]

How many hours in the Viennese worker's day or week could be devoted to leisure activities? It is a difficult question to answer. Although the eight-hour day became law in 1919, the average five-and-a-half-day week was at best limited to forty-eight hours.[7] Travel time to and from work further reduced free time on weekdays, as did the overtime frequently necessary to supplement low wages. The largest block of free time was during the one-

and-a-half-day weekend, but with great gender differences resulting from women's frequent triple burden of work, housework, and child rearing. Younger workers generally had fewer domestic responsibilities and greater opportunity for leisure. The onset of the depression in 1930 and the rapid increase of unemployment confounded the previous time budgets of workers. Employers used the crisis to extend the workweek, speed up work processes, and renege on paid vacations.[8] Even the growing army of unemployed responded variously to their forced freedom. Young workers, including married couples without children, functioned in the part-time gray market to improvise their survival and used their newly found free time creatively, while family heads suffered from a loss of identity and self-respect.[9]

When one considers these variations, it becomes clear that the question posed above can be answered only in general terms: there was an absolute gain in workers' leisure time, but it was modified by the conditions described. Most important may have been the attitude, widespread in the 1920s, that the worker had a right to free time, and that he toiled in order to win as large a portion of it as possible.

Commercial Culture

One of the most important commercial institutions in the life of the Viennese working class was the *Gasthaus* or *Wirtshaus*. These establishments provided meals and snacks, alcoholic beverages, lodgings, and meeting rooms. On Saturday night and Sundays popular singers entertained and *Schrammeln* quartets (two violins, a bass, and a clarinet) played dance music. In the last quarter of the nineteenth century, when half the Viennese work force was still artisanal and half the workers did not live in their own homes, *Gasthäuser* were a virtual home away from home and the most important center of worker sociability.[10] As communication centers they also had a sexual, political, and acculturating function: on weekends male journeymen and female workers were able to enjoy the erotic ambience; journeymen established bases there for their particular trade *(Gesellenherbergen)*, turning them into hiring halls, a locale for the settlement of conflicts at the workplace, and strike centers; and the majority of journeymen, who were of Bohemian and Moravian origin and spoke Czech at the workplace, attempted assimilation by speaking German.

By 1919 the *Gasthaus* had lost many of its preindustrial social and political functions. The enormous increase in marriages and more stable worker families, and the maintenance of wartime rent control by the municipal administration, made the surrogate domicile function of the *Gasthaus* obsolete. Similarly, the full-scale development of trade unions and the Socialist party largely replaced the *Gasthaus* as a primitive center of worker economic and political activities. Yet *Gasthäuser* continued to play a significant role in the working-class communities of Vienna. Their proximity to the growing number of industrial enterprises made them the lunchtime center of male worker communication.[11] *Gasthäuser* also dotted the streets of working-class neighborhoods, where they continued to provide weekend entertainment for both sexes and a weekday meeting place for male workers.[12] Despite the Socialist party's attempt to locate party activities either in worker centers *(Arbeiterheime)* or in special meeting rooms in the larger municipal housing projects, many party locals continued to meet in the neighborhood *Gasthaus* of worker choice.[13]

On the whole, however, the function of the *Gasthaus* was reduced to a place where entertainment and sociability had to be paid for. No doubt their number declined between 1870 and the early 1930s, as Joseph Ehmer demonstrates.[14] But their number remained considerable in the postwar period (3,713 in 1931) and showed a remarkable stability during the economic crisis of 1930–33, when reduced worker budgets made commercial entertainments more of a luxury.[15] This suggests that, as one of the oldest forms of commercial culture, they were able to maintain a significant place in worker lives despite the alternative offered by the Socialist party culture and the competition of mass-cultural entertainment.

Another long-standing and particularly Viennese institution of worker relaxation was the *Heurigen*. The winegrowing areas to the north and south of Vienna, virtually encroaching on the city limits, were dotted with small vintners who sold their new wine to a mixed public of Viennese seeking cool, convivial surroundings in the summer months. Serving a popular clientele, these small establishments were located in or near the vineyards and consisted of little more than a few shade trees and some rough wooden tables and benches. For the price of several quarter liters of wine, workers and their families could enjoy a weekend evening meal, usually consisting of cold snacks brought from home. Singing and dancing, occasionally accompanied by an instrument, provided the improvised entertainment for adults; the children played in the surrounding fields. The atmosphere was one of high spirits occasioned by a sense of liberation from the oppressive atmosphere of the workplace and crowded domiciles. Most of the *Heurigen* could be reached by public transportation followed by a short walk.

No doubt the *Heurigen* had its heyday before World War I, but the institution seems to have continued to draw the public in the interwar period. Unfortunately, this popular site of entertainment remains entirely unstudied, making it difficult to ascertain the extent of its continuing popularity. There is some evidence to suggest that the *Nobelheurigen*—large establishments with hot buffets, music, and formal entertainment, flower and toy sellers, and taxis on demand, serving the Viennese *Bürger* and foreign tourists—largely went bankrupt during the depression; but the small enterprises, which workers had frequented by tradition, seem to have survived.[16]

The largest domain of popular entertainment in Vienna was the Prater, a former royal hunting preserve donated to the public in the 1880s. This huge green space contained lawns and shaded walks, ponds for rowing, bicycle lanes and bridle paths, soccer fields, an amusement park with the famous *Riesenrad* (Ferris wheel), numerous *Gasthäuser* and snack stands, and (by the early 1930s) a stadium accommodating 60,000 spectators and a swimming pool and sunning lawns with a capacity of many hundreds.[17] From its earliest days as a public facility, the Prater was a focal point of mass entertainment and activity. An exhibit of 1895 entitled "Venice in Vienna," emulating the city of canals and harboring restaurants and theaters, attracted 2 million visitors in its first year. The Olympia Arena, opened in 1902, was the largest open-air theater in Europe, with 4,000 seats. The first May Day procession of workers in 1890 along the Ringstrasse terminated with a convocation in the Prater.

The Prater continued as a center of spectacles and crowds in the 1920s and 1930s: a folk costume parade of 1921 attracted 400,000 spectators; the flower parade of 1925, with 400 floats, was viewed by half a million; in 1928, the tenth German Song Society Festival assembled 40,000 singers; and the biannual *Wiener Messe* (trade fair) lured hundreds of thousands to see the latest products of domestic and foreign industry and technology. The Viennese who flocked to the Prater on a typical Sunday afternoon entered the grounds free of charge and also paid nothing as spectators of passing parades or the general scene. Almost everything else came at a price, from the beer, soft drinks, and sausages consumed for small change by the great multitude to the costlier rides in the amusement park, rented rowboats, *Gasthäuser*, circuses, and variety theaters. As worker budgets rarely allowed for more than five Schillings a week for nonessential items, workers no doubt consumed the commercial offerings sparingly.[18] Even so, workers contributed to and imbibed the holiday atmosphere of the crowd—an ambience produced in large measure by the commercial cultural products for sale.[19]

The Prater also housed a large number of the most sophisticated commercial entertainments: circuses and variety theaters.[20] Although both of these had a long history as distinct entertainments before the war, they increasingly borrowed from each other in the postwar period and attempted to modernize in order to compete with newer commercial competitors such as film. *Variétés* introduced "living photographs" (Biotophone) and Bioscope film shorts; circuses sought new audiences by featuring boxing matches.[21] Both struggled in the first postwar years with inflation, later with the municipal entertainment luxury tax, and after 1930 with the depression. Their attempt to recapture their previous standing as leading commercial forms of popular entertainment were largely a rear-guard action against new technological commercial culture—the mass media. Yet if we consider the number of spectators that continued to be drawn to circuses and *Variétés* in the interwar years, their role as a popular leisure activity among workers needs further consideration.

Animal acts continued to be the distinguishing characteristic of circuses, making them a favorite as family entertainment. Although only two permanent circus structures, the Zentral and Renz, remained in or near the Prater, numerous foreign circuses pitched their huge tents there for engagements each season. In 1931 the largest of these had a now almost unbelievable number of spectators: Circus Kludsky drew 10,000, Zirkus Gleich 12,000, and Zirkus Krone 13,000 per performance.[22] Even smaller circuses with a tent capacity of 5,000 totaled an audience of 300,000 in

1927.[23] Less celebrated as purveyors of spectacles but equally important as popular entertainment were circuses outside the Prater. Tenting on open spaces in the part of the city with large working-class populations, these circuses drew steady audiences at very modest prices.[24]

Both the price structure and total audience size are difficult to ascertain. Cheaper seats in even the largest circuses appear to have been within the 50 Groschen to 1 Schilling range, and were 50 percent cheaper in the small local ones. Approximations of attendance are even more difficult. By using the Krone's ratio of tent capacity to seasonal audience for the largest circuses (10,000 to 13,000 seats), we arrive at an annual audience of more than 600,000 for each of these. By adding the several larger and smaller circuses in the Prater to the small local ones (several hundred seats), one might arrive at a weekly—that is weekend—total of 80,000 to 100,000 circusgoers.[25] Though these figures suggest that circuses continued to play a significant role among popular entertainments in Vienna, even weathering the financial stresses after 1930, their decline is also palpable.[26]

The Viennese *Varieté* theater, comparable to the American vaudeville, the English music hall, and the French *café concert,* shared with its foreign counterparts the reputation of being a serious competitor to the elite "legitimate" theater, and was without rival in attracting a large and varied public. From the second half of the nineteenth century to the end of the war, *Varieté* enjoyed an enormous popularity among the lower middle and working class. Though the dominant elite culture clung to the opera house, theater, and concert hall, even it emulated and borrowed from the *Varieté* (operettas, for instance) or created versions of it sufficiently elevated and exclusive to be acceptable.[27]

From its beginnings the Viennese *Varieté* thrived on its versatility as a mixed art form, offering short numbers of great variety to please audiences of diverse taste: scenes from operettas, skits, popular songs, dance, comedians, and above all orchestral music.[28] After the war the *Varietés* were unsettled by economic instability and threatened by the new mass media; they attempted to hold their audiences by adding acrobats, jugglers, strongmen, clowns, and clairvoyants as well as Duncanesque dance and the American Charleston and Black Bottom to their offerings. In keeping with a changed climate in public morals after the war, partial nudity increasingly became a top drawing card on *Varieté* stages.[29]

The largest *Varietés* in and outside the Prater featured leading stars of the Viennese stage, screen, and concert hall as well as international stars like the American Josephine Baker, the Parisian Mistinguett, and the Tiller Girls of Berlin.[30] Given the prices of these first-run establishments (more than one Schilling minimum), the traditional public frequented cheaper and more congenial *Varietés* such as the Leicht-Varieté in the Prater. There, for little more than 50 Groschen, one could enjoy an excellent program with the same distinguished stars (who considered performance before a popular audience a confirmation of success), plus a plate of kolbas, potatoes, and gravy.[31]

By 1933 nine of the large, successful *Varieté* theaters employing some

300 artists were still playing to full houses. But competition from mass culture had taken its toll: some of the most famous *Varietés* had been converted into movie houses in the previous five years (Colosseum, Apollo, Johann Strauss, Neues Orpheum, Lustspieltheater).[32] The *Varieté* as truly popular entertainment survived best in the small wooden structures dispersed throughout the working-class districts of Favoriten, Ottakring, and Brigittenau.[33] With 200 to 300 seats, and prices in the 50 Groschen to one Schilling range, they offered alternating programs of theater, revues, peasant comedies, and *Varieté* to a loyal public. To these must be added numerous *Gasthäuser* which provided *Varieté* on a much reduced scale as an entertainment supplementary to the food and drink consumed by their customers. The smallest establishments with less than fifty seats, offering *Varieté* of sorts, were excluded from the municipal luxury tax and license requirement and could offer more or less amateur entertainment at very low prices.[34] Like the circus, *Varieté* was a declining form of commercial entertainment that struggled to maintain its past popularity. Given the diversity of establishments, the size of its public is difficult to estimate. It is doubtful whether weekend audiences ever exceeded 20–30,000. In all likelihood working-class consumers formed less than half that number.

Two other commercial entertainments attractive to workers deserve mention: dancing and balls. In the postwar period the spontaneous dancing at *Gasthäuser* and *Heurigen* was increasingly superseded among the young by dancing at commercial dancing schools. These establishments provided not only instruction but also a place for the initiated to take their pleasure by exercising their skill.[35] A Sunday afternoon enjoyed by young workers in this ambience of proper dress and comportment brought them into contact with an unfamiliar but attractive world. No doubt dancing schools also played an important role in providing a setting for the courtship of young workers. The nature of this individualistic entertainment particularly disturbed Socialist party educators and youth leaders. Their denunciation of couple dancing as senseless pleasure-seeking and sensuality (a sign of false consciousness), and their championing of folk dancing—in which the circle symbolized equality, purity of spirit, and the collectivity—will be discussed later. Related occasions for young workers to enjoy couple dancing were the public balls organized—usually annually around nationally observed holidays—by various trades (laundresses, tramwaymen, firemen, seamstresses, etc.). At these festivities dancing to live bands, food, alcohol, prizes, and special events were a great attraction, drawing large numbers of participants.

Noncommercial Leisure-Time Activities

The noncommercial leisure-time activities of Viennese workers are the most elusive of the various cultural forms under consideration. There is no way to measure the extent to which such activities took up the workers' precious free time and competed with commercial and mass culture as well as Socialist party culture. We know of their existence from contemporary oral histories as well as questionnaires and social studies of the time. They must, if by nothing more than enumeration, enter our calculation of how much of the workers' time was available for the new proletarian counterculture the SDAP attempted to create.

The immediate locale of childhood recreation was the street or the empty spaces of the working-class neighborhood.[36] As we will see in the later discussion of the family and sexuality (chapter 6), urban niches played an important role in the socialization of working-class children. Through play in its broadest sense the authorities and boundaries of the community were apprehended, social and sexual distinctions among the players were drawn and observed, and a domain of selfhood distinct from the adult world was established. In ascertaining the social aspect of street play, two of its intrinsic aspects should not be overlooked: its liberating character (from parental authority and the depressing confinement of crowded domiciles), and its spontaneous organizational possibilities (single-sex and mixed-sex groupings, proximate and distant locales, cooperative and competitive forms.)[37] Seen in this light, children's street play was a liberating contrast not only to the authority imposed by school and parents, but also to such municipal facilities as after-school centers and socialist institutions for children like the Rote Falken, whose structures were predetermined and predicated on discipline.

Among adult Viennese workers the single most preferred noncommercial form of recreation was rambling and hiking. The strongly expressed desire to escape from the dehumanization of the workplace and the confinement of domiciles into nature led thousands of workers to explore the vast expanse of the Wienerwald bordering the city. Sunday was the preferred day for such excursions, for then even married women with children were able to wrest several hours of recreation from their domestic responsibilities.[38] The virtual paean to nature found in contemporary studies and recent reminiscences seems to imply more than the desire for recreation away from the confines of urban life. It suggests the momentary escape from all the restraints of everyday life into a different and regenerative world where a restrictive order did not exist and individual inclinations could be followed.[39]

Swimming was another popular leisure-time activity. It was pursued at various commercial beaches along the lower Danube or at the growing number of municipal swimming pools, both of which served the tens of thousands gripped by a virtual swimming mania. The passion for swimming among Viennese workers—virtually absent in the large cities of other countries—was made possible and encouraged by the numerous swimming pools created under the auspices of municipal socialism.[40] But young workers were especially drawn to the rugged beaches of the Lobau along the upper Danube, a terrain known colloquially as the "proletarian Riviera." Here, swimming was free not only of paid admissions but also of imposed rules and structures.[41] Beginning on Saturday afternoons swarms of young workers

Kongressbad, one of the largest swimming pools in Europe (measuring 100 × 20 meters), with 450,000 paying bathers in 1930 (VGA)

arrived with friends, carrying provisions, and often spent the night under the stars. Oral histories and memoirists speak of the unique sense of liberation they experienced there from the cares of the workweek and from the regimentation of everyday life, and marvel at the mixture of spontaneity, individualism, and fellowship which prevailed.[42] It is not surprising that many young workers, too poor to travel elsewhere during their week of statutory vacation, regarded the Lobau as their Riviera.

Swimming and spontaneous youth culture along the Lobau (VGA)

Soccer was another noncommercial form of recreation among young male workers. It required little or no organization beyond a sufficient number of friends to constitute two teams. Its locale was the neighborhood spaces familiar from childhood as well as open terrain along the Danube and in the Prater.[43] What made this unorganized sport attractive to young workers was the opportunity to associate spontaneously with one's friends and the sense of freedom imparted by the ability to shape and control all aspects of the activity. Unorganized soccer was by no means a substitute for organized sports offered by the Socialist party or for spectator sports, whose popularity was steadily increasing. But it remained one of the increasing number of recreational choices available to young male workers in their leisure time. In the realm of sports as in other recreational forms, commercial, noncommercial, mass cultural, and party cultural activities coexisted and vied with one another—a simultaneity important in understanding the extent to which workers were accessible to the proletarian counterculture the SDAP was creating for them.[44]

Finally, a frequently overlooked noncommercial leisure activity which combined recreation with the quality of the workers' living standard were the *Schrebergärten* or small garden plots on the periphery of the city. Worker families strove to attain one of these allotments leased by the municipality for a nominal fee and also by private entrepreneurs. In their "garden" the worker family grew vegetables and fruits to be eaten in the summer but also preserved for later consumption, raised rabbits, but most of all enjoyed the freedom of being in touch with nature on their own plot and usually in the proximity of friends and neighbors from the city. By 1932 some 14,000 of these *Schrebergärten* existed.[45] A number of these included primitive overnight habitations which had been illegally converted from tool sheds. Especially during the depression, these garden plots became an essential supplemental source of food for many families.

No doubt other minor forms of commercial and noncommercial culture existed which have not been accounted for here. What emerges from the forms considered is the wide variety of leisure-time activities, other than those proffered by the Socialist party, available to and engaged in by Viennese workers. Though the difficulty in measuring the extent of worker participation in these has not been overcome, it is fair to say that they represented a significant competition to SDAP attempts to create a proletarian counterculture. The latter faced an even more serious challenge from mass culture, whose unique power to create needs and satisfy consumers made it a powerful adversary. All this competition among cultural forms was made possible, in the 1920s and early 1930s, by the shortened workweek and attendant growth of leisure time of the working class.[46]

Popular Culture Condemned

The Socialist party viewed popular culture, the wide array of commercial and noncommercial leisure-time activities in which the workers participated, as an obstacle to be overcome, if the party's "civilizing" mission was to succeed. As we have seen (chapter 4), the party succeeded in attracting only a relatively limited number of workers to its cultural program. To attain its goal of incorporating the large mass of workers into its comprehensive proletarian counterculture, to create "neue Menschen," it was forced to deal with existing cultural practices of the working class. In confronting the workers' widespread consumption of commercial culture, party publications assumed the same tutelary role they had pursued in condemning the symbolic capitalist content of workers' homes.[47] But in the realm of culture the tone of criticism was less vehement about how workers ought to change, and more realistic about the strength of cultural bonds in the workers' world. This does not mean that the party had no dogma on the subject of entertainment; it seemed, however, less certain that its dogma could win it the converts it desired.

In this regard, it is interesting to follow the course of the socialists' argument in the pages of *Sozialistische Erziehung,* a journal addressed to youth leaders and educators. In its supplement, *Die-Praxis,* leaders and Socialist Worker Youth (SAJ) members aired practical problems of the movement. One of the first in a series of commentaries on entertainment attempts to draw a distinction between bourgeois and proletarian forms.[48] Bourgeois entertainment is characterized as upholding the social order and as deflecting workers from their political and economic mission. The form of proletarian entertainment, the argument continues, is less important than its limits; it may include the Charleston and Black Bottom as well as folk dances. It is the intention and extent which differentiate bourgeois from proletarian forms. The former is pleasure-seeking for its own sake; the latter keeps entertainment in bounds as a form of rest and relaxation, so as to gather new strength for the serious work and struggles of the movement.

This position is criticized in the following month for speaking only about the need for socialist youth to keep entertainment within bounds without dealing with the contents or bias of entertainment.[49] The bourgeois entertainment industry—particularly the Prater but also *Varietés* and operettas—creates illusions to make working-class youth forget the class struggle and the fact that the social order can be changed. Does working-class youth have the right to waste its time, to relax? The answer is no. The individual has the duty to use his leisure time to read, learn, harden his body for struggle through sport, and to participate in party activities.

At the suggestion of the editors, the next contributor returns to the subject of dancing raised earlier.[50] She pictures the typical commercial dancing school as a den of eroticism: stale air, a small dance floor, closely pressed bodies undulating to music which inflames the senses, flirtations. But even folk dances can be erotic, she admits. What distinguishes the latter from the former is a "social and hygienic" quality. In the magic circle of the folk dance there are no fixed partners—a sense of equality prevails. The folk dance is also distinguished by a healthy casualness of dress in comparison to the unhealthy dress codes of urban dancing establishments, including tightly fitting dresses, stiff collars, and ties. Socialist youth choose the folk dance because it corresponds to their ideology of freedom. But this claim is denied by another discussant who insists that folk dancing has failed to hold the interest of youth even in the SAJ, and that commercial dancing schools are here to stay.[51] If the party wants to attract proletarian youth, it must compete with social dancing entertainments of its own and use them as a means of agitation. Only then will it be able to educate those who come. A subsequent critic finds such views antiquated and unrealistic.[52] People are attracted to the SAJ because it provides conviviality: the opposite sex, sports, and trips. For the overwhelming majority, conviction, current politics, and the desire for knowledge and education are the last reasons for joining.

The final word is spoken by a party elder who has little patience with the ambiguities and confusions of the previous young discussants.[53] He rejects the suggestion that dancing can be used as a means of attracting the unorganized proletarian youth, because social dancing simply cannot be separated from its bourgeois milieu: immoral smoking, drinking, clothing mania, flirtatiousness, and mock gallantry. The values of abstinence from smoking and drinking—central to the creation of "neue Menschen"—cannot be compromised even in building a mass organization. Besides, he adds, the SAJ could never compete in attracting seventeen-year-olds who have so far kept their distance from the party, by offering dancing cleansed of its bourgeois detritus. Any attempt in that direction would not only fail but also undermine the existing cohesiveness of SAJ youth and functionaries.

It comes as no surprise that the SDAP expected a great deal from the Viennese working masses in whose name it spoke. The somewhat ragged discussion above illustrates how the message of self-denial had reached certain echelons of the SAJ. At the same time it suggests that the normal desire for a leisure time of joy and pleasure among the young continued to be associated with aspects of the popular culture. The contradiction between an outright condemnation of commercial leisure, and yearning for the same on the part of the discussants, could not be resolved by the party leadership. It seems that there was a good deal of wishful thinking in the belief that the socialist youth were such single-minded and future-oriented zealots as to deny themselves pleasures outside those deemed to serve the collective good.[54] Here, as in other aspects of the socialists' cultural program, there is considerable ignorance of both the existing sociocultural context in which workers were socialized and of the psychological mechanisms constituting and shaping behavior. Despite the SDAP's alleged adherence to Alfred Adler's ego psychology, there is little evidence of it, save a somewhat vulgar environmentalist behaviorism, in the party's approach to the workers themselves.[55]

The application of party dogma on popular culture to young workers struggling with problems of identity and caught up in the confusing desires for conformity and selfhood is but another example of a persistent socialist insensitivity to the workers as they really were. In attempting to combat the corrupting influence of popular culture, the socialists stressed self-denial, abstention, and postponement of gratification. In its place they offered regimented collective experiences whose ability to "substitute for everything else" continued to be questioned even by those already integrated in the movement. Simple denunciations of commercial culture without a compensation for its social and psychological qualities was not very effective. Yet when confronted with the far greater enticing powers of mass culture, socialist leaders could find no other response.

The Cinema: A Dream Factory?

Of the major forms of mass culture available after the war—film, radio, spectator sports—the film was clearly the most popular. In Austria as in other countries, the nascent prewar film industry had developed rapidly during the war, when its propagandistic possibilities were discovered and exploited.[56] Austrian film production reached its zenith in the expansionist climate of postwar inflation. At that time the predominant production company, Sascha-Film, founded by Alexander Kolowrat in 1918, was joined by more than a half dozen others financed by various banks, and sizable film studios were built. During this brief golden age of the Viennese/Austrian film, an average of forty films a year were produced,[57] including spectacles involving huge and complicated sets and casts of hundreds. A remarkable array of talent was active in Vienna at this time, including the directors Michael Curtiz (Kertesz), Alexander Korda, Fritz Lang, G. W. Pabst, Otto Preminger, Billy Wilder, Fred Zinnemann, Robert Viennae, and Jacques Feyder, as well as such actors as Conrad Veidt, Walter Slezak, Paul Lukacs, Fritz Kortner, Marlene Dietrich, Greta Garbo, and Peter Lorre, to mention but a few.

By 1925 economic stabilization had brought about the virtual collapse of Austrian film production, with all major companies except Sascha-Film disappearing from the scene, while directors, cameramen, and actors left for Berlin or Hollywood, and the annual number of films produced shrank to nine.[58] The stagnation in domestic film production continued for the next decade, and the demands of film distributors and exhibitors were met by American, German, and other foreign imports.[59] Neither the introduction of import quotas nor the lifting of national censorship in 1926 succeeded in significantly increasing the quantity or improving the quality of Austrian films.

The troubled development of the Austrian film industry and its struggle against foreign competition were not unique. European cinema on the whole underwent a similar crisis but, unlike the Austrian, it recovered to produce a distinct and internationally marketable film.[60] In retrospect, it is difficult to explain why film production in Austria declined to sheer mediocrity. The collapse of major companies left a vacuum into which the large number of talented resident cineasts might have moved. That is exactly what happened in France, where many films were created as single collective efforts, where financing had to be improvised imaginatively, and where the talent which fled in the mid-1920s was encouraged to return. Actually, the possibilities for creating artistic films with a serious social content should have been greater in Austria than in France. Both the SDAP and the trade unions were engaged in developing and supporting a working-class culture; in France there was no such development and possibility for support.[61]

As we shall see later, the Austrian socialist leaders' attitude toward film ranged from ambiguous to negative. To a large extent they judged this new medium with the yardstick of the old, elite culture in which they had been raised (see chapter 4) and regarded it as the newest and most threatening form of cheap entertainment. To say that the socialists, in neglecting the production of films, missed an opportunity to strengthen their own cultural program by attuning it to popular tastes is not to minimize the serious obstacles which would have been faced and the risk involved.[62]

The health of the Austrian film industry did not affect the size of the audience which flocked to see films not only for their content but as technical and artistic novelties. By the end of the war there were already many movie theaters in Vienna, and their number, seating capacity, and frequency of performance grew at a remarkable pace in the following decade and a half.[63] The following table compares three years for which the most reliable information about movie theater admissions in Vienna is available.[64]

	Yearly	Weekly	Daily
1926	12.42 million	238,846	34,120
1928	29.39 million	565,234	80,747
1933	28.03 million	539,163	77,023

By 1926, which was an off year because of the reorganization of the film industry, there were already 170 movie theaters in Vienna, 160 of which had daily showings. By 1933 the number of theaters had grown to 179, with 99 percent showing sound films, and 7 with over 1,000 seats. As the above figures represent averages on a weekly and daily basis, certain corrections must be made so as to arrive at attendance on days when Viennese workers were at leisure. According to theater owners and managers, the most popular days were Saturday and Sunday, followed by Friday and Monday, with a sharp drop in ticket sales in midweek. Attendance was reduced in general during the five warm-weather months.[65] If we adjust the above figures in the light of this information, the number of weekend admissions would more than double the daily averages, yielding a total of 350,000 filmgoers.

By whatever yardstick we use—560,000 weekly filmgoers or 350,000 on weekends—the film had become the most popular entertainment among the Viennese by the late 1920s. How many of these were workers? Unfortunately, the official film statistics lack the necessary refinement, and an answer can be given only by approximation. Accordingly, if we use SDAP and trade union membership as well as the social composition of the SDAP as a frame of reference, about 50 percent of weekly film audiences (280,000) and 60 percent of weekend filmgoers (210,000) were workers.[66]

What made the film such a popular form of entertainment?[67] A study of Viennese theater and film audiences conducted in 1936 throws some light on this subject.[68] The cheap admission price and informality were mentioned by an overwhelming majority of moviegoers. There was a marked preference for sentimental filmscripts with melodramatic obstacles leading to a Hollywood-style "happy end." Attendance in the company of others was preferred, because it heightened the experience and, in the presence of the opposite sex in the darkened theater, aroused erotic feelings. For younger viewers the film provided an opportunity to widen their horizons, introducing them to strange new worlds. More mature viewers sought relaxation. The choice of films by working-class viewers was strongly influenced by advertising; comedies were popular; educational subjects without plot were shunned; music and favorite actors were sought after; and the plot as well as its plausibility were of relative unimportance.

The young were the most committed film viewers, as a study of 1933 demonstrates.[69] More than 95 percent of the 10,000 surveyed were enthusiastic, with working-class children ranking highest in positive responses. In Vienna, movie theater attendance of these increased steadily with age: 8–13-year-olds attended at least two times a month; nearly 40 percent of 14–16-year-olds attended 4 to 7 times a month; and nearly 8 percent of the latter group attended more than 8 times. Interestingly, frequency of attendance by children of the unemployed was not reduced.[70] More than half the working-class children went to the movies with friends; the rest divide equally between those going alone and those going with parents and other adults. The reasons given for liking films were not too different from those of adults: films were entertaining, educational, cheap, exciting, romantic, realistic, offered actors as role models, presented the world of now.

What are we to make of the fact that Viennese workers, young and adult, flocked to the movies; that they preferred comedies, excitement, and happy ends; that they chose films on the basis of actors rather than plots; and that the plausibility of the screenplay was of little importance to them? Was the nearly unanimous judgment of socialist educators, culture experts, and critics correct in regarding films as an opiate of the working class, created by capitalism to seduce workers from their true goal? Virtually from its beginnings down to the present, the film industry has been characterized as a "dream factory" creating illusions in place of reality.[71] Is it wrong to have dreams or to share in them? Don't other forms of entertainment and art also present illusions and special perspectives on reality? The socialists never asked themselves these questions or attempted to answer them. They viewed film as a degenerate art form judged by the yardstick of linear elite culture (mainly drama and the novel).

What may well have drawn the workers to the cinema, in addition to the need to relax and socialize in informal surroundings, or even the need to escape from the ever-present reality of their ongoing struggle in the workplace and domestic sphere, was the ability of film to incorporate and even surpass the more traditional and still consumed forms of commercial culture: more spectacular representations of exotic places and people, sharper images of daily life, a greater immediacy of feeling, and a broader scope for empathy than the circus or *Varieté* could provide. What may have "seduced" workers to pay frequent visits to the cinema is a new kind of seeing, and later seeing and hearing, which allowed them to perceive the subjects of entertainment in a new and dynamic way. Even bad and trivial films contained that novel charm, which may well account for the success of some of the most soporific examples.

But even the very best examples of cinematic art, with the greatest complexity and suggestiveness, such as *Der blaue Engel,* could be enjoyed in some of its dimensions by workers with unsophisticated taste. The same was not true for examples of elite art, such as Thomas Mann's novel *The Magic Mountain,* whose prolix style and complicated plotting precluded the possibility of being enjoyed on a simpler level. The visual dynamics of film, allowing for growth of perception with experience, lent a democratic aspect to cinema which very few contemporaries were able to appreciate.[72]

Was the quality of films viewed by Viennese workers really as bad as socialist critics made it out to be? Was it really an ocean of kitsch in which an occasional pearl came to light? It is well to remember that there is a wide spectrum of quality in all art and entertainment forms. It is remarkable that hardly any of the cinema's most outspoken critics ever asked how many of the plays brokered by the Kunststelle or the numerous books serialized in socialist publications were kitsch as they defined it.

About 400 to 500 films were exhibited in Viennese cinemas each year.[73] Of these, no doubt, the majority were of limited artistic or intellectual merit—light comedies, historical pageants, musicals, adventures in bizarre settings, pedestrian tragedies, and tales of miraculous salvation or success— most often routinely but sometimes well crafted, featuring well-known stars and negligible screenplays which frequently contained imaginative scenes. Some 20 percent perhaps were downright kitsch—renditions of the "fallen women," the "charming crown prince," "happy peasants," "flowers of the harem," and "jungle adventures"—generally of low technical quality and geared to diverting the most passive viewers.

The weekly diet of the film public, however, was enriched by the best films on the international market at the time, and they were frequently being shown at ten or more theaters at once. A very incomplete sampling of these would include *Anna Christie, Der letzte Mann, Sacco und Vanzetti, Metropolis, Berlin Alexanderplatz, Emil und die Detektive, Grand Hotel, The Hunchback of Notre Dame, Marius, Resurrection, An American Tragedy, Charley's Aunt, Gas, Der Hauptmann von Köpenick, Huckleberry Finn, Kameradschaft, A nous la liberté,* and *Twenty-Four Hours.* The mix in quality of films shown in Vienna appears to have been about the same as in Berlin, Paris, London, and New York.[74] A look at the films with the leading box office sales for 1930/1931 confirms the distribution of quality films available to Viennese filmgoers. Among the twelve films with the largest ticket sales were *Dreyfus, Der blaue Engel,* and *Atlantic.*[75]

Domestic and imported expressionist spectacles and fantasies *(Sodom und Gemorrha, Die Sklavenkönigin, Das Kabinett des Doktor Caligari, Der Golem)* were dominant during the postwar inflationary period. With the return to economic stability, such flights of fancy gave way to more sober subjects and forms of presentation. This search for a new sobriety and objectivity, called "neue Sachlichkeit," originated in painting and literature but was quickly adopted by other arts including film.[76] There too it took the form of a search for the plague spots and social problems of society exposed through an intensified realism that threw the dark side of contemporary life into sharp focus. The evils and temptations of the city, especially licentiousness, were a running thread through many films of the mid- to late 1920s. The formula on which such films as *Café Elektric* (1927) and *Gefährdete Mädchen* (1928) were based included a presentation of the sinful temptations of the city, which plunge the individual from the good life into an abyss of degeneracy from which only a pure love brings rescue.[77] Political subjects as well *(Oberst Redl,* 1925, and *Die Brandstifter Europas,* 1926) took the form of exposés, but without sentimental endings. The blending of serious social subjects with erotic representations particularly aroused the ire of socialist culture experts, for whom they epitomized not only the kitsch of capitalist film production, but also the danger to the working class and its youth of being manipulated and corrupted.

The popularity of film as an entertainment form with a potential for influencing the masses of workers was recognized in socialist publications in the first postwar years. But film was a mass product of capitalism and as such, by definition, of poor quality, arousing the public's worst instincts, wasting its time, trading in cheap illusions, and robbing it of hard-earned money. The SDAP would have to intercede, to turn the cinema into an institution of enlightenment and good entertainment.[78] These early journalistic interventions suffered from the ambiguity of recognizing the importance of film in general while denouncing films specifically as worthless trash. At the time, film commentaries were written by theater critics and feature writers, who judged films with the yardstick of high culture.

Condemnation brought the socialists no closer to gaining any influence over the films viewed by the workers. In 1924 a cinema conference was initiated by the municipality in the hope of developing a program for film reform.[79] There, socialist party functionaries and cineasts clashed on a variety of subjects. The suggestion that the sound film would one day surpass the silent in importance was dismissed out of hand. If that happens, a party speaker replied, film would become nothing more than an inferior substitute for theater and opera. That the pictorial aspects of film should receive more attention and the literary qualities be deemphasized was treated as passé, even though the socialist writers of film criticism present practiced just the opposite. Fritz Lang's suggestion that it was the function of film to replace theater and opera was answered with the derisive remark that the ability of directors to exercise artistic judgment was questionable. Nor was Béla Balazs spared when he demanded that critics focus their attention on the visual. Until films become true art, plots will continue to be the critics' main interest, he was told.

Although the conference exposed the fundamental difference between cinema practitioners and socialist critics and functionaries, it did mark a watershed in the SDAP's actual intervention. It had been suggested earlier that the party create a viewers' organization, produce its own films, and establish a leasing company, all of which could assure that films would be made and shown that corresponded to the party's aims.[80] Plans were set in motion to realize the last of these proposals. But the conference did not greatly alter the tenor or improve the quality of film reviews.

Although *Die Arbeiter-Zeitung* and *Bildungsarbeit* began to carry regular reviews in 1924, and *Das kleine Blatt* and other socialist publications followed suit late in the decade, films continued to be judged on their literary merits above all. One leafs in vain through the weekly column of Fritz Rosenfeld, the socialists' best reviewer, in search of a deviation from potted reviews that use pejoratives to judge films on the basis of their content, with grudging asides about good acting or well-rendered scenery.[81] The cinema as a visual experience, as a distinct art form present even in average films, is missing or at best measured against the deficiencies of the plot. Only in the Soviet film, brought to Vienna at the behest of distributors and theater owners, does Rosenfeld combine aesthetics and content in laudatory reviews.[82]

Even the best films of the period are compared unfavorably to the literature on which they are based. *Der blaue Engel* is a case in point. Johann Hirsch, in a long review for *Das kleine Blatt,* the widest-read socialist daily, considered the film a degraded version of Heinrich Mann's novel.[83] The right-wing views of Alfred Hugenberg (owner of UFA) had prevailed, Hirsch charged, in reducing the novel's critical sharpness. Although he praises the cinematic force of the film, he entirely misses the sadomasochistic destruction of the central character, representing a way of life; the presence of observant but passive bystanders during acts of brutality; and the wolf-pack character of the young students—all of which reflected reality in Weimar Germany far better than the novel's attack on authoritarianism in imperial Germany. Hirsch simply fails to see, or discounts, these new dimensions. He concludes that one will be able to see this film with pleasure, adding: "But one will benefit intellectually only afterward when one reads *Professor Unrat,* the novel of Heinrich Mann . . . and learns from it what aspects of German literature may be included in the German film and what the masters of the film industry exclude."

While socialist film reviewing appeared to be stuck in a predictable groove, a lively exposition of fundamental problems concerning the working class's approach to cinema appeared in leading publications—*Der Jugendliche Arbeiter, Bildungsarbeit, Der Kampf*—and was aired on the radio. A dominant theme of this discussion, repeated year after year by the SDAP's film expert Rosenberg, was the immense power of film as entertainment in comparison to books, theaters, and other cultural forms. But films represented the values of the bourgeoisie and used subtle means to make working-class audiences believe in the immutability of the present social order. The first step in turning the film into a weapon of the working class was to demystify it by exposing its class bias.[84]

The question was how? If the workers escaped from the hardships of daily life into the cinema's make-believe world of class deception, what steps should the SDAP take to alter that—to bring its own cultural politics to bear?[85] Logically, the party ought to produce films of its own but, Rosenfeld insisted, that was impossible because of the cost. He argued that control of movie theaters and, through them, of distribution was feasible. Such socialist cinemas could draw upon German socialist films, Russian films, and artistic films on the international market to provide a rich diet for proletarian audiences. The profits from such an enterprise could be used in conjunction with similar socialist cinema chains in other countries to create an international socialist production company whose films would have a sound and secure market in the cinema chains.[86]

Rosenfeld's proposal was imaginative and far-reaching; furthermore, as we shall see, some practical steps in that direction had already been taken by the party. Rosenfeld failed to mention that the difficulties in establishing a socialist film policy stemmed from differences between a small group of younger, more politically daring functionaries of the Bildungszentrale, struggling to create the SDAP's cultural program, and the older, more cautious party leaders who held the purse strings. The conflict was fought out behind the scenes; when it surfaced from time to time, it revealed how divided the perceptions regarding the potential of film really were.

An article by a young Bildungszentrale functionary, for example, accused the party of not having responded sufficiently to the potential of film.[87] For the younger generation the turbulence of the war and postwar world was captured by the film and neglected by the theater. The film was not only developing into a worthy art form but also becoming the perfect reflection of the rapid tempo of contemporary life, and thus of the experience of the masses. Party leaders, he charged, who were socialized thirty years ago, clung to theater and concert hall as the pillars of culture and looked down on film, by which they felt threatened, as a degrading commercial product. Until now, he concluded, the party's approach to the film had been half-hearted; the time had come to undertake the struggle for an Austrian socialist film.[88]

It is difficult to explain the continued demand until 1933 by socialist film critics and members of the Bildungszentrale that the SDAP intervene in the distribution and exhibition of films, when the party had initiated just such a program. One can only conclude that these persistent critical voices were directed at the party's efforts themselves. How far had the party gone in seeking to influence commercial film viewed by working-class audiences, and what opportunities had it missed?

The Viennese party leadership commissioned several election campaign films as early as 1923.[89] The experiment proved so successful that additional films were produced for the election of 1927. These were exhibited in party-owned theaters as well as a number of others rented specifically for special showings. An open-air projection in the Naschmarkt (central market) was said to have reached an audience of 10,000.[90] In 1930 the party began the production and distribution of eight-millimeter films, which could be shown anywhere a simple screen could be mounted. All these films had an educational aim and propagandistic focus on such subjects as May Day celebrations, the accomplishments of the socialist municipal administration, and the working-class Olympics of 1931. Lest only the already converted be reached, the party also entered the arena of commercial entertainment films.

In May 1926 the Kino-Betriebsgesellschaft m.b.H (Kiba) was founded by the Arbeiterbank[91] with the blessing of the SDAP secretary Robert Danneberg, the city councillor for finances Hugo Breitner, and the socialist mayor Karl Seitz.[92] The intended purpose of Kiba was to organize and increase socialist movie theaters and to supply them with worthy films. The time for such an enterprise was propitious, because a new Viennese cinema law taking effect that year, recalling all former theater licenses and issuing new ones, made it possible for Kiba to buy or lease additional theaters. Despite this unique opportunity, neither the Arbeiterbank nor the party leaders showed much interest in expansion.[93]

Only a year later, when the brothers Edmund and Philip Hamber, owners of the production/exhibition firm Oela and the distribution company Allianz, began to take a leading role in Kiba, did it begin to grow.[94] As manager of Kiba, Edmund Hamber converted five movie houses into first-run sound theaters. By 1931 Kiba managed nine theaters in Vienna; by 1932, twelve. Beginning in 1930, this expansion was coupled with profits sufficient to cover Kiba's losses during its first three years. The ambitions of the Hamber brothers went much further. In 1930 their Allianz and two further distribution firms were associated with Kiba, which also purchased the extensive Vita film studio at the end of 1931.[95] The following year the Hambers engaged in a public controversy with Breitner over the increase of luxury taxes for Kiba theaters.[96] An appeal by the Hambers to the party leadership fell on deaf ears, and a personal scandal involving the brothers served as ground for their dismissal. In view of Kiba's overextension, particularly through the purchase of the Vita studio, and a general loss of confidence, both the Arbeiterbank and party leadership were anxious to get rid of Kiba. Negotiations were begun with a number of prospective buyers, principal of which was UFA, the bête noire of socialist criticism of reactionary capitalist film politics. But Kiba remained, unsold, only to meet the fate of other party enterprises after February 1934.

As a business Kiba was remarkably successful. By 1932 the Hamber brothers had turned a simple movie theater association into a complex theater chain and distribution enterprise which supplied 15 percent of Viennese cinemas with films.⁹⁷ Did the quality of films distributed and shown by Kiba reflect the long-expressed aim of the SDAP for a socially conscious film as an antidote to the typical illusions of the capitalist dream factory? They did not. From the beginning Edmund Hamber retained control over film programming, and his only concern was to show a profit. From the late 1920s on, members of the Bildungszentrale and socialist film critics complained about Kiba's unsocialist and antisocialist programming. Rosenfeld particularly lambasted the Kiba management for exhibiting films in its theaters far worse—more reactionary, more trivial, and less artistic—than could be found in commercial theaters.⁹⁸ Edmund Hamber defended his programming policy before the SDAP executive: the Viennese public was not interested in political films; Russian films played to smaller audiences; the middle-class film audience in theaters supplied by Kiba had to be taken into consideration as well as the workers. Apparently, the party executive agreed with Hamber.⁹⁹ Julius Deutsch was sent to make it clear to Rosenfeld how much money was at stake; film criticism of Kiba programs in *Die Arbeiter-Zeitung* was put into other hands.¹⁰⁰

Kiba was a commerical success to the end; as a socialist cultural experiment it was a failure from the beginning. In 1925 the Bildungszentrale had petitioned the party executive to create a Volks-Kino-Verband, an association of all the socialist-owned and controlled theaters, with an eye toward improved block programming, only to be told that the undertaking was too risky. Rosenfeld's suggestion that an association of working-class film viewers be formed also failed to get serious consideration. And the opportunity to buy or lease additional theaters, made possible by the cinema law of 1926, was allowed to pass, as was the possibility of cooperating in film production with the German Prometheus company.¹⁰¹ Instead the SDAP executive, socialist municipal officials, trade unions, and the Arbeiterbank gambled on the creation of Kiba. Having turned down all other suggestions for implementing a socialist film policy as too risky, they took the biggest risk of all by putting themselves into the hands of the Hamber brothers, two skillful practitioners in a new, volatile, and ruthless industry.¹⁰²

The failure of Kiba to bring a socialist influence to bear on the mass media revealed a fundamental split between functionaries of the Bildungszentrale, who wanted the distribution and exhibition of films to reflect the party's demand for socially relevant and artistically well-made films, and party bosses in the executive and municipal government, who were content with commercial success. The socialist notables were trapped by the narrowness of their cultural perspective. On the one hand, they wanted to raise films to the level of elite culture; on the other, they wanted it to be functional—to educate the workers and aid them in the ongoing class struggle.¹⁰³ Film made them uncomfortable. Its novelty and experimental aspects, its lack of a long and venerable tradition, made it as unreliable as modern art in the eyes of party doyens, who clung to the established elite culture they had been taught to respect.

The socialists were not alone in failing to understand the unique qualities of film as an art form and mass entertainment.[104] They seemed to have comprehended its magnetic attraction of audiences, particularly of workers, and its ability to compete with and surpass other forms of commercial culture. Quite correctly, they saw it as a dangerous competitor to their project to create "neue Menschen," in that it kept workers away from the party's planned programs and distracted them from participating in the class struggle in their leisure time. They, like virtually everyone else, underestimated the attractiveness of films as very special visual and aural entertainments. Their judgment of the film as a pseudoliterary medium of banal themes and illogical plots simply missed its power to stimulate the imagination through immediacy, discontinuities, motion, superimposition, simultaneity, and abrupt transitions. All these special qualities belonged to a new realm of entertainment which did not abide by the classical canons the socialists were accustomed to, and by which they judged and hoped to transform what they considered mere kitsch.

It would be unfair to castigate the socialist leaders for failing to be avantgarde in their appreciation of film. (Their choice of architecture for municipal housing signaled their limited tolerance for artistic innovation.) But their limited insight into workers' lives and needs is another matter. It seems to have occurred to none of them that the party's cultural program was politically overloading the workers, providing them with a diet of commitment which made no allowance for other human needs. Since the SDAP looked down upon pleasure for its own sake as baggage to be shed on the way to "neue Menschen," the workers, who still very much needed such pleasure, turned to mass culture where it was to be found—in the marketplace. One cannot help but remark certain similarities between the socialists' functionalist approach to leisure and cinema in particular—the denial of pleasure for its own sake—and the position of the Catholic church on that subject. By and large the church treated cinema as a corruptor of morals, and films as the purveyors of immoral kitsch. It sought to protect the faithful from such corruption by publishing lists of approved films suitable for Catholic filmgoers.[105]

Radio: Pulpit of the People?

From its origins as a mass medium in Austria, radio was a public enterprise providing entertainment and information but without the commercial imperative of other mass cultural forms.[106] Unlike the privately owned film industry, whose market was international, radio was largely aimed at the national public. That limited focus, together with radio's unique ability to penetrate the private sphere of a growing listening public, made it part of the political battleground between the socialists and their Christian Social and other political opponents. As we shall see, the socialists' single-minded concern with the content of radio programming—as part of the *Kulturkampf*—led them to overlook the special qualities of radio to entertain

through distraction. Consequently, they overestimated the kind of cultural/
educational impact they might achieve, because they viewed the working-
class listener as passive and malleable in either a good or a bad direction.

Radio as a mass medium began with the constitution of Ravag (Österreichische Radio-Verkehrs A.G.) in 1924 as a mixed public and private
enterprise comprising the national government, the Viennese municipal
government, a bank, and several industrial companies.[107] In the negotiations leading up to the founding of Ravag, the SDAP and Chancellor Ignaz
Seipel had come to an agreement whereby the dominance of any political
party was to be avoided. Indeed, the socialists had one out of five seats on
the executive committee, by which only unanimous decisions could be
taken, and ample representation on the advisory council (seven out of
twenty-four) which supervised but did not control the programming.[108]
Even so, as actual practice soon demonstrated, the socialists were outnumbered on the program subcommittee (two out of six) by a coalition of conservative opponents. Moreover, Ravag was managed by Anton Rintelen,
governor of the province of Steiermark and president of the Steirerbank,
serving as Ravag's president, and by Oskar Czeija, a subordinate of Rintelin,
serving as director. These men viewed radio as an industry like any other
and were concerned with the profits to be gained from the listeners' fees of
a growing audience. They were certainly out of tune with the socialists' educational intentions on business and ideological grounds.[109] The same may be
said for the various program heads—music, literature, lecture and science,
news—whose antisocialist orientation was never in doubt.[110]

The structure of Ravag, in which the socialists maintained a veto power
on the executive committee but their opponents held control over the
implementing positions, was not auspicious for elevating the culture of
workers by providing them with social and political information and the best
in elite art. In all the later socialist criticism of Ravag's programming, no
mention was made of two principal concessions which the SDAP had agreed
to in the initial negotiations. The first was to exclude all political, religious,
and sexual subjects in the interest of "neutral" programming.[111] The second
made the national news service the only source of the twice-a-day news
broadcasts. In the first three years of operation, Ravag abided by the neutrality agreement above and tended to present bland information (crime,
weather, stock market, sports) to the exclusion of anything controversial.[112]
But the potential for a slanted politicized news remained in the hands of the
SDAP's opponents, who dominated the national government throughout
the period.

These early skirmishes as well as later struggles for control of Ravag had
little impact on the enthusiasm of the Austrian public for this technological
marvel, which brought entertainment directly to their homes. The number
of licensed listeners soared from 83,000 at the end of the first year to
508,000 in 1933.[113] Until 1928 two-thirds of the listeners were from Vienna;
thereafter Viennese preponderance declined to 55 percent by 1933. By that
time half the households in Vienna were radio listeners, as compared to one-

quarter of households in the provinces. During the first ten years of Ravag, Austria emerged as a principal radio listener country in Europe, surpassing Germany and not far behind Britain in the ratio of listeners to population.[114] Viennese enthusiam for the "ether waves" can in part be explained by the low license fee of two Schillings per month and the fact that initially radio receivers were largely crystal sets constructed by the listeners. The introduction of relatively expensive tube radios, which gradually replaced the crystal set, did not diminish the number of licensed listeners. The annual increase of listeners flattened out only with the onset of the depression in 1931.[115]

Neither of the political camps locked in combat over control of Ravag and its programming understood the fascination with radio of the Viennese in general and the workers in particular. When the results of a large national survey of listeners' attitudes became available in 1931, the findings were misinterpreted by both sides. The attitude prevailed that the listening public was purely passive. Listener reception was equated with perception, without recognition of the listeners' power to transform conveyed values to suit their experience and needs.

The struggle within and about Ravag continued. During the period of agreed-upon neutrality, from 1924 to 1928, the socialists praised educational programs such as language courses, theater previews, and readings and discussions by writers especially of naturalist and expressionist works. They criticized the programming in general for failing to present Vienna as a center of the music world, and for operating at the cultural level of *Heu-*

Listening to the radio with a crystal set also became a communal experience.
(Bilderarchiv, Die Wiener Stadt- und Landesbibliothek)

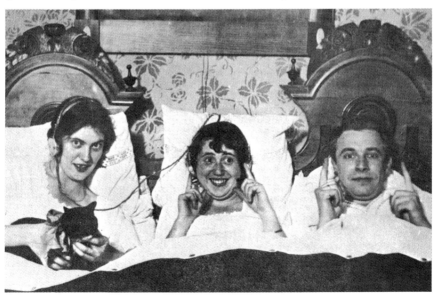

rigen entertainment. Their strongest complaints were directed at the clerical, Christian Social, and monarchist slant of musical programs and fairy tales.[116] Socialist criticism was accompanied by the only instance of direct pressure by the SDAP-dominated municipal government. Early in 1925 Finance Councillor Breitner announced the extension of the existing luxury tax on Ravag income at the rate of 30 percent. A reduction or suspension of the tax could be discussed, he maintained, only when the quality of the programs had been substantially improved.[117]

Although the political right had succeeded in keeping Ravag committed to light entertainment and free of controversial subjects, it began to question the value of neutrality in the aftermath of the bloody July days surrounding the burning of the Palace of Justice.[118] Radio technicians had joined the general strike of July 15–16, and Ravag had remained silent. When broadcasting was resumed the next day, the conservative and Catholic press was even more outraged, because a resumé of events was delivered by Councillor Breitner on behalf of Mayor Seitz.[119]

By this time it had become clear that neither political camp was satisfied with neutralism: the socialists because Ravag served as a conduit of bourgeois values, which distracted workers from the class struggle; the conservatives because Ravag failed to sufficiently express a Christian and German worldview, and remained under threat of political intervention by the socialist-dominated Viennese government. By tacit agreement between the two camps a struggle of ideas was to find a place in Ravag programming, but each side had its own conception of what such pluralism meant.[120]

The working out of positions was spearheaded by various radio clubs—worker, Catholic, and bourgeois—which had been organized at the time of Ravag's creation. Worker radio enthusiasts consisting of hobbyists and listeners organized their club in 1924, which counted 3,000 members by the end of that year.[121] Although the SDAP included this club among its cultural organizations, it had been formed without the initiative or assistance of the party and freely admitted party members, the unaffiliated, communists, and employees. Until its demise in 1932 (with 18,000 members),[122] it pursued goals independent of and often at odds with the party.

In the conflict within Ravag over the nature of pluralism, the Workers' Radio Club proposed that the SDAP representatives demand an independent and substantial worker programming as well as more worker representatives on the program subcommittee, based on the overwhelming preponderance of Viennese and worker listeners. This strong stand (often repeated and expanded to include the hiring of radio-competent artists, announcers, and writers) found little sympathy among established party officials sitting on Ravag committees, who claimed that their power to maneuver and bargain was being undermined. The "compromise" offered to and accepted by the SDAP as "worker radio" was the "Workers' Hour," a half-hour program made available to the Chamber of Workers and Employees with equal time allotted to the Chambers of Commerce and Agriculture. But even this "Workers' Hour" was severely restricted by the Ravag administration, so that a program about the reasons for the law against night work for bakers and another about school reform were canceled out of hand.[123]

The behind-the-scenes struggle over program revision had obviously been hard, for the Ravag administration made some attempt until 1931 to give more air time to worker subjects and activities (but always providing equal time to Catholic and conservative interests). No doubt it was in part guided by the business reality that the majority of the paying radio audience was in Vienna. Such concessions included celebrations of May Day and Republic Day (November 12); a series on the cultural importance of Vienna; an extension of educational lectures; and direct transmission of worker symphony concerts and the International Worker Olympics.[124]

On the surface these concessions in the name of pluralism might suggest that Ravag was moving in a direction that would make the eventual realization of socialist goals possible. But quite the opposite was true. In the conservative camp the issue was not cultural programming with political or social overtones, but radio as an instrument of political power. The lesson of July 1927, when Ravag went off the air and returned with socialist municipal officials having their say, was not lost on Christian Social politicians or the Ravag administration. To forestall a repetition of Viennese control over radio in case of emergency, the Ravag administration used a major part of its annual profits to build provincial stations to transmit national broadcasts, but also capable of broadcasting independently of Vienna.[125]

By 1931 the pressures of right-wing organizations on Ravag had become ominous. Following an appeal by Cardinal Piffl, thousands of new members streamed into the Catholic radio club. The protofascist Heimwehr also extended its influence over Ravag, with 10,000 of its members joining the conservative radio club.[126] Moreover, conservative and right-wing politicians were given more and more opportunities to speak on the radio, while socialist politicians were excluded.[127] When Ravag refused to air a speech by the German Nazi leader Georg Strasser in June 1932, Austrian Nazis attacked the Ravag headquarters.[128] The struggle over Ravag had become (or, apparently, always had been) a fight for political control which closely paralleled the battle for power in the national political arena.[129]

This excursion through the thickets of Ravag struggles reveals the extent to which the SDAP, virtually until the end, misinterpreted their opponents' ultimate aims. For the opponents, "keeping the Marxists in Vienna in check" meant not just foiling their educational and culturally ennobled programs for the workers. It meant bringing the powerful direct political power of radio under their control. Therefore the socialists' campaign for their kind of programming could be waged only on the level of national politics and won only if they succeeded there. They confined their efforts to appeals within Ravag and refused to go beyond them to fight for their case.

Surely the bargaining for the establishment of Ravag was badly carried out, giving the SDAP inadequate representation on the governing bodies.[130] Nor did the socialists dare to use the power of the municipal government they controlled to force an accommodation that would serve their cultural interests. These would have been served best by a second program within Ravag, a demand made repeatedly within the Workers' Radio Club but rejected by SDAP representatives in Ravag as too uncompromising. Least of all did they consider mobilizing the mass of worker listeners for the direct economic action of membership resignation. Only in 1933, when such a tactic had lost most of its power, did the SDAP form an organization which sponsored mass resignations on a selective basis (5,000 at a time).[131]

When one considers that the socialists' mission was to create a proletarian counterculture, one is struck by their blindness to and disregard for the needs and interests of the working-class radio audience. Bowing to continued SDAP pressure, Ravag finally carried out a substantial survey of radio listeners in 1931.[132] It was conducted by the economic/psychological research center of the University of Vienna under the direction of Paul Lazersfeld and was based on 110,312 listener survey responses. A breakdown of the findings gave a clear picture of the composition and program preferences of Viennese listeners. Among the most interesting general information was the following: that nearly half were workers and employees; that three-quarters used tube radios with a speaker; that the most popular listening time was between 7:00 and 10:00 p.m.; and that the largest age group of listeners (46 percent) was thirty-one to fifty years old.

Far more revealing were the listeners' program preferences and desires. They were asked to rank program types offered by Ravag on a scale of most- to least-preferred items. The most positive response (in descending order) was received by *Varieté* evenings, light entertainment concerts, comedies, Vienna evenings, and dialect plays. The most negative response (in descending order) was received by chamber music, choir music, literary readings, symphony concerts, and lectures about music.[133] The sixteen items in the negative category contained virtually every program preferred by the SDAP for its worker listeners. What kind of programs did the listeners want to hear? The answers did not depart significantly from the evaluation of Ravag offerings: an overwhelming preference for light and cheerful entertainment in music and literature and for lectures on unusual and "sensational" subjects.[134] There was only a marginal difference in the response of men and women to the above.[135]

The official Ravag reaction was to praise the survey and to carry out one important change by broadcasting lighter entertainment at the most popular listening time, and more serious programs in the late evening hours. For the rest, Ravag claimed to be offering such a wide choice that all listener interests could be satisfied, if individuals learned to be selective.[136] The socialist press treated the report of survey findings in a bland way. The briefest resumé covered the main points, glossing over listener preferences, but offered no commentary.[137] It seems as if the SDAP was concerned with the size of its public but not with what it wanted to hear. If it had really wanted to formulate its radio policy on the basis of what workers liked and expected, it could have commissioned its own survey long before 1931.

But the socialists had rather fixed ideas about the cultural needs of the workers, and a functional approach to how radio might be used to further these ends. The socialists, like virtually everyone else at the time, had little understanding of radio as a mass medium. Throughout the period broadcasting was at the lowest level of radio's potential and was put in the service of simply transmitting other cultural forms for a mass audience. The actualizing power of live, simultaneous broadcasts of events, giving the listener a sense of presence, was hardly ever achieved, because the talent employed by Ravag came from traditional cultural sectors and lacked "radio imagination."[138]

The socialists' desire to monopolize the workers' leisure time, to uplift them from mere pleasure for its own sake to a nobler cultural level, was based on an ignorance of the workers' actual work experience and leisure needs. It would seem that radio was popular among the workers because it provided an easily accessible means of physical and mental release from the tensions accumulated at the workplace. Returning from the monotony and drone of sewing machines or the piercing sound of drill presses, a worker may have needed and found relaxation and distraction in "light" programs. The SDAP discounted this need for relaxation, because it feared that the passive worker would fall prey to the evil messages of his enemies. The party's view of the worker was static, just as its understanding of radio as a mass medium was undialectical, leading to an equation of transmission, reception, and perception.[139] To be sure, radio insinuated itself into the workers' private sphere, but they retained the power to select and filter and, ultimately, to turn the radio on and off.

Spectator Sports: Gladiators of Capitalism?

A survey of mass-cultural competitors to the Socialist party culture would be incomplete without some assessment of the claims of spectator sports on the leisure time of workers. It is important to remember that organized sports were a centerpiece of party culture, with ASKÖ claiming a membership of 110,000 in Vienna by 1931. Although boxing drew periodic crowds of several thousand for matches held in the Prater, it was soccer, the most indigenous popular sport, in which the development of a mass cultural character was most pronounced.[140]

In the enjoyment of soccer by male Viennese workers we can observe the transition from noncommercial to commerical, to mass-spectator participation: from the street play of children and urban niche play of young workers to the formation of dozens of local clubs and the development of district teams; from the creation of professional clubs organized after 1925 into two metropolitan leagues, to national select teams competing in European cup games. Mass spectator soccer coexisted with soccer as a popular performance sport and drew upon the latter for its enthusiastic fans. Both emphasized superior collective effort and individual skill, elevating these to a physical/aesthetic level, and thereby enhanced the excitement and loyalty of team-oriented fans. In no other mass cultural form (film or radio, for example) was there such a relationship between the passive and the active participant.

It raises the question of why workers ceased to be satisfied with participation alone. What was there in soccer as a mass sport that attracted tens of thousands as spectators? Julius Deutsch, an important sports functionary of the SDAP, allegedly had the answer.[141] The worker-spectator was a victim of his own desire for cheap distractions, which in the guise of political neutrality estranged him from his own class. Capitalist sports, Deutsch intoned, seduced the spectator with the achievements of stars; socialist sport aimed collectively to develop the physical competence and grace of the worker's body. In other words, the low motives of mass sport were posed against the lofty goals of the party's sports activities; the primitive enjoyments of sports fans against the higher strivings of the collectivity. Aside from the fact that Deutsch neglected the overdisciplined and almost militaristic quality of ASKÖ organization and activities, he simply sidestepped the question of mass sport popularity.

Hendrik deMan's earlier critical observations about the needs and motives of worker spectators came much closer to the mark. The workers, he observed, attained a heightened sense of self from the alternative tension created by the sports contest.[142] Present-day sports historians have suggested that the accumulated emotions and aggressions of the monotonous workday could no longer be compensated for by sport participation, which no longer sufficed to ameliorate feelings of social inferiority and give expression to the natural desire for personal recognition.[143]

Beginning with the late 1920s, both socialist and mass spectator sports were given enthusiastic coverage in the popular party publications *Der Kuckuck* and *Das kleine Blatt*. This is hardly surprising when we consider that, from the early 1920s on, Vienna was a soccer city. As early as 1926, international games with 40,000 spectators were high points of the season. The stadium at the Hohe Warte in the northern outskirts of the city, with an overflow capacity of up to 70,000 spectators, was one of the largest in Europe.[144] In 1931 the municipality built a modern stadium in the Prater with a capacity of 60,000. Although the socialist city fathers had created this facility to house party sport activities as well as socialist mass festivals, soccer matches were a constant attraction. In the early 1930s, with two large stadiums and numerous district soccer fields, weekend crowds of 150,000 to 200,000 were unexceptional.[145] To these must be added the tens of thousands of radio fans who tuned in to occasional direct broadcasts of important games.

In view of such large attendance figures, the drawing power of mass spectator sports can hardly be disputed. It remains to place these in perspective in relation to noncommercial and commerical sports and the great variety of other leisure-time activities in which Viennese workers participated so enthusiastically and in such large numbers.

The SDAP's desire to permeate the workers' private sphere and to fill that with a tight network of party-organized and party-directed cultural activities

Soccer spectators (Bilderarchiv, Die Wiener Stadt- und Landesbibliothek)

was intended to compensate for the party's inability to alter power relations at the workplace or in the national political arena. But the SDAP's cultural program also had the loftier goal of transforming the workers into "neue Menschen." This utopian goal was based on the premise that in their present state workers lived at a brutish level and lacked the class consciousness necessary to resist the manipulations of the capitalist power structure and bourgeois culture to which they were exposed.[146] In the eyes of socialist leaders, therefore, a total transformation of the workers' private sphere was necessary—one that expunged their present life-styles and excluded all cultural forms and influences that did not emanate from the party. More than forty socialist organizations were expected to satisfy all the workers' cultural needs.

A closer look at the cultural environment in Vienna has revealed that the SDAP's expectations were both dogmatic and unrealistic. The texture of cultural life in which the workers participated was both complex and rich. Vienna offered a full and varied menu of cultural outlets—noncommercial, commercial, and mass—which coexisted and were enjoyed by the working population. The number of Viennese who participated in these is remarkable: on a typical weekend every man, woman, and child would seem to have partaken. Yet this seems unlikely. How then can we account for the masses who streamed into the *Gasthäuser, Heurigen, Varietés,* circuses, the Prater, movie theaters, sports stadiums, and other attractions every weekend? It would seem that many workers engaged in more than one cultural activity in their leisure time.

That supposition, however, is inconsistent with the existing information on household budgets and the proportion of worker wages available for nonessential expenditures. One explanation of this discrepancy might be that the recorded budgets disguised leisure-time expenditures under essential categories. Another would point to the large number of free or very inexpensive cultural activities available (hiking, swimming, local *Varieté*, movie, and sports admissions). Household budget statistics also fail to take account of the considerably larger proportion of wages available for leisure expenditures by younger workers, particularly those who were single and married couples without children.[147]

The strident tone of the socialists' criticism of nonparty culture in their various publications makes it clear that the SDAP was aware that its own program had not yet engaged the majority of workers. The party's attempt to reach out to these was marred by the uncritical approach to its own cultural program and its inability to understand why workers were attracted to commercial and mass culture. Party leaders were oblivious to the rigid structures of their organizations (the uniforms, drill, and military language of their sports organizations, for instance) and the ideological overloading of their cultural demands on the workers. The "other" culture was generally dismissed as degraded and degrading: the *Heurigen*-Prater commercial culture as an outdated form of cheap amusements; and mass culture as trash and kitsch which, unless ennobled, put the workers at the mercy of capitalism.[148]

Mass culture drew the socialists' strongest fire because they understood it least. With few exceptions they failed to appreciate film and radio as unique artistic forms. In this they were not alone, mirroring attitudes prevalent not only in the dominant culture but even among film and radio program creators themselves. The socialists were suspicious and even fearful of the powerful combination of images in motion, for instance, which had a vocabulary and structure of their own, and which they found irrational if they did not follow the linear forms of high culture they were accustomed to. They also found it particularly incomprehensible that audiences might get enjoyment from the technical novelty of film and radio unrelated to content or any hidden or obvious messages they might contain.

The socialists' denunciation of the mass media's bad taste reveals an inability to appreciate what one might call the democratic aspect of these leisure activities attractive to working-class audiences. The enjoyment of film and radio involved a very low level of social differentiation. The darkened movie theater provided a sense of anonymity in which clothes, personal appearance, and comportment did not matter, and stood in stark contrast to the theater and opera, where seeing others and being seen were part of the ritual. Both film and radio provided a realm of choice allowing for individual taste and at the same time a sense of equality, because the menu of choices before everyone was the same. Many films were playing at the same time. Friends, newspapers, or previous films with the same actors could determine the choice. Or no apparent choice needed to be made beyond immediate impulse or convenience. Radio could be kept playing as a background distraction to which the listener tuned in and out, or specific programs could be selected. In either case, the menu of choices was predetermined and the same for everyone, control was private and complete, and the social context was the domicile. Such flexibility of choice was a liberating experience, in contrast to the discipline and restrictions of the workplace. Moreover, the price differentials in movie theaters were much smaller than in opera or theater, and radio fees were the same for everyone. Enjoyment of radio and film required little effort and at the same time gave a sense of belonging, of being equal to others.

Were the audiences of the mass media really as passive as the socialists believed—ready victims of bad taste and hostile ideologies? Until recently one simply assumed that audience reception and perception were the same. This went for mass entertainment as well as mass consumption products. But the process of perception appears to be very complex, involving the personality of viewers as differentiated in various ways: by age, gender, formal and informal education, political experience, and so on. Studies of American film audiences in the early silent era have revealed that there was considerable verbal and physical audience reaction to what appeared on the screen.[149] In a similar way, fascination with a particular radio program, following it weekly and with commentary and discussion or later reference among peers, as well as leaving the radio on and using it as a kind of environmental background, are examples of interaction with the mass media.

One should not castigate the socialists for being no more sophisticated about mass culture than anyone else at the time, or for failing to recognize that the relationship between production, consumption, and use was not a linear process open to simplistic corrections based on party dogma. But their lack of psychological insight into the life-styles and daily pressures on the working class for whom they claimed to speak, and their unsubtlety in arguing against the "other" cultures and for their own, suggest why their proletarian counterculture failed to attract a majority of workers. Unfortunately for the SDAP, the purveyors of mass culture were far more adept in attracting their audience.

CHAPTER 6

The Worker Family: Invasions of the Private Sphere

As we have seen, the SDAP attempted to create a climate in the public sphere of Vienna conducive to the implantation and acceptance of its cultural program. Broad reforms in housing, health, social welfare, and education aimed to improve the workers' quality of life. A complex network of party organizations assumed the difficult task of creating a proletarian subculture capable of carrying out the transformation of workers into "neue Menschen." The SDAP leaders realized from the beginning that the effectiveness of both municipal socialism and their comprehensive cultural program depended on reaching into the workers' private sphere. The worker family, therefore, received particular attention as the most fundamental agency for influencing and shaping changes in behavior and consciousness. It was not so well understood that the family, as a center of private life and repository of accepted habits and practices, commanded powers to resist intrusions into its realm of activity and control.[1]

The nuclear family model became pervasive in Vienna and other European cities only between the late nineteenth century and the early 1920s, as an adaptation to changing productive techniques of high industrial capitalism, which demanded a stable worker existence and assured reproduction of labor. This "closed" form of family socialization gradually replaced previous "open" forms such as concubinage and other types of unregulated associations with a high degree of illegitimate births.[2] The emerging disciplined and orderly worker family approximated the model proposed by middle-class social reformers of the late nineteenth century. They had hoped to integrate the worker family into the norms of bourgeois life and thereby to assure stability and peace in the social order. Whether the stable and orderly worker family would provide the basis for proletarian embourgeoisement or for greater class consciousness and participation in working-class organizations remained an open question.[3]

The Austrian socialist leaders clearly aimed at the second of these two possibilities and sought to strengthen the formal structure and shape the values of the nuclear working-class family.[4] In following this strategy, the Austromarxists appeared to reject the accepted Marxist canon, which anticipated the dissolution of the family under capitalism and its replacement by communal forms of social organization.[5] In the Viennese context after 1919, their embracing of the nuclear family model was pragmatic; it also brought them dangerously close to the interventionist position of their bourgeois and clerical opponents. The bourgeois regarded the family as the foundation of social stability and conformity; the clericals considered it the primary spiritual unit of Christian morality. As we shall see, the socialists' intervention in the workers' private sphere, like the presumption of their opponents, assumed that the worker family was a passive entity. That view yielded a paternalism with a social purpose. As one recent commentator has observed: "Whilst the old order and its father figures had fallen from power, the social democratic leaders came forward in the chaos of the first postwar years as new father-figures."[6]

Men were not a direct subject of the SDAP's attempt to transform the working-class family. They were, however, always in the background as beneficiaries of the more orderly, relaxing, and peaceful home environment to be created. The task of building this domestic haven was placed on the shoulders of women, whose nature, appearance, and responsibilities were to be altered and enhanced, and whose role as wives and mothers was to be redefined. In its attempt to transform the worker family, the SDAP paid particular attention to the rearing of children and to the later organized life of youth, in the belief that the aspired-to goal of "neue Menschen" depended on the next generation. If the socialists' goal had an ethical/social purpose, their means of attaining it was largely limited to forms of discipline, self-denial, and the postponement of gratification.

The "New Woman" and the "Triple Burden"

What actual place was accorded to women in the cultural experiment to transform working-class life? Socialist party publications were silent or at best obtuse on the subject of women per se or of female consciousness and identity. This subject was generally subsumed under various higher social goals: the creation of *ordentliche*[7] (orderly, decent, respectable, and disciplined) worker families; the need for rational and controlled reproduction, leading to a "healthy" new generation; and the desire to make a varied party life central to the lives of workers. Since female workers accounted for almost 40 percent of the total labor force, and since 80 percent of married women were in some way employed,[8] the party literature devoted considerable space to the plight of women compelled to bear the triple burden of work, household, and child rearing.[9] In attempting to rescue working-class women from this plight, the socialist reformers hypostatized the "new woman" as the female part of the "neue Menschen" they were in the process of creating. As we shall see, here as elsewhere in the socialists' transformational program, the reformers failed to distinguish between their modest initiatives and their expectations—a confusion of the present with the future that resulted in a mythic substitution for reality.

What picture of the "new woman" did the socialist literature project for its readers? Her physical appearance was youthful, with a slender *garçon* figure made supple by sports, with bobbed hair and unrestraining garments bespeaking an active life; her temperament was fearless, open, and relaxed. To her husband she was a comrade; for her children she was a friend.[10] The working-class woman of yesterday—careworn in appearance, imprisoned by her clothes, unapproachable by those who needed her[11]—was to be abolished by waving a magic wand.[12] This image, like many other aspects of the socialists' program, was adapted from middle-class attempts to redefine the role of females in society. A seminal work in this liberation movement was the widely translated and somewhat scandalous *La Garçonne* by Victor Margueritte, which featured an independent, self-assured, and worldly woman who fought against the double standard and demanded the right to sexual experimentation. In the year of its publication (1922) it sold 300,000 copies; by 1929 sales reached over a million in France alone.[13] In the 1920s the *garçonne*/flapper became a widespread female role model in the industrialized world.[14]

How was the transformation envisaged by the SDAP to be accomplished? One standard answer was the equalization of female and male wages, making it possible for women to turn over housework and child care to paid, trained help.[15] The most common advice for reducing the triple burden was the rationalization of housework. This fascination with rationalization in the domestic sphere echoed the latest developments in the scientific management of industrial production and reflected the emphasis on science and efficiency of the home economics movement in the United States. It went hand in hand with the house-proudness exemplified by cleanliness and neatness, an aesthetic of simplicity and functionality, and the demand for formal training in efficient housework.[16] The SDAP's conception of domestic rationalization sought mainly to lighten the burden of each woman in her home, thereby contradicting the demand for professionalization of housework.

The party offered working-class women a variety of practical advice. They were encouraged to provide themselves with electric hot plates and irons, sewing machines, and vacuum cleaners.[17] When the costliness of these implements was remembered, the suggestion was made that women forgo the "luxury of personal presents such as jewelry and dresses" in favor of these labor-saving devices.[18] Otherwise, women were advised to purchase and use these machines collectively.[19]

Rationalization of the household was the keynote, and popular weeklies like *Der Kuckuck* and *Die Unzufriedene* provided a steady stream of labor- and money-saving tips for the simplification of housework.[20] One of the most influential pamphleteers of the period turned her ingenuity to simplifying the elaborate Sunday lunch—the bane of working-class women.[21] Accord-

The New Woman

Utilitarian clothing contrasted with bourgeois finery. *(Die Unzufriedene)*

Loose-fitting clothing, sensible shoes, and bobbed hair (VGA)

Literate and intellectually curious (VGA)

A body made supple by sports *(Der Kuckuck)*

ing to her formula, soup was to be abandoned in favor of cold canapés (with sardines, capers, or olives!); the main dish and baked dessert were to be prepared on the previous afternoon; and the accumulated mound of dishes, if neither husband nor children were inclined to wash them, should be left for Monday. Thus the harassed housewife was "liberated" on Sunday afternoon.

On occasion, the subject of the sexual division of labor in the household was raised but never really explored.[22] Instead, the socialist reformers offered the by-now-familiar nostrums: equal pay, a shortened workday, extension of the social support system (nurseries, kindergartens, youth centers) and collective facilities, and trained, paid houseworkers. The apparent object of these measures was to reduce the triple burden of working women and to make it possible for them to "participate in the working-class movement and to remain intellectually sharp by reading 'sensible' periodicals and books."[23] But there were other goals set for the free time to be won for women. Time and again the literature applauded the opportunity thus created for women to devote themselves emotionally to husband and children.[24]

It seems that the time gained by women through the rationalization of housework was not to be at their own disposal. The socialist reformers had already allocated it: husbands ground down by conditions at work were to be weaned from the *Gasthaus* and tied to the home with tenderness and understanding. Marriage itself was to be altered by these opportunities for freedom. Helene Bauer saw that old institution being transformed into "an erotic-comradely relationship of equals," as women gained status through their work.[25] Her excessive optimism about the liberating power of work for women led to a sharp critique. Bauer's notion, it was argued, might apply to a few bourgeois women, but for proletarian women work remained a burden rather than a sign of progress in status.[26]

However the visions of the "new woman" were formulated, and no matter how many new creative attributes the image was endowed with, the emphasis in the end was always on woman's role as mother. The sculptures of women selected by the city fathers for public places such as municipal housing invariably depicted the static, ample, nurturing mother rather than the dynamic, *garçon*-figured new woman.[27] Repeatedly motherhood was invoked as woman's "most noble" calling.[28] The whole subject was subsumed under the rubric "population politics," denoting a eugenic approach to the creation of a healthy and supportable new generation (see below).

SDAP publications laid great stress on the healthy female body as the means to a "natural" beauty. Central to good personal hygiene was the daily bath and rubdown, including those parts below the navel.[29] The use of cosmetics was discouraged, save for homeopathic remedies for less than glowing facial skin. After 1930, with the onset of the economic crisis and increased competition for jobs, periodicals made concessions to the use of commercial cosmetics to enhance the appearance of female job seekers.[30]

In addition to personal hygiene, fresh air and especially physical exercise were obligatory for the new woman. Working women of childbearing age were encouraged to participate in the SDAP sports program or at the very least to do ten minutes of calisthenics on their own before going to work.³¹ Thereby they were to be assured of two benefits: to become healthy and strong in preparation for motherhood, and to retain their youth and beauty just like bourgeois women.³² The pregnant and postpartum working woman received similar advice. If she took reasonable care of her body during these critical periods, she would retain her charm and attractiveness in the eyes of her husband.³³ Such care included avoidance of heavy work by those pregnant; avoidance of all work and excitement during the first weeks of the childbed period; and massages by a professional, or at least calisthenics for several months, for nursing mothers. That economic circumstances or financial resources might play a role in making such care possible was raised but never explored.

What was the everyday life of Viennese working-class women really like? To what extent were they in a position to be transformed into the new woman? In attempting to answer these questions, a useful point of departure is the excellent survey studies of industrial workers and homeworkers carried out by Käthe Leichter at the time.³⁴ According to the census of 1934, 41.5 percent of Viennese women over fifteen were in full employment.³⁵ Of these, 46.5 percent were workers, 24.4 percent were employees, and 13.2 percent were domestics.³⁶ It seems reasonable to look for answers to the above questions among workers in industry, where the larger context, contact with trade union and party, and accessibility to new ideas were most likely to lead to the conflict and gradual blending of traditional values and changing circumstances.³⁷

As we follow the industrial working woman through her normal day and extrapolate her experience for the week, month, and year, it becomes apparent that the socialist reformers' rendition of the triple burden treated it far too lightly and schematically. Working hours for most women began at 7:00 a.m. and ended at 5:00 p.m., but with the inclusion of travel time, this made for an absence from home of eleven to twelve hours a day.³⁸ But, considering their household obligations, the workday began between 5:00 and 6:00 a.m. and lasted until 10:00 or 11:00 p.m., making for a total workday of sixteen to eighteen hours.³⁹ Almost half of the women and three-quarters of those married did all the housework; those receiving assistance relied overwhelmingly on mothers and mothers-in-law.⁴⁰

Conditions in the homes of these female workers were not more promising for the rationalization of housework. In managing their household 18 percent had gas, electricity, and running water; but an equal number had none of these (though more than a third had electricity and water). Almost half of the women workers, and even a quarter of those married, did not have a home of their own but lived with parents or as subtenants. Bedrooms were shared with two or more persons by more than half, and with three or more persons by more than a third of the women.⁴¹ Even those who were

Female factory workers *(Der Kuckuck)*

fortunate enough to live in the new municipal housing (10.8 percent) generally shared their bedrooms with husbands and children because of the limited space (38 to 48 square meters) in these apartments.[42] Latter-day oral histories have added interesting details to this picture of crowding. It was common for young married couples to wait five to six years for an apartment of their own and to live cheek by jowl with parents and younger siblings. It was not uncommon for children to share their parents' bed or the bed of the same-sexed parent.[43]

The triple burden of many working women included child care, which further occupied their time and drained their energy. Some mothers complained about available kindergartens because they accepted children only at age four; many opened their doors only at 8:00 a.m., one hour after the adult workday had begun; some served no lunchtime meal; and most had long and frequent holiday periods or closed abruptly because of childhood diseases. In many cases the small fees charged by kindergartens and after-school centers were beyond the means of the family.[44] Was the triple burden lightened on weekends? Three-quarters of the sample and four-fifths of the married women devoted Saturday afternoon (the morning was a workday) to housework. Only Sunday afternoon was available to most women as a time for rest and/or recreation; one-third of those married and two-thirds of those single had Sunday morning free.[45]

It is difficult to see how under such conditions women should have considered their work as an enhancement of their status; indeed, they did not.

Aside from the apparent additional hardship, work created conflicts in female identity based on narrowly defined gender role models. The delineation of women's place in the home was strongly reinforced in the postwar period by an emphasis on creating a "real home" for the workers, one that was "neat and clean."[46] This was made possible in part by the stabilization of domiciles through rent control and in part through the building initiatives of the municipality. The goal of the "proper" home was not only a response to influences from the dominant bourgeois culture; it was in every way promoted by the schools and social welfare agencies of the municipality.[47] Such increased valuation of home and domesticity led to an increase in the variety of housework.[48] It also strengthened existing patterns of gender-role definition by which females were associated with home and household at an early age.[49] Schoolgirls might have resisted such expectations of domesticity in the hope of finding employment, but their limited prospects—dressmaker and nanny—reflected interests associated with female activities such as needlework and taking care of small children.[50] Besides, their opportunity for developing skills through apprenticeship was very limited.[51]

Conflicts of identity were no doubt reinforced by the realities of the labor market, where women were given the most menial positions, were the first to be fired, and received wages that were only 50–65 percent of male wages for equal work.[52] Although women were protected by law from night shifts, heavy physical labor, and dangerous occupations, labor inspectors reported frequent breaches of the rules.[53] The condition of female homeworkers was far worse[54]: their wages were 50 percent less than those of women in industry; they had no collective wage contracts; they suffered from intermittent unemployment; their living quarters were among the smallest and most densely populated, and served as workrooms in addition to their many other functions. Domestic workers were the most exploited and least protected of all. Although a law of 1920 regulated hours of work, wages, time off, and vacations, working conditions remained largely unsupervised.[55]

The trade unions did little to alter the impression that women were an unwanted presence at the workplace. Lip service was given to equal pay for equal work at trade union congresses,[56] but on the shop floor the attitude prevailed that women took away men's jobs.[57] There was a widespread attack on married working women as "double earners" which the trade unions appear to have abetted.[58] This lack of support is astounding when one considers that the working women of Vienna supplied 26.4 percent of the trade union membership.[59] That the trade unions made little effort to integrate women workers or to accord them positions in their organizations commensurate with their numbers can be adduced from the low percentage of female shop stewards,[60] the male orientation of trade union papers, and the underrepresentation of women trade unionists at general congresses.[61] It is small wonder, then, that trade unionism for women workers remained a formality, something expected of them, and that only 21.7 percent of the women trade unionists in Leichter's study ever attended union meetings and only 3 percent read the union papers.[62]

Why, then, did women work in factories? Leichter concludes that it was out of pure economic necessity.[63] Would they have continued to work, if their husbands or fathers had been able to support them? Eighty-five percent answered no.[64] The imperatives for such a choice are not difficult to understand. A retreat from work into the household was the only way open for women workers to reduce the triple burden. Neither the city fathers nor the socialist reformers had been able to create sufficient and appropriate social services to reduce their labor in the domestic sphere, nor had they seriously broached the traditional sexual division of labor there, which would have made a greater difference than all the labor-saving devices and rationalization schemes. Yet there are indications that women derived certain psychological benefits from work outside the home in the form of female solidarity.[65]

If we look at the bare facts offered in Leichter's study, we need hardly wonder that working women were light years removed from that attractive image of the new woman projected in the socialist literature. How could a working woman transform her body into the figure of a *garçon*, when her diet consisted largely of bread, starchy grains, and fat; when coffee was her mainstay morning, noon, and night; and sugar was the cheapest source of calories?[66] What time or energy was there in the working woman's day for sports, meetings, cinema, concerts, theaters, or even reading?[67] Given the stress of meeting her daily responsibilities, what opportunity was there for her to be "fearless, open, and relaxed," to become "a comrade to her husband and friend to her children"?

A word or two about the great variety of helpful hints to the housewife/mother served up by the socialist reformers should suffice. Labor-saving implements were simply beyond the means of all but a few working women. Nor did they have the time to organize the collective use, or even the small change necessary for the collective purchase, of the same. How could the typical kitchen-and-room or room-and-a-half worker apartment be rationalized, given the density of habitation, multiple use of all space, and frequent absence of basic amenities? The problem was not further or better organization but more space. For most women, the daily ablutions called for would have meant stripping in the kitchen in full view of children and other adults, not to speak of bringing cold water from the hallway tap. The exercise forced upon most working women consisted of household preparations after rising and a brisk walk to the workplace. Calisthenics, even for ten minutes, was for women with much more leisure in their daily routines. Working women who became pregnant surely did not have to be told what was best for them and the infant to come. It was not out of ignorance that they cut short their legally guaranteed lying-in and postpartum leaves but out of fear of losing their jobs.[68] Moreover, how could the average pregnant or nursing working woman avoid heavy work, as she was counseled to by the reformers, or afford massages to revitalize her body? The suggested time-saving Sunday lunch with olive canapés deserves no comment.

It might be appropriate at this point to ask whom these socialist messages might have reached and served. In Vienna, certainly the SDAP functionaries and cadres, of whom there were more than 20,000 by 1931, and functionaries in the trade unions and Chamber of Workers and Employees.[69] Beyond them, workers in safe and well-paid employment in the public and municipal sectors, who had already reached a lower-middle-class living standard.[70] But the rank-and-file female worker in industry and especially the homeworkers and domestics were by and large beyond reach.

Surely the socialist reformers were well-meaning, especially in the light of their many initiatives in bettering the lives of Austrian workers. Why, then, were their transformational plans for women so unrealistic, so blind to the actual life-styles and deprivations of working women? In the cultural laboratory of Vienna a fundamental and perhaps unbridgeable distance between leaders and masses came to light.[71] The average worker and the higher functionary inhabited two different worlds between which there was little contact. Female leaders in parliament all belonged to an older generation whose age in 1930 ranged from fifty-one to sixty-seven.[72] But even younger female leaders who had professions outside politics and were more perceptive about social realities in general, such as Käthe Leichter or Marianne Pollak, were middle-class professionals viewing the working-class world from afar. It was all well and good for Therese Schlesinger, Anna Boschek, Adelheid Popp, Gabriele Proft, and even Marianne Pollak and Käthe Leichter to exhort working-class women to rationalize housework. These leaders knew next to nothing about a burden they were able to turn over to hired help.[73]

The same applies to the other nostrums they offered to their readers. Most spoke in imperatives—"society should," "the municipality is obligated"—and in their self-congratulatory formulations often neglected what actually might have been done.[74] Moreover, the socialist reformers tended to view the existing worker subcultures as barbaric and to deny their positive aspects. Their attempts at superimposing the new woman over working-class reality may have appealed to an elite of functionaries and privileged workers but could not find resonance among ordinary proletarian women.

Sexuality: Repression and Expression

The study of the place of sexuality in working-class culture is still in its infancy.[75] The most noteworthy investigations stop at the threshold of the twentieth century, on the eve of full-fledged industrialization. The sparseness of data on all periods has made sexuality particularly difficult to investigate. The reticence of both memoirists and oral history subjects to reveal the most intimate aspects of their private lives has forced historians to work with very fragmentary evidence and to rely on inference and contextual reconstructions. If this is so, then why study the subject at all? Because our understanding of working-class culture would be very incomplete without some glimpses of intimate life.[76] There, at the core of the workers' private sphere, lie the emotional resources that have made it possible for them to express their selfhood at the workplace and to respond to the most oppressive aspects of wage labor.[77] There, also, are embedded a variety of subcultural forms and associations which, acting as both disabilities and strengths, have made it possible for workers to resist exploitative manipulations and reformist schemes for self- and collective improvement.

In its attempt at a total transformation of workers' culture, the SDAP erased the boundaries between the public and private spheres and between the social and the sexual. As one typical programmatic statement put it: "Sexual relations meet a physiological and psychological need, whose satisfaction has social consequences. For that reason sexual activity is not simply a private matter."[78] The attempt by Socialist party leaders and municipal functionaries to incorporate sexuality in their cultural experiment affords us some insight into the relationship between working-class life and a party's attempt to direct and reshape it: into the tenacious imperviousness of older subcultural forms to reform efforts which refused to take them into account; into the seemingly unbridgeable social and cultural distance between socialist leaders and the rank and file; into the bourgeois and liberal origins of socialist reform efforts; and into the ultimate similarity of socialist political and cultural goals, both of which could be attained only by struggle and direct confrontation with the political-cultural opponent.

What place was accorded to sexuality in the transformation of working-class life? From the beginning of the Viennese experiment, sexuality received considerable attention from the socialist reformers. But with very few exceptions they were primarily concerned with the social effect of sexuality on the party, on the worker family, and particularly on the next generation, which as "neue Menschen" was expected to make socialist culture a reality. Sexuality was viewed as having a social utility, especially in uplifting the moral standards of worker families. The end product was to be the *ordentliche* (orderly, decent, respectable) family; sexuality would have to be shaped and constrained to accomplish that end.[79] Sexuality played a similar socially practical role in the extensive discussions of "population politics," from which the need for rational and controlled reproduction leading to a healthy new generation emerged.

But sexuality was not only viewed as a means to social ends. Much of the socialists' concern with the subject centered on its possible negative powers, which threatened to distract workers from the programs, organizations, and activities being created by the SDAP, and to lead them into private spheres that were at best neutral in relation to collective culture. Seen in that light, sexuality was to be sublimated so as to make the workers' "marriage" to the party possible. The socialist reformers showed little concern for sex as a source of pleasure and as a normal and important part of everyday life. In its "pure" form sex was an embarrassment and was treated obliquely. Most frequently it was dealt with in party publications as a problem of moral control in which women and youth were the primary subjects of concern.

Population Politics

Socialist periodical and pamphlet literature was obsessed with the dangers of prostitution, to which it claimed working-class females were exposed. In part these fears were only a continuation of a major preoccupation of middle-class reformers of the 1880s and 1890s.[80] They also reflected the popular and totally unreliable tracts on the dangers and evils of prostitution and pervasiveness of venereal disease circulating in the 1920s.[81] But even the increase of prostitution during wartime and the immediate postwar period, a further source of socialist anxiety, seems to have been exaggerated. In 1920 there were eight thousand arrests for prostitution in Vienna, and 40 percent of those arrested were from the middle class.[82] A later survey of venereologists, gynecologists, and alienists pointed to the marked decline of prostitution in postwar Vienna.[83] Attempts to lay the ghost of decadence to rest and to situate sexual waywardness within the context of social neglect in general were rare and reached only a select audience of specialists.[84] For popular consumption the socialists provided cautionary and moralizing articles and tracts.

Socialist party sermonizing against sexual decadence began early. An article of November 1919, in the biweekly for female SDAP members, took the reader rapidly through the general evils of capitalism to the dangers of female promiscuity, leading ultimately to prostitution.[85] The two, although sequential, were different, the anonymous author insisted. Promiscuous women engaged in sexual intercourse for its own sake because they desired men, whereas prostitutes sold themselves for money. But "mothers" did neither, because they desired only one specific man and engaged in intercourse only for the purpose of procreation. Far more influential were the marriage pamphlets of Johann Ferch, a popular socialist writer of romantic fiction and founder of the Union Against Forced Motherhood. One of his frequently reprinted pamphlets is a treasure trove of male middle-class attitudes on marriage and sexuality.[86] He begins by proclaiming that the place of sex in love and marriage has been exaggerated, fulminates against those who call for sexual freedom for young women as despoilers of the ideal of homemaking, catalogs the evils of premarital sex, and warns that casual love will make a person incapable of the true love on which marriage is built. Nor does Ferch neglect the double standard: the horror a man might feel upon discovering that his "true love" had been possessed by other men. In view of the present difficulties in founding a home, the prerequisite of a happy marriage, he advises women to be on their guard against sexual desire, which generally stems from men.

Both of these patronizing warnings are reminiscent of nineteenth-century moralizing in Germany, France, England, and Austria. A greater irony lies in the fact that these socialist sermons—restricting the conjugal act to procreation by married women and denouncing premarital sex as the act of fallen angels—might have come directly from the pastoral letters of Austrian bishops in 1919 and virtually every year thereafter.[87]

The above examples are typical of the verbal sublimation served up in the party literature. One further illustration is necessary to demonstrate the predominant eugenic strain in virtually all discussions of the sexual question. A physician writing in *Die Unzufriedene*, the SDAP's popular weekly aimed at unaffiliated women, praised the virtues of marriage and building a home, but strongly urged women not to succumb to the prevalent notion of love without marriage.[88] Sexual relations before the age of twenty were particularly dangerous, the doctor insisted, because the female sexual organs, as yet immature, would be permanently damaged and future offspring might be harmed. Moreover, no woman should enter the state of matrimony without obtaining a certificate of health from her prospective spouse, since the well-being of the next generation was at stake.[89]

In a later attack on sexual abstinence literature, Wilhelm Reich singled out the harmfulness of designating an arbitrary age—twenty or even twenty-four—as medically appropriate for the onset of sexual intercourse. In his experience as a sex counselor, he maintained, those who had not made the transition from masturbation to intercourse by the age of twenty experienced difficulties in doing so later.[90]

As early as 1922, under the guidance of the anatomist Julius Tandler, the municipal council created a marriage consultation clinic to certify to the health of prospective sexual partners. Its director, the gynecologist Karl Kautsky, Jr., made it clear that the central purpose of the clinic was to prevent the inheritance of disabilities and to improve the "quality" of the population. In his publicity for the clinic Kautsky assured the public that no contraceptives would be provided. By 1927 Tandler and Kautsky were forced to admit that the clinic was a failure for lack of clients. In Germany, by contrast, one hundred such clinics had been created by 1928, providing information on contraception and sexual technique.[91]

The subject of sexual promiscuity was also aired in the more scientific setting of an international congress of the World League for Sexual Reform in 1931. There Tandler, a socialist member of the municipal council and head of its Public Welfare Office, presented the official SDAP view.[92] Sexual problems arising from sexual pathology, he asserted, were one of the principal sources of moral decay and social disintegration. The chief cause of this misery, he insisted, was the overcrowding of habitations; therefore the basis of sexual reform must be a public program to create new housing for the working class.[93] In the ensuing discussion Tandler was criticized for linking sexuality essentially with procreation, for failing to recognize it as a special condition of human existence, and for avoiding the reality that promiscuity in the working class had its origins in the repression of women by men.[94] A further interpellation challenged the right of society to punish the sexual transgressions of youth, so long as no one assumed responsibility for sex education.[95]

The most pointed attack on Tandler's traditionalist position came from Wilhelm Reich.[96] In the average Viennese worker domicile, he claimed, four persons share a single room and as a consequence the sexual act is performed fully clothed, in fear of disturbance, or with an indifference to others present. This condition, he added, leads not to promiscuity but to the general impoverishment of sexuality in the working class. Youth, who cohabit in the woods during fair weather, are driven into dark doorways in other seasons. Therefore it was the sexual repression (created and maintained by capitalism) of the workers, and not their license, which was the true problem to be confronted.

Although Tandler and Reich (and other critics) agreed that decent housing would have a positive effect on the sexual life of workers, the two sides disagreed fundamentally about the essence of worker sexuality. For Tandler it was reproduction in the stable surroundings of municipal housing and under the influence of the party's institutions, which would guard against moral decay and assure the creation of respectable worker families. Reich argued for a sexual expressiveness—even for youth and the unmarried— that in the eyes of the party leadership bordered on anarchic permissiveness and rationalized the very license the party hoped to stamp out. SDAP leaders opted for Tandler's more orderly position, and the pages of party publications, particularly those meant for mass circulation, generally remained closed to opposing points of view.[97]

The most personally and socially dangerous aspect of sexuality was not promiscuity (an arbitrary designation at best) but its consequences in illegitimate children, unwanted births, and illegally terminated pregnancies. Abortion, according to paragraph 144 of the penal code, was punishable by substantial prison terms.[98] However, in the absence of sex education in the schools and of readily available and inexpensive birth control devices (both of which were fought vigorously by the Catholic church), it was also a prevalent form of birth control in the working class. How widespread the practice actually was is difficult to say. One medical source estimates an abortion rate of 20 to 40 percent of all pregnancies.[99] Considering that abortion was illegal and therefore outside the reach of official numerical notation, the upward reaches of that range seem appropriate for the working class. At any rate, the incidence of abortion was high enough to make it a major issue of social and political controversy.

Throughout the period the SDAP occupied a series of ambiguous positions on the abortion question. In 1920 a National Conference of SDAP Women demanded the revision of paragraph 144 to allow for abortion in the first trimester of pregnancy. At the end of the year Adelheid Popp presented that proposal for reform to parliament, where the Christian Social party prevented it from being formally discussed.[100] In internal party discussions during the following years the trimester model was maintained, but arguments in its support were largely eugenic. The demand by the gynecologist Karl Kautsky that each medical intervention be duly reported to the police to underline the seriousness of the act undertaken raised doubts about the conviction of the reformers.[101]

Socialist reservations about the trimester plan or, more specifically, about allowing women a major say in determining the need for abortion, became clear at a conference of SDAP physicians on abortion and population politics held in May 1924.[102] The conference adopted the position of Tandler, popularly called "the medical Pope of social democracy," that under no circumstances could abortion be performed on demand, because society had to keep control of sexual reproduction and could not become dependent on the needs of individuals. In his influential writings Tandler had insisted that life is created at the moment of conception and that society and not the mother has juridical control over the embryo.[103] Tandler proposed three criteria to determine the justification for abortion: medical, eugenic, and social. The first two were to be determined by a panel of physicians. As concerns the social criteria, a panel consisting of a judge, a physician, a woman, a lawyer representing the embryo, and a representative of society were to determine each case on its merits.[104]

The SDAP program of 1926, which outlined the party's position on a variety of social and cultural questions, also included a special section on birth control. Although the party refused to adopt Tandler's complicated and restrictive plan, the determination of the need for abortion was implicitly left in the hands of experts. The program recommended abortion in a public hospital if the birth might affect the health of the mother, produce an abnormal child, or endanger the mother's economic existence or that of her children.[105] In the years that followed, the SDAP never went beyond this position.[106]

The abortion platform at the Linz party congress of 1926 had been managed by the male leaders without apparent opposition.[107] But at the National Women's Conference preceding the congress there had been a broader spectrum of views, mainly because more radically feminist delegates from Styria challenged the conservative positions of the party doyennes.[108] The radical Styrians rejected the right of men to place limitations on the need for or right to abortions and argued for women's right to control their own bodies. This challenge to the official SDAP position was easily voted down by the female leaders, whose views on abortion were of a piece. They ranged from Therese Schlesinger, who favored Tandler's plan of determining social criteria through panels, to Adelheid Popp and Gabriele Proft, who cautioned against going too far beyond the present mentality of working-class women, to Emmy Freundlich, who insisted that childbearing was a female duty and proposed that men should have a say on prospective abortions.[109] The common ground among leading female socialists appears to have been the fear of bearing the eugenically unfit, who would become a burden to society.[110] On the subject of eugenics the socialists had unexpected bedfellows among racists, anti-Semites, and National Socialists, who demanded that paragraph 144 be used to weed out the unfit and to promote the motherhood of healthy women.[111]

Female party leaders continued to argue for reform of the law in public and in print, but usually did so by intoning the importance of motherhood in allowing pregnancies to be terminated for social reasons.[112] It is difficult to explain why the position of prominent female socialists on the abortion question, though more differentiated than that of male leaders, should have failed to go much beyond the official party view. Schlesinger offers some insights into the difficulties experienced by female socialists within the inner precincts of the SDAP. In order to function in a party of men, she suggests, women had to accept the view that female oppression was a condition of capitalism, had to be on the defensive against charges of putting gender issues before the "important" questions of the party, and ultimately had to internalize what was expected of them.[113]

The SDAP's equivocation on paragraph 144 can in part be explained by its principled stand on birth control. It called for the creation of public birth control clinics and the dispensing of contraceptives through the public health service.[114] These demands had been made repeatedly in the past, especially by socialist women, who regarded these measures as essential in reducing the need for abortion and in making the separation of pleasure from procreation in intercourse.[115] But neither the SDAP nor the municipality developed a strong and comprehensive plan to turn such programmatic exhortations into reality. This is not to say that scattered attempts were not made by or with the blessing of the above institutions, but they lacked the single-minded commitment with which programs in housing, welfare, and health were undertaken. The municipality did create thirty-six mothers' consultation clinics throughout Vienna, but their emphasis was on the problems of childbirth, female pre- and postpartum health, and infant care.[116]

The problem of inadequate information about birth control and the inaccessibility of contraceptives remained largely unsolved. A few consultation clinics were created for married couples and the Association for Birth Control did sponsor lectures on sexuality, as did women's groups of the party's cultural associations.[117] But such efforts hardly touched more than a very small proportion of the working class. Discussions about and advertisements for contraception did creep into party publications. This was particularly true of *Die Unzufriedene,* which gave very serious attention to women's issues (such as abolishing the law which still permitted "reasonable" physical punishment of wives by husbands). But for every article in the party literature touching on sexual questions, there were scores dealing with proletarian motherhood. The latter was equated with a healthy sex life in which erotic pleasure appeared to play no role. Instead, the negative consequences of sexuality in unwanted births and abortions ran as a danger signal through the popular party publications.[118]

The important beginning made by Wilhelm Reich in opening six sexual consultation clinics for workers and employees in 1929 received no support from the party or municipality. The clinics were staffed by psychoanalysts and midwives and were open two hours a day for consultation by youth as well as married and unmarried clients.[119] Reich's extremely sober pamphlets on sexuality, written in a style that the average reader could understand and dealing with coitus and contraception in a clear and supportive manner, were not published by the party publishing house.[120] Nor did the party make printed material on birth control or sexual practices available to the working-class public. Neither the popular sex manual *Die vollkomene Ehe* by Hendrik van de Velde, the frank guide for youth *Bub und Mädl* by the sex reformer Max Hodann, nor the numerous other versions of practical enlightenment available in Germany were published by the SDAP or otherwise made available at prices affordable by workers.[121] Various marginal petitioners for party or municipal action were too dispersed to make the public health service carry out the frequently demanded and promised free dispensing of pessaries and other contraceptives.

In looking back on the SDAP's efforts regarding abortion reform and birth control, one cannot escape the impression that the party gave lip service to the second in order to avoid having to confront the first. Both these issues were of great importance to workers. The party failed to provide assistance in this aspect of private life not because it shunned intervention there but because it feared the emotionally charged atmosphere surrounding sexuality as a public issue.

The SDAP does deserve credit for having made the issues of birth control and abortion part of its official program. At the same time it raises the question of why the party was so cautious in its proposals. The leaders explained their moderate approach as a means of disarming the political opposition in parliament. The Christian Socials had no program of their own save the absolute rejection of any modification of paragraph 144 and the support of steady population growth. They used this simple negative weapon to ward off all socialist attempts from 1920 to 1932 to restructure the abortion law.[122] In the face of such determined and successful opposition in parliament, was moderation the best course, and was parliament the only or primary arena for waging a campaign for reform? By keeping the struggle confined within strict legislative bounds, the SDAP prevented any expression of public sentiment, any mobilization of action from below by diverse groups in society among whom abortion was practiced in constant fear. In Germany in 1931, a massive mobilization of the public attempted to abolish the restrictive antiabortion laws.[123] Nothing comparable took place in Austria.

This failure to encourage initiatives from below points to one of the cardinal weaknesses of the SDAP. The highly bureaucratized and paternalist party saw no need for rank-and-file initiatives other than symbolic mass celebrations which it organized and controlled. The fear of losing control appears to have been uppermost in the minds of the leaders; mass mobilization threatened the legality to which they were committed above all else. The socialist workers' culture, which the party was attempting to implant in

Vienna, also served to enhance the passivity of the rank and file. After all, what need was there for popular expression on abortion (or on other issues), when the party claimed to be taking care of all of the workers' needs and problems through its network of social and cultural organizations?

Such criticism must not overlook the fact that the SDAP had genuine reasons for fearing the opposition of the Christian Socials on the abortion and birth control issues. In no European country did the Catholic church advance more conservative views or play a more direct political role.[124] Every attempt by the socialists to reduce the public influence of the church, such as the abolition of compulsory religious instruction in the schools, resulted in a bitter struggle in parliament with the Christian Socials, and in the streets with a host of Catholic Action groups.[125] Since the church equated morality with Christianity, of which it was the sole guardian and only spokesman, it fought most vigorously any attempts to tamper with what it defined loosely as moral conduct. It equated abortion with murder and threatened transgressors with excommunication, denounced the artificial restriction of the number of children in families as blasphemy, opposed coeducation and sex education in the schools as invitations to lust, and blamed all these "signs of modern degeneracy" on socialist immorality.[126]

Several pastoral letters were specifically addressed to the moral conduct of girls and women in the form of admonitions.[127] Girls were to be segregated and closely guarded during gymnastics and swimming; during medical examinations in the schools their modesty was to be assured by female physicians; and they were to be restrained from mixed activities such as hiking and dancing or else be closely chaperoned. Women, as guardians of "pure morality," were cautioned against wearing revealing modern clothing and instructed to bathe only at sex-segregated pools and beaches. Attempts by the church to convert its moral dicta into secular law were narrowly prevented by the SDAP on constitutional grounds.[128]

During the 1920s several incidents brought the *Kulturkampf* to the point of explosion, all of them involving sexuality and public morality. The first of these revolved around the Vienna premiere in 1921 of Arthur Schnitzler's play *Der Reigen* and pitted the Christian Social federal government against the socialist Viennese municipal and provincial government.[129] At issue was the central theme of the play in ten scenes linking heterosexual couples in an unending chain of coital relations. These, cutting across class lines, represented a seamless web of lies and desire, deceit and misery, calculation and feeling. The Christian Social press waged a campaign of denunciation against Schnitzler and his socialist supporters in vicious anti-Semitic epithets, and various Catholic Action groups prepared for physical intervention to prevent the play's performance beyond a trial period. The SDAP chose to fight the issue of pornography and censorship on the constitutional grounds that the Viennese government was legally empowered to make a decision in the case. The party refused to defend the artistic merits of Schnitzler's reflections on sexuality or to answer the Christian Social charges of sexual degeneracy. The mayor of Vienna, Jakob Reumann, explained the party's failure to take a position on the sexual question by saying that the morality of Viennese workers was not affected by the play because they did not go to see it.[130]

Two other scandals brought the issue of sexuality before the public.[131] In 1924 the Catholic church unleashed a poisonous campaign against the writer Hugo Bettauer, whose street novels were popular in the Viennese worker libraries and whose sexual reform magazines reached a circulation of 60,000. In both he advocated sexual freedom and experimentation and championed women's rights. To the Catholic religious and lay leadership Bettauer represented an arch fiend, sexual demon, and embodiment of pornography, a blot on public morality that had to be erased. Federal Chancellor Ignaz Seipel demanded that the municipal government purify itself of this evil by exercising censorship. Again the socialists on the municipal council resorted to constitutional arguments and refused to take up the question of sexual morality. The storm whipped up by Catholic Action groups subsided by the end of the year, but it had created a climate of hate and violence in which the murder of Bettauer by a right-wing fanatic in the following spring was a natural consequence.

In 1928, posters of the scantily clad body of the black Paris-based American dancer Josephine Baker appeared in Vienna, announcing her schedule of performances.[132] For the Catholic leadership, public exposure to pornography and degeneracy was again being flaunted. Chancellor Seipel instructed Catholic Action to mobilize the Viennese population to finally rid the city of the socialist government which tolerated such filth. The socialists fought off Christian Social demands for censorship and a special antipornography law with the by-now-routine constitutional weapons. At the same time the SDAP daily, *Die Arbeiter-Zeitung*, characterized Baker's performance as "the coupling of a naked woman with naked money."

These brief summaries of Catholic-socialist confrontations over sexuality and morality cannot re-create the violent emotional climate in which they took place. They should make clear, however, that struggles on the cultural front were at heart political in nature. Every attempt by the socialists to introduce new laws, practices, or facilities in the cultural and social realm was met by the church and its party with a violent onslaught in parliament and the street. The Catholic offensive against any change in an Austria it considered to be Christian and capitalist by definition meant that any part of the socialists' cultural and social program—not just abortion and birth control—would meet potent resistance. If the socialist leaders understood this reality, they did not act on it. By limiting their defense to the legal realm and the protection of the constitution, while their opponents organized pressure groups and fomented street actions, the socialists remained in a purely defensive position. Abortion and birth control were issues which might have brought the SDAP much public support not only from party members or workers but even from Catholics of the opposing party, who also suffered from the restrictions of paragraph 144.

Youth: Abstinence, Discipline, and Sublimation

Just as the SDAP and the municipality limited their concern for adult worker sexuality to the question of healthy reproduction, in which marital motherhood received high marks and the sexuality of others was left to the realm of sublimation and repression, their concern with the sexuality of youth almost exclusively sought its postponement. Youth was clearly the most important group for the socialist reformers, who viewed it as the standard bearer of the desired transformation leading to "neue Menschen" and an alternate proletarian culture. The concerted party and municipal effort, initiated shortly after the war, to prepare children and youth for the "true path" met with considerable difficulties. In addition to the home, the traditional place for the socialization of working-class children and youth was the street—in reality, a shorthand term for a variety of urban niches free of adult supervision and control. From the beginning of the republic, socialist reformers characterized the street as the locale of disorder, promiscuity, and decadence from which the young had to be rescued.[133] The rough and ready activities in which groups of children engaged in their "territories" was believed to lead to criminality. For boys this meant stealing; for girls, the far more irremediable drift toward prostitution.[134]

"Neue Menschen": the generation of fulfillment (VGA)

Who was responsible for this wild and dangerous socialization leading to promiscuities of various kinds? The capitalist system, of course; but beyond that, the more tractable proletarian family was identified.[135] It lacked the capacity for the proper rearing of the young, it was argued, because of inadequate living space, the high rate of divorce, the frequency with which both parents worked outside the home, and the prevalence of adult authoritarianism. In short, the disorderly worker family could not be entrusted with the rearing of orderly socialists of the future.

The SDAP made its program for children and youth the centerpiece of its countercultural efforts. It created a series of lockstepped organizations and activities for every age group. The Kinderfreunde (for ages six to ten), a prewar parents' association concerned with imparting a proletarian consciousness to their children, was incorporated into the SDAP structure in 1922.[136] Its main function was to counteract the prevalent street socialization of working-class children with various supplementary educational facilities and programs. The focus of its activities was in after-school centers in which homework was to be done under supervision and community responsibility was to be instilled. The strict discipline and regimentation practiced at these centers discouraged many children, and particularly the older boys, from attending them.[137]

The Rote Falken (Red Falcons, for ages ten to fourteen), created by the SDAP in 1926, were an adaptation of the folkloric Wandervogel, the regimented Boy Scouts, and the politicized Soviet Russian Pioneers. Dressed in green hiking shirts and shorts or short skirts with red neckerchiefs and subject to twelve commandments, they were most akin to the Scouts. But their goal was an ethical socialism to be attained by a strict proletarian discipline, personal purity, and obedience to leaders. Falcons were expected to reject bourgeois values and habits, especially drinking, smoking, and trashy entertainment. By 1932 the organization in Vienna numbered about 6,000.[138]

The Socialist Worker Youth (SAJ) was most important as a transmission belt for young people (ages fourteen to twenty-one) into the SDAP and its Schutzbund.[139] Abstinence, sexual sublimation, and puritanism were guiding principles in the ethical preparation of youthful members. Their activity was restricted to sports and education. With the exception of electoral canvassing, they were prevented from engaging in direct political activities on the grounds of political immaturity.[140] The party leadership feared that its control over the SAJ would be weakened by political experience, which would incline the youth toward radicalism. The Vienna organization numbered about 10,500 in 1932.[141] It is remarkable how much the SDAP expected from its youth and how little confidence it had in it. The SAJ program seemed to be directed toward the postponement of adulthood. That orientation might have been appropriate for middle-class youth absorbed by formal education. It was unrealistic for working-class youth who by and large worked as full adults from their fourteenth year.

The significant socialization of the next generation was controlled by the party, which made available the experts necessary for the task. The only

Socialist Worker Youth: a pure mind in a disciplined body *(Der Kuckuck)*

function reserved for parents in the home was to offer the emotional attachment and tenderness which biological ties alone could provide.[142] The SDAP's whole effort was revolutionary in an unintended way: it brought both sexes together in myriad activities and even encouraged the "natural" association of the sexes in a manner previously unknown in Catholic Austria.[143] Understandably, such proximity raised the sexual question.

Socialist spokesmen had a ready answer, one that generally avoided the questions which arose in practice from the daily experience of their charges. The party, they maintained, through its age-linked organizations, was helping youth to experience the wonderful state of comradeship, the main way station on the road to becoming more fully realized socialists.[144] But the subject of sexuality could not be treated only in the form of commandments, such as that of the Rote Falken to be "constantly pure in thought, word, and deed."[145] The discussions of the unmentionable in the party literature very much followed the pattern of writings on sexuality in general: idealization, emphasis on sublimation through sports and exercise, and an insistence on the postponement of gratification until physical and political maturity.[146] A closer look at a small selection of these advisory and cautionary tracts should suffice to illustrate the SDAP's general approach.

A particularly moralizing form of presenting the subject of sexuality to youth was the paternal heart-to-heart talk. The prominent youth leader Otto Kanitz, speaking to the multitude of proletarian fourteen-year-old boys, allows that sex drives will soon announce themselves with great force.[147] But he cautions that the normal yearning for a woman and child is premature for them ("Just as the unripe seed of an apple can't produce a tree"). Kanitz continues with this nature analogy to warn against "casting the seed in the wrong place" (the diseased prostitute) and commends purity until maturity for the sake of the health of the children to come. In a similar sermon for girls, Gerta Morberger assumes the tone of friend and comrade. What do socialist girls expect in their relationship with the opposite sex?[148] They are aware that in these chaotic times the temptation exists for fleeting relationships. But they want neither the old restrictive marriage nor the new immorality of going from partner to partner. They will bring socialist lifestyles to their union with a life companion to further the full development of both.

Sexuality as a reward for youth who had matured into responsible young socialist activists was a common theme. Joseph Luitpold Stern, a principal leader in the socialist culture movement, insists that the sexual question for youth must be confronted by the party in an open and matter-of-fact way.[149] But he goes no further in his honesty than to contrast the puppy love of the seventeen-year-old, which must be contained by wholesome activity in the movement, with the love of mature comrades, which combines sex and the great ideas of humanity. Other writers are more direct on the need for postponement of gratification. Kanitz proposed that every socialist youth group be addressed by a socialist physician once or twice a year about human sexuality, conception, and venereal disease.[150] The purpose of such enlightenment was to ensure the wholesomeness of youth and to make it aware of the complexity of love, wherein the sexual step was to be the last one.

Some went so far as to propose that youth be sensitized to "the finest expressions of the erotic."[151] But all they could advise was to put off the physical union of a couple until the final culmination of the relationship—a crescendo in which physical, emotional, and intellectual qualities would merge. Marianne Pollak offered a socialist romance in which the youthful confusions of a seventeen-year-old girl are all resolved by the SAJ.[152] There she enters a platonic relationship with a dedicated young socialist who helps her to find a vocation as a teacher of young children, achieve a sense of community, and abandon her previous desire for frivolous love.

The most active and organizationally integrated working-class youth were bombarded with admonitions against sexual experimentation and urgings toward abstinence from alcohol, tobacco, and sex. But, as the following example illustrates, this led to conflicts between party admonitions and older social practices in the working class and added to the confusion of youth on sexual matters.[153] In the autobiographical sketch of his transformation from village poverty and working-class unawareness to socialism, the later socialist leader Joseph Buttinger recounts the problems surrounding his sexual initiation.[154] At little more than sixteen he fell in love with a girl his own age. But they did not begin sexual relations until four to six months later, because he was under the influence of the party's puritanical teach-

part on context and inference in place of direct evidence.[165] It is based on in-depth oral history interviews, memoirs, contemporary empirical studies of schoolchildren, surveys of female factory workers, and pedagogical tracts.

In attempting to trace the nature and development of sexual knowledge, initiation, and practice in the Viennese working class, we must return to the localities of proletarian socialization: the family and the street, which played a dominant role, regardless of stratification, within the working class.[166] Common denominators of family life experienced by proletarian children and youth were scant means and resources, overcrowded living spaces, and restraint and control exercised by paternal/maternal authority. A family life of scarcity prepared children for life in the wider economic and ideological system of domination and subordination and set the pattern of the gender-specific roles of working-class men and women.

Where and how did sexuality enter in? In the typical worker domicile all family members slept in the one crowded bedroom; family life was conducted in the kitchen. This remained true even among young couples who were fortunate enough to acquire an apartment in the new municipal housing, where parents and children still shared the bedroom, and family life shifted from the kitchen (now too small) to the living room.[167] In the tenements, where 90 percent of worker families lived, beds were shared by same-sexed children, and the youngest child typically slept in the double bed between the parents (sometimes to the age of ten).[168] Despite a decided avoidance of nudity by parents, under such conditions of overcrowding early confrontations with sexuality appear to have been unavoidable.[169]

Notwithstanding the reluctance of oral history subjects to remember or speak about sexual matters or their reticence about having seen or known anything, sexuality seems to have been encountered in the following ways: through direct observation of the "primal act"; in a vaguer sense through wakefulness resulting from adult night-time traffic in the bedroom; through homeoerotic experience with sibling bed partners; and through the experience of older siblings and neighbors in the dense network of tenement life. Sexual knowledge thus acquired was not likely to be discussed in the proletarian family. The subject was taboo, and children were expected to practice a ritual blindness or ignorance of the subject—expected not to see their parents' half-exposed bodies during ablutions or to express anything but ignorance about their mother's pregnancy and the arrival of a new sibling.

Further light is shed on socialization in the family and sexuality by two contemporary studies of pre- and postpubescent working-class schoolgirls. Margarete Rada, a star graduate of the Psychological Institute of the University of Vienna codirected by Charlotte Bühler, presents a picture of early maturation in which girls of limited intellectual knowledge or interest revealed an unexpected sophistication about sexual matters.[170] On the whole the girls were well informed about menstruation, pregnancy, intercourse, illegitimacy, paragraph 144, blood tests to determine paternity, and pre- and extramarital relations. What was the source of this knowledge?

Without exception, Rada maintains, it came from the daily experience in the home: the parental act in the shared bedroom, the birth of younger siblings, the "misfortune" of an older sister, discussions among female family members and neighbors. For a small percentage of girls, this knowledge was supplemented by sexual experimentation.

For the girls in her charge, Rada observes, sexuality was not something mysterious and forbidden to be whispered about. It was talked about only rarely and then without any sense of reserve or shame (as one would find among middle-class girls of that age). In short, for working-class girls sexuality was a matter-of-fact part of their daily lives, which was a cumulative part of their experience from the earliest years. One other finding deserves mention. Away from school the girls had little supervision, with both parents frequently working and nearly half of the girls spending even Sundays on their own.[171] But, as Sieder makes clear, even in homes where there was parental supervision—the control of homework by the father, for instance—the common determinants of scarcity and crowding prevailed.[172]

Hildegard Hetzer seconds the findings of Rada in her study of working-class children and youth.[173] But she draws a distinction between the cared-for and uncared-for. It is among the latter that she finds not only ready conversance with sexual subjects but also a variety of sexual experiences including intercourse (fourteen- to sixteen-year-old girls). The cared-for, she claims, have more self-control over their drives and are more given to intellectual and cultural pursuits.[174] But her categories are vague: by "cared-for" she means bourgeois or skilled elite workers; by "uncared-for" she suggests the working class as a whole. Both Hetzer and Rada, following the lead of their mentor, Charlotte Bühler, regard sexual precocity—early knowledge and early confrontation with practice—as the cause of intellectual and cultural impoverishment and of generally low expectations among working-class youth and, most important, as the source of "uncontrolled" sexual expression.[175] At the same time both observed that these same youth lived up to their responsibilities at work (school-leaving age was fourteen) and at home. Both singled out precocious sexuality as a social disability on the one hand, yet demonstrated on the other that the girls gave no special importance to sexuality in their conversations or interactions but integrated the subject into their daily lives.

Could it be that Hetzer and Rada failed to see that working-class childhood and youth demanded a precocity in all things because adulthood, or at least its heaviest responsibilities, came so early? They did admit that control over drives by the cared-for (who could continue their studies) also went with childishness and dependency. One gets the impression that these two studies created a problem viewed out of context for which only the "idealism" of the SDAP and its programs offered the solution. Another way of looking at conditions in the proletarian home would have been to confront the general deprivation among working-class youth, for which neither psychological theories nor socialist play groups and youth organizations could offer an alternative.

In their emphasis on the proletarian home as the source for sexual knowledge of their charges, Rada and Hetzer slighted the at least equally important influence of the street. Despite middle-class alarmist denunciations of the street as the source of criminality, and despite the socialists' organizing efforts to bring working-class children under the protection and control of experts and the party, the street remained a principal place of socialization.[176] In the proximate street or lot (as distinct from distant unsupervised urban niches), where girls could readily be recalled by their mothers for domestic chores, girls and boys mixed freely in a rich combination of traditional and improvised play.[177] It was their territory, their property, in which they learned about the ruling class through its agents; about differences between children of their own class and those of better-off workers, who had to disobey parental commandments to participate; and about the survival strategies necessary for their adult working lives.[178] Rough horseplay and repeated association also, it would seem, was a physical learning experience for both sexes. It should be seen in combination with the long-standing observations made about the home, especially for the pubescent, as part of the prelude experience of an adult life. What the socialist reformers regarded as a seductive stimulation of sexual drives, the children of the street experienced as living and growing up.

Young people continually dipped back in later years into their storehouse of street experiences, or these simply flowed along with maturation. As unemployment in the early 1930s affected close to 50 percent of young workers between sixteen and twenty-five, many improvised strategies for survival based on the "street wisdom" learned earlier.[179] Along the banks of the Danube and especially the Lobau (dubbed the "proletarian Riviera"), where sun, water, and nature were free, colonies of young workers sprang up. They lived there in the fair seasons and followed a great variety of cultural, athletic, and political interests, found occasional work in the gray market, and made plans for the future. Sexual relations there were as unsupervised as was the rest of life.

What was the normal age for workers to become sexually active? No one can give an answer based on empirical evidence. But it certainly was not the prescriptive twenty, twenty-two, or twenty-four years of age that socialist reformers were so fond of quoting as appropriate to the health and development of female sexual organs.[180] We have already learned about Buttinger's initiation at seventeen; his later experience is instructive as well.[181] As a socialist youth leader in St. Veit in Carinthia, he fell in love with a girl fourteen years old. The relationship was kept secret for a year because it would have seemed "unnatural." She was unusually mature, having gone to work immediately after leaving school. All the same, Buttinger hints that the community was alarmed though not outraged when it learned about the liaison. Two important insights can be gained from this case. First, that an open sexual relationship at the age of fourteen or fifteen was out of the ordinary (though his initiation at seventeen raised no eyebrows). Second, that even though Buttinger was an exemplary product of the socialist reform movement and as serious and dedicated a young socialist as the party could hope for, his most private life was governed by traditional norms rather than by party strictures, or the latter at most led to personal conflicts.

For most working-class youth, being sexually active also brought uncertainties and problems in its wake. In discussing the fifty most common questions and problems brought to the sexual consultation clinics, Reich mentions menstruation, premature ejaculation, birth control, failure to achieve satisfaction, frequence of coitus, age of initiation, and forced abstinence.[182] Victor Frankel, a physician active in the municipal youth consultation clinics, recounts that of two thousand clients seen during a three-year period, one-third suffered from sexual problems.[183]

Lest Buttinger's case appear as an aberration, we must look further for clues to sexuality within socialist organizations themselves. Notwithstanding the socialists' stress on a higher kind of coeducational purity in organizations like the Rote Falken (age ten to fourteen), Sieder's oral-history subjects recount that, because of the lack of space and privacy in their worker domiciles to meet the needs of friendship, love, and sex, the Rote Falken and Socialist Worker Youth (age fourteen to twenty-one) were considered attractive places to meet the opposite sex.[184] It was precisely for this reason that working-class parents were cautious about letting their daughters participate in mixed groups, expecially on overnight trips.[185] This caution was not simply a matter of hypocrisy on the part of parents, who knew full well that their daughters could not be protected from a variety of sexual experiences at the workplace. It must be seen as part of the sense of respectability present in virtually all sectors of the working class that may be summarized as the ability to manage the private life of the family within the restraints placed on it by society and circumstances.[186]

The coeducational Socialist Worker Youth was given the key role in the socialists' project of weaning malleable youth—the socialists of tomorrow—from the authority of working-class parents.[187] In the right-wing press the organization was slandered as a hotbed of sexual permissiveness. In reality the youth leaders (or "chairmen") preached abstinence from alcohol, tobacco, and kitsch as well as sexual purity, and discouraged individual relationships as disruptive of collective experience. Bimonthly lectures on sexual problems were aimed at reducing the interference of sex with organizational aims and work. Memoirs and oral histories suggest, however, that sexual expressions among the socialist youth found an outlet despite the proscriptions of the puritanical leaders.

In the numerous summer camps sponsored by the SDAP, erotic experiences during the seminude morning ablutions of both sexes or in innocuous folk dancing, for instance, did take place.[188] In the tent settlements of unemployed youth in the Lobau and among those able to join them only on weekends, there was a tremendous variety of spontaneous activities. But, as one youth leader recalls, "in between, boys and their girlfriends disappeared for a while behind the bushes."[189] When couples were formed, however, they tended to leave the SAJ. Whatever sexual activity prevailed in the organization, it was restricted to the out-of-doors during the temperate seasons. According to Safrian and Sieder, the first genital encounters of SAJ took place after the age of eighteen.[190] Even then, fear of pregnancy was ever present among the young women, giving a sense of reality to the dangers of sexuality preached by the socialist leaders.[191]

No doubt the socialists' introduction of sex education in their organizations is praiseworthy, but the content and purpose of their actual efforts raise serious doubts about whether their intentions were to enlighten their young members. In a seminal statement on sexual education, Otto Kanitz made it clear that, aside from imparting knowledge about reproduction, sex education should prepare the young for the necessary subordination of their sex drive to the laws of socialist ethics.[192] A former member of the Cooperative of Socialist Teachers explains what this meant in practice: fourteen-year-olds were impressed with the responsibility to be assumed in later sexual relations, and older youth were lectured on new theories of childhood sexuality, the Oedipus complex, and sexual taboos; but "in general, there was little advance beyond the birds and the bees."[193] Similarly, in a pamphlet intended for socialist teachers and youth leaders, the child psychologist Joseph Friedjung did not go beyond the time-worn animal analogies in recommending what teachers should tell their pupils. He was extremely vague about what might be said to adolescents other than the usual cautions about prostitution and the responsibilities of parenthood.[194]

In the SDAP's approach to the workers' intimate sphere as in other aspects of the party's cultural program, there is little evidence of the frequently claimed close relationship to Adlerian individual psychology (see also chapter 4).[195] No doubt the idea of human malleability subject to intervention and improvement was attractive to socialist leaders and experts embarked on creating "neue Menschen." But if one looks over the papers delivered at the Congress of Marxist Individual Psychology held in Vienna in 1927, one is hard put to find descriptions of actual application and practice.[196] Instead, there is a pervasive confidence that individual psychology will solve the sexual problems of working-class men and women. Such bald assertions are seconded by intellectual obfuscation that can hardly be taken for Adlerian practice. In a pamphlet on marriage directed at teachers, Sophie Lazersfeld (a leading individual psychologist) discusses the ambivalence of women overcome by the choice between being a comrade or madonna to their mates and the dangers for women of becoming Galatea to Pygmalion.[197] As interesting as such subjects might have been in their own right, particularly to a small group of educated female professionals, one finds it difficult to relate them to the education of working-class women, the realities of working-class marriages, and the narrow gender roles assigned to women.

What clues are there to the sex life of adult Viennese workers? In view of the early sexual maturation and onset of adult responsibilities, it is not surprising that sexual intercourse and cohabitation before marriage (often for many years) seems to have been widely practiced.[198] From the point of view of socialist reformers, this behavior was exemplary of the "disorderly living" among workers they aimed to correct. In working-class neighborhoods it was accepted as part of the courtship pattern leading to marriage. A number of oral-history subjects reported that their parents gave their tacit consent to their sexual relationship by allowing the couple to live on the premises. The choice of marriage partners was largely in the hands of the young people. Among the desirable characteristics looked for in their prospective mate, women often mentioned safe employment and fidelity.[199] It was customary for courting couples to get married when the woman became pregnant. The ceremony itself and the wedding night seldom attained the importance given to them by the middle class.[200]

The best source of indirect information about sexuality can be found in studies of the birth rate in the working class.[201] In the generation of women born after 1900, the majority had only one child and virtually none more than two.[202] This feat was accomplished—without the assistance of the municipality or socialist reformers—by the couples and especially by the women themselves. It stemmed from the recognition by workers that their aspirations to or maintainance of a higher standard of living depended on a smaller family size.[203] Moreover, Viennese proletarian women, most of whom were employed for wages, apparently recognized that the only possible reduction in the triple burden of work, housework, and child care could be achieved through reducing the number of children.[204]

It is in the domain of birth control, where proletarian couples needed the most assistance, that the SDAP failed them most abysmally. The methods of contraception available to workers were primitive, unreliable, impoverishing of the coital act, and dangerous. Workers who had served in the army had experienced commercial sex and become acquainted with condoms. But there is little evidence that these or other rubber contraceptives were used, partly because of inconvenience (the absence of privacy to apply these implements) but mainly because of cost.[205] Coitus interruptus is the formal technique most frequently mentioned in memoirs and oral histories.[206] Not only was this form of prevention unreliable, it also depended on the skill and good will of the male partner. Control and reduction of births probably depended at least as much on abortions resorted to by women regardless of the danger to their health and of falling foul of the law. It was a method totally controlled by the women themselves. In a number of cases of abortions by married women that came to court, the procedure had been carried out without the husband even knowing that his wife was pregnant.[207]

Though abortion was practiced widely as a form of birth control, it was a threat to women's health and a breach of the law. But there are indications that paragraph 144 was fully subscribed to mainly by the Catholic clergy and diehard leaders of the Christian Social party. Both the number and notoriety of abortion trials appear to have declined sharply in the fifteen years after the war.[208] In the sixteen case files for the 1921–32 that I found in the municipal archives, none of the women was actually punished for having attempted or carried out an abortion. In all of them actual sentences were reduced by the judges (through a provision for leniency) to one-year probation, giving child care, health, work, and family obligations as the reasons.[209] Only midwives with prior arrest or conviction records were given prison terms of two to nine months. The Viennese judges seem to have balked at punishing women who had undertaken these desperate acts.[210] The humane considerations of these pillars of society suggests that the SDAP missed an important opportunity to launch a broad-based campaign against paragraph 144.

Frequently a crisis developed in the sexual relations of couples after the desired number of children (one or two) had been born. Then wives faced the cruel choice of denying conjugal rights to their husbands or having recourse to the "maker of angels."[211] The following case is typical of this dilemma and throws a clear light on the context of worker sexual practice.[212] Josepha P., born in 1902, was a domestic servant who saved for her dowry and looked for a husband. She rejected one handsome suitor because he was too sexually forward, as she wanted to avoid the fate of her mother, who gave birth to six children. Josepha chose an older man in the hope of having fewer children. Her relationship was consummated before her marriage, after which she moved into her mother-in-law's two-room flat. The couple slept in the bedroom, which they were forced to share with the husband's brother, and two children were born there during their four-year residence. After the birth of her second child, Josepha refused to have intercourse with her husband; she succeeded only partly since a third child was born.[213] Clearly, living in cramped quarters with parents and in-laws, and the resulting lack of privacy, put limits on sexual expressiveness and the use of contraceptives.[214]

Under the typical conditions of working-class life, not the promiscuity feared by socialist reformers and city fathers, but a sexual life of deprivation seems to have been the norm. One can only speculate through indirect evidence about the quality of such sexuality. Leichter's survey of female industrial workers makes clear that the triple burden on women was not only physically but also psychologically debilitating.[215] A total workday of some sixteen hours for the majority of women and a life in general characterized by haste and devoid of all privacy left little time and energy for conjugal intimacy. To these restraints must be added the stresses of unemployment, which rose astronomically during the depression and frequently made women the sole breadwinners. Several studies reveal that rising unemployment had an adverse effect on the greater freedom of male workers and seriously impaired their sense of identity and of daily rhythm, leading to general psychological depression.[216]

Admittedly, one does have to read between the lines of such evidence, but it suggests that worker sexuality and the general conditions of their lives were closely interdependent. From this distant vantage point it appears that changes in the sexual lives of workers had to occur at the workplace before the bedroom. The orderly and decent worker family the socialists attempted to create depended on an improvement in the quality of life per se and on a basic change in gender relations, rather than on moral uplift. No reform effort could succeed, it would seem, that did not begin to accept the workers' primary subcultures as a point of departure and allow initiatives for change to spring from them.[217]

Virtually every feature of the SDAP and municipal leaders' attempt to harness sexuality to a new socialist workers' culture was centered on controlling and redefining the role of girls and women. In their view of females as creatures of instinct, these leaders mirrored the attitudes of middle-class reformers and moralists of previous generations.[218] The main function assigned to women in the desired orderly worker families was to act as affective centers.[219] As Otto Bauer put it, the wife was to organize the home in such a way that her husband would find it a place of peace, privacy, and comfort.[220] The role assigned to women within the SDAP was equally circumscribed and tertiary. With the exception of a small number of very visible functionaries, the typical female activist was relegated to social welfare activities closely resembling middle-class sociability, but at a tangent to the real (political) life of the party.[221] To a large extent, the working-class woman's "marriage to the party" was a means of reinforcing and enhancing the role she was expected to play in the family.

What explanation is there for such a restricted role assignment—for the apprehensions expressed by male leaders about female sexuality in a constant harping on promiscuity, and for their failure to deal with actual problems of women such as birth control? In part the answer may be found in the middle-class values and attitudes of the principal male socialist leaders, which at times bordered on misogyny. A few illustrations should suffice. In his capacity as a senior member of the Viennese medical faculty, Julius Tandler opposed the admission of female students and made it difficult for them during examinations by asking questions about male genitalia to embarrass them. As a powerful city councillor, he refused financial support to coeducational summer camps and forbad kindergarten teachers to wear dresses that revealed knees and calves.[222] Otto Kanitz's views on sexuality were colored by his forced childhood conversion to Catholicism and influenced by the Catholic educator Friedrich Wilhelm Forster. Kanitz spoke of the imperative of creating an "ethical will" based on purity and self-denial and even sought to popularize Marxist thought in the form of catechisms.[223] Otto Bauer lived like a liberal *Bürger* of his time. He married a woman ten years older than he and had a mistress ten years younger. Friedrich Adler practiced the double standard by "rescuing" young women—sexual affairs which he described in detail to his estranged wife.[224]

It is not surprising that none of these men advanced positions on female sexuality that would have brought them into conflict with the dominant culture. One is reminded, by contrast, of Léon Blum's treatise on marriage of 1907, in which he viewed monogamy as the culmination of sexual experimentation and experience by both men and women, and of the fact that he allowed the book to be reprinted in 1937, when he was prime minister.[225]

The limitations of the socialists' transformational plans on the sexual question may be viewed from another perspective. From the beginning of the republic the SDAP engaged in the class struggle on the political *and* cultural front. After 1927, as the extraparliamentary intentions of the political opposition became more apparent and Bauer's conception of the "balance of class forces" more unrealistic, the SDAP increasingly retreated to the cultural realm in the enclave of Vienna. Even when the signs of a final political confrontation became unavoidable, Bauer and other leaders refused to confront their enemies. The rising in February 1934 was a tragic postscript in which the workers suddenly became their own leaders in a doomed resistance.[226]

The same paralysis of will on the cultural front may help to explain the socialists' timid position on matters like birth control and abortion. To have taken a forceful stand through the municipal council on these and related issues would have led to a confrontation with the Catholic church, which was also the power behind the political opposition. As we have seen, the SDAP refused to take a position outside constitutional legality on the Schnitzler, Bettauer, and Baker sex scandals. In fighting against paragraph 144 in parliament, it relied on the most innocuous formulations in the hope of gaining a compromise from an uncompromising opponent. The party was not prepared to mobilize the larger public on issues that clearly had wide appeal. No doubt the leaders were correct in assuming that going beyond the "legal" ground of parliamentary and party action would bring on a civil war on the cultural front leading to a final showdown.[227] That risk the socialists were not prepared to take.

Some nagging questions remain. Was the purpose of the socialists' moralizing a means of dispelling the popular middle-class view that workers were promiscuous savages, was it rhetorical, or was it part of a strategy to uplift the working class?[228] If anything, the socialists' emphasis on the dangers of promiscuity and degeneracy among workers reinforced the hostile image of disorderly living among them. The campaign to create orderly, decent worker families through procedures outlined by the party demeaned the existing family structures and sexual behavior in the working-class community. It constituted a refusal to accept the fact that norms and standards of behavior and a sense of respectability already existed there. In this refusal to see the value of existing subcultures, the Austrian socialists were no different from reformers elsewhere in denigrating the conditions which were to be transformed by institutional intervention. They did believe that sexuality had a place in the creation of a total socialist culture based on the worker family, and asserted that sexual mores and practices were as mutable as any other aspect of the workers' lives. But socialist means of uplifting the workers—of making them "neue Menschen"—did not go beyond condemnation of existing life-styles, admonitions, and directives formulated by leaders at a great distance from the arenas of working-class life.

CHAPTER 7

Conclusion

> Against the idea of force, the force of ideas.
> Austromarxist aphorism

In the course of my examination I may have overemphasized the exaggerations and contradictions in the Austrian socialists' claims and accomplishments and the limited scope of their venture. That these were embedded in the plans and development of the cultural experiment has, I believe, been demonstrated in detail. We ought to remind ourselves that despite serious practical failings, fundamental flaws in conception, and far-reaching dangers, red Vienna succeeded as no other metropolis had in improvising and innovating social reforms and cultural activities for its working class within the political limits of a polity hostile to such efforts.

The SDAP's cultural preparation for a socialist future in the present was unique. It went beyond traditional social democratic reform legislation in seeking to encompass the Viennese working class through an intricate network of party cultural organizations and activities that had both an educational content and symbolic force. While critical of the conception and execution of this program, one still marvels at the daring vision, for instance, of the public housing palaces as total worker environments containing laundries, bathhouses, kindergartens, libraries, meeting rooms, swimming pools, cooperative stores, youth and mothers' consultation clinics, and much more. The purpose of these enclaves was to provide the workers with a setting for the "political culture"—the Austromarxist special formula leading to the maturation of the working class—through which the consciousness necessary for the creation of "neue Menschen" could be instilled. One also cannot but be impressed by the forty or so cultural organizations striving to engage and elevate the workers during their newly gained leisure time and thereby to bind them to the party. The combination of this comprehensive cultural network and the SDAP's political and trade union strength were formidable. But, as I have attempted to illustrate, the potential of the Viennese cultural experiment was not realized. The obstacles faced and created by the socialists in carrying out their project were both political and cultural.

Political Limits

The SDAP emerged as a mass party after the collapse of the old regime at the end of the war. Its future power and role were largely shaped during 1918–19, when politics were in flux and the nature of the new republic was yet to be determined. The most dynamic political force at the time was the workers' councils, a partly organized mass movement of workers and demobilized soldiers with revolutionary aims that went beyond establishing a republic on a capitalist base. The councils were influenced by the Bolshevik Revolution and the brief revolutionary regimes in Bavaria and Hungary. In Vienna, particularly, where a fledgling Communist party was newly active, the councils threatened to become the directing force of the masses of workers, who were largely politically unorganized.[1] The SDAP, thus threatened from the left, adroitly maneuvered the Vienna workers council into accepting the principle of proletarian democracy, which allowed the socialists to bind the communists and other radical groups to decisions by majority rule. By the autumn of 1919, with the counterrevolution successful in Hungary, the councils were pushed to the sidelines. The Constituent Assembly, which had been elected in February, was able to function and secure the new republic without further threat of an alternative source of power.

The socialists had succeeded in keeping revolution from the gates of Vienna. Had they gained by it, and if so, what? The SDAP succeeded in making itself the sole spokesman for the workers of Austria, able to formulate its programs and to navigate the parliamentary waters without being seriously challenged by the Communist party (KPÖ), which remained a sect throughout the period.[2] But the withering away of the workers councils also had its costs for the socialists. Until the adoption of the constitution in 1920, the councils served as a powerful reminder to the Christian Social and Pan-German parties of a revolution that threatened to create a social order totally unacceptable to them. The SDAP had gained 43.4% of the seats in the Constitutional Assembly, compared to 54.7% for the combined opposition, yet remained the dominant partner in the coalition formed with the Christian Socials. This short-lived advantage must be attributed to the councils' demonstrative presence in Vienna.

No doubt the SDAP played this radical card to its advantage in pushing through fundamental social and economic legislation at the time. But should it have pressed for more—demanded structural reforms and constitutional guarantees that would have served its long-range socialist goals and put the republic on a sounder foundation? There is no easy answer. On the whole, the SDAP leadership was unprepared for the rapidly moving events surrounding the collapse of the old order and the party's emerging central role in establishing the republic. One must sympathize with their inability to see all the advantages the crisis might bring them; they were not above the battle.

In my view the socialist leaders missed two important possibilities during those transitional months. First, the nationalization of some primary industries and banks and the legal implantation of the factory councils as agencies of codetermination so as to give the SDAP a strategic base in the economy—a form of industrial democracy without which there could be no real change in the condition of the working class. Second, curtailment of the power of the Catholic church, going beyond Glöckel's reactivation of the law prohibiting religious practices in the schools. A much clearer and more decisive separation of church and state was necessary—removing all religious influence from the schools and prohibiting members of the clergy from holding public office, for instance—in order to reduce the ideological/psychological advantage of the Christian Socials and thereby make them just a political adversary. The importance of these steps was not easy to see, nor would they have been easy to carry out. But they were necessary, even if accomplished only in part, both for the socialists' own cultural program and the survival of the republic. As it turned out, quick and decisive action was essential; by 1920 the future alignment of political forces into unequal camps was already determined.

But Otto Bauer and the SDAP directorate were not favored with our powers of hindsight. In the midst of the struggles to establish the republic, they sought a path for Austrian socialism different from both German social democratic reformism and Bolshevism. Its main contours were conceived to be the electoral conquest of political power and the creation of a far-reaching cultural program transforming the working masses and preparing them for the democratic attainment of socialism. Such a class struggle on the two fronts of politics and culture was predicated on the neutral position of the republican state, with constitutional guarantees of political pluralism and democratic procedures. But the neutrality of the state was illusory; in reality the disposition of power within it, beginning with the breakup of the coalition government in 1920, already pitted the national government controlled by Christian Socials and various allies against socialist-ruled Vienna.

Bauer's optimistic assumption that the SDAP would come to power at the polls was contradicted by municipal and especially national elections from 1919 to 1930, in which the socialists' votes and mandates increased only slightly.[3] The SDAP's ability to attract voters outside its ranks appears to have reached its limits between 1930 and 1932.[4] In Vienna especially, it seems that the SDAP had already attracted all possible support from the liberal middle class in the Jewish and Czech communities (which had no real alternatives) and from a small proportion of the lower middle class attracted by rent control and various municipal reform measures. The SDAP's further movement in the direction of a "people's party" was inhibited by its well-advertised program of culture for the working class alone.

The socialists' aspiration to power through the polls was illusory in another sense. What if the magical 51% had actually been reached? Would the socialists then have been able to use their democratically gained right to nationalize the means of production, as the Linz Program of 1926 proclaimed? Given the deep political and ideological divisions in Austria, such a step would certainly have led to civil war, for what amounts to a social revolution cannot be undertaken with a slim parliamentary majority, especially when the defeated minority is prepared to defend its contrary interests with force. That was the conclusion arrived at by Léon Blum, head of the French Poplular Front government in 1936, when a general strike signaled popular support far greater than the slim majority by which the Blum government attained power. Considering the actual disposition of political forces and interests in the population, Blum concluded, the socialists could exercise power in alliance with others but could not conquer it, in the sense of "transforming capitalism into socialism."[5]

The claims to transformational power on the basis of a slim electoral majority in a society divided into two bitterly opposed camps flirted with civil war, which Bauer insisted was to be avoided at virtually any cost. His concept of the "balance of class forces"—the Austromarxist leitmotiv of the First Republic—was a theoretical device to keep the republic intact while the socialists sought to increase their electorate and prepared the workers culturally for their future mission. But, as Bauer's critics had pointed out in 1924, one could hardly speak of a balance of class power when the capitalist system and its social order remained in control, or even of a long-range equilibrium, since the conservative opposition had regrouped after its brief disarray immediately after the war.[6] But neither Bauer's nor the party's position was altered by these critiques. Consequently, the cultural project rested on the illusion that a supportive political context would develop in tandem with it. The police violence on July 15, 1927, during the workers' attack on the Palace of Justice, shook the confidence of many socialists in the neutrality of the state. Karl Renner suggested that the SDAP enter into a coalition with the Christian Socials in order to safeguard the republic. It was rejected by Bauer out of hand as requiring the abandonment of the entire cultural effort in Vienna.[7]

After 1927, as the political fortunes of the SDAP grew dimmer, the cultural project, originally conceived as a major weapon in the armory of class struggle, more and more became a substitute for politics. The cultural experiment and Vienna became synonymous, as the capital-enclave increasingly assumed a defensive position in a hostile country. Toward the end, the socialists' projected relationship between politics and culture became reversed. Instead of providing a protected environment for culture, politics depended more and more on cultural expression. The 100,000 workers, representing the gamut of cultural organizations, parading through the streets of Vienna during the International Worker Olympics in 1931, and the more than 200,000 spectators at the worker festival in the new stadium celebrating the symbolic fall of capitalism, were neither culture nor politics, but a metaphor for both. Opponents of the SDAP (Christian Socials, Pan-

Germans, Heimwehr, and Nazis) were impervious to the SDAP's symbols of strength and rejected its emblems of republicanism: democracy and neutrality of the state.[8] They fought with raised visors in all arenas to liquidate the socialist enemy.

Cultural Limits

The specific shortcomings of the socialists' cultural project have been thoroughly discussed in the body of this work. There are, however, a number of questions raised by the experiment which deserve further comment. What is a "socialist" culture and who determines its content? What dangers are inherent in the discipline by which such a cultural program is implemented and in the orderliness toward which it strives? In what sense was the Austromarxist experiment both a model and a dead end?

Leaders of the SDAP were vague about the socialist content or quality of the varied facets of their cultural program. To be sure, the workers were to be educated and thereby brought to a higher level of consciousness and provided with the facilities and necessities of a more dignified life. It remained unclear how such an individual and collective improvement was different from old-fashioned liberal ideals which had become platitudinous. Similarly, more and better housing, kindergartens, and libraries were certainly desirable, but did these improvements differ from the goals of the reformist socialism the Austromarxists were committed to surpassing? The socialists essence sought after by the SDAP becomes most illusive in the arts.[9] Did a symphony by Beethoven become socialist if it was played by the Workers' Symphony Orchestra for an audience of workers, as was suggested at the time? What made Jack London an acceptable author and Karl May a purveyor of kitsch in the eyes of socialist culture experts? Why were Käthe Kollwitz's paintings of lower-class misery preferred to a madonna by Raphael? What were the criteria used to determine appropriateness or desirability from a socialist perspective of home furnishings, decorations, books, dress, radio programs, and so on?

I am not defending the impoverished taste prevalent in worker subcultures. The knickknacks, antimacassars, framed proverbs, and other items which adorned workers' homes speak for a widespread deprivation of taste, lack of opportunity to come in contact with the wider world, and constrained household budgets. But, however poor and deficient the workers' taste was, it could not simply be commanded away by people outside the subcultures and replaced by items alien to them. The education of taste has been found to be a slow and long-term process.

All questions about the socialist content or essence of culture lead us back to the valuers who determined what should be included and excluded in the menu of culture set before the workers. It was a small elite of party leaders and directors of cultural programs who made such decisions. Their values and tastes reflected their own generally middle-class and German-*Bil*-

dung-oriented socialization, or an adaptation to the values emanating from it, rather than the vague ideal of a socialist culture. One of the most important demonstrations of the Viennese experiment is that a cultural program projected from above onto a population below is destined to remain the expression of an elite. The good intentions of leaders are not sufficient to make a cultural program truly popular and acceptable. It would seem that only by a gradual process of negotiation between cultural innovators or reformers and the existing subcultures they hope to elevate, could there be a move toward the kind of transformation the Austromarxists envisaged.

Not only was the content of the socialists' project problematic; the means by which it was carried out, and the demands it made on those for whom it was intended, also help to explain its limited success at the time and are revealing about cultural experiments in general. The socialist leaders appeared as authority figures in the workers' world in two related guises: as city fathers and as oligarchs of the SDAP. Furthermore, their cultural program was paternalistic, especially in its demand for discipline both in the organizations themselves and on the part of the workers undergoing "civilization." This method of control clearly contradicted the Austromarxists' long-range goal of creating self-confident and assertive workers. The call for an orderly and respectable worker family (aside from echoing what middle-class critics had demanded for some time) was ultimately aimed at the diversity of life-styles in working-class communities. I have no intention of romanticizing worker subcultures by overlooking the forms of dissonance to be found there. But these were far outweighed by codes of coping with the hardships of life in a respectable manner. No matter how well meant, the socialists' constant reiteration of the need for betterment not only demeaned the workers for what they were, but also, indirectly, suggested conformism of a new kind. In the socialists' struggle to save the workers from the effects of commercial and mass culture, they seem not to have realized that they were equally guilty of turning the workers into passive consumers, albeit of their own brand of cultural products.

The Viennese cultural experiment remains the clearest example of the possibilities and limits of providing a foretaste of the socialist utopia in the present, of devising and implanting a unique proletarian culture in a society that has not experienced a fundamental revolution.[10] As a model it reveals the fragility of such an enterprise, because of its clear dependence on political power. It also demonstrates that culture per se cannot compensate for the economic deprivations and general hardships of life. The lofty goal of education for knowledge was circumscribed by the means of its implementation, in the relationship between subject and object, leaders and masses. Perhaps the most significant legacy of the Vienna experience is the concrete challenge it presents to the once fashionable interpretation of Gramsci's hegemony theory, which suggests that the workers could create a counter-hegemonic culture before they succeeded in capturing state power.[11]

In interwar Europe red Vienna's experiment to create a working-class culture without revolution represented a countermodel to Stalinist totalitarian culture.¹² Today both examples, culture without power and power without culture, have apparently reached a dead end. The Austrian experiment was not to be replicated, not even in Austria after World War II and the attainment of state independence in 1955. A new "social partnership" (between the trade unions, owners' associations, and political parties) has largely depoliticized national life. In the emerging welfare state there seems to be little need or desire for a compensatory culture. The current flight into private life and private culture raises a cardinal question: Is there no halfway house between a party culture totally initiated and controlled from above (of which the Russian and East European examples currently unraveling are the models), and the totally privatized individual hedonism manipulated by the market, which creates desires and demands in order to satisfy them at a price and gives satisfactions that are as brief as this season's styles and fads?

Despite all shortcomings, the brief Vienna experiment has left a lasting afterglow of nostalgia for its promise, which could not be fulfilled.[13] The desire to create "a revolution in the soul of man" reached far beyond traditional socialist aspirations into the realm of human yearning for a future in which individual development and the community become a harmonious whole. As the Viennese worker song put it so well:

> Wir sind das Bauvolk der kommenden Welt,
> Wir sind der Sämann, die Saat und das Feld.
> Wir sind die Schnitter der kommenden Mahd.
> Wir sind die Zukunft, und wir sind die Tat.

> We are the builders of the future world:
> We are the field, the sower and seed,
> We are the reapers of harvests to come,
> We are the future and the deed.

Notes

Chapter 1

1. By the early 1930s the Heimwehr was a formidable paramilitary organization with a small delegation in parliament *(Heimatblock)* on which Chancellor Engelbert Dollfuss depended in dismantling the republican structure. Its politics was a mixture of nostalgic monarchism, pan-Germanism, and fascism given coherence by a violent antisocialism. Under the leadership of Prince Ernst Rüdiger von Starhemberg, it pursued a putschist policy and played a vital role in the February events. See C. Earl Edmondson, "The Heimwehr and February 1934: Reflections and Questions," in Anson Rabinbach, ed., *The Austrian Socialist Experiment: Social Democracy and Austromarxism, 1918–1934* (Boulder, Colo., 1985); idem, *The Heimwehr and Austrian Politics, 1918–1936* (Athens, Ga., 1978); F. L. Carsten, *Fascist Movements in Austria* (London, 1977); and Ludwig Jedlicka, "The Austrian Heimwehr," in Walter Laqueur and George L. Mosse, eds., *International Fascism, 1920–1945* (New York, 1966).

2. *Le Populaire,* Feb. 14 and 15, 1934. In the latter issue the Trade Union International (IFTU), meeting in Paris, issued a statement of solidarity with the Austrian workers.

3. *Daily Herald,* Feb. 14, 1934. In the same issue the TUC announced the creation of a Fund to Aid Austrian Workers and on the following day reported that substantial monies already had been pledged.

4. *Le Populaire,* Feb. 16, 1934. All translations are my own unless otherwise noted.

5. *Daily Herald,* Feb. 17, 1934.

6. See Marcel Cachin on Feb. 16 and 19, 1934. His claim that the policy of accommodation had led both the German and Austrian Socialist parties to defeat had a certain credibility. He of course failed to mention the considerable responsibility of the German Communist party for the collapse of the German left.

7. See *Rundschau über Politik, Wirtschaft und Arbeiterbewegung,* March 1, 1934, 663.

8. See James Donnadieu, *Le Temps,* Feb. 14, 1934. He also blamed the Heimwehr, but considered the socialists more responsible for the crisis.

9. See Wladimir D'Ormesson, *Le Figaro,* Feb. 17, 1934. Austrian socialism in its Viennese enclave, he claimed, suffered from exhaustion: an inability to further animate the masses save for "dangerous incitements to class warfare."

10. Feb. 13–17, 1934.

11. See Feb. 13–16, 1934, and especially Feb. 15 and 12.

12. Most important in the large literature on the February risings are Erich Fröschl and Helge Zoitl, eds., *Februar 1934: Ursachen, Fakten, Folgen* (Vienna, 1985); Joseph Weidenholzer, "Bedeutung und Hintergrund des 12. Februar 1934," in *"Es wird nicht mehr Verhandelt": Der 12. Februar 1934 in Oberösterreich* (Linz, 1984); Gerhard Botz, "Strategies of Political Violence: Chance Events and Structural Effects as Casual Factors in the February Rising of the Austrian Social Democrats," in Rabinbach, *Austrian Socialist Experiment;* Ludwig Jedlicka and Rudolf Neck, eds., *Das Jahr 1934: 12. Februar* (Vienna, 1975); Kurt Peball, *Die Kämpfe in Wien im Februar 1934* (Vienna, 1974); and Norbert Leser, "12 Thesen zum 12. Februar 1934," in Internationale Tagung der Historiker der Arbeiterbewegung, *ITH-Tagungsbericht* (Vienna, 1976), IX.

13. For the disorientation of leaderless workers at the district level, see Hans Safrian, "Mobilisierte Basis ohne Waffen: Militanz und Resignation im Februar 1934 am Beispiel der oberen und Unteren Leopoldstadt," in Helmut Konrad and Wolfgang Maderthaner, eds., *Neuere Studien zur Arbeitergeschichte* (Vienna, 1984), II.

14. Harold Laski called Vienna "the best governed city in the world." *Daily Herald*, Feb. 17, 1934, 10. The *New York Times* lauded "Vienna under progressive socialist rule for its public housing and social services." Feb. 13, 1934, 2.

15. See Abbot Gleason, Peter Kenez, and Richard Stites, eds., *Bolshevik Culture: Experiment and Order in the Russian Revolution* (Bloomington, Ind., 1985).

16. See Otto Bauer, "Gleichgewicht der Klassenkräfte," *Der Kampf* 17:2 (Feb. 1924), 56–67; idem, *Die österreichische Revolution* (Vienna, 1923).

17. See Giacomo Marramao, "Zum Problem der Demokratie in der politischen Theorie Otto Bauers," in D. Albers et al., eds., *Otto Bauer und der "dritte" Weg* (Frankfurt, 1979).

18. There has been an unfortunate tendency among some younger Austrian historians to romanticize the SDAP and its cultural efforts. In part this was an overraction to the virtual silence about working-class history in the two decades following World War II. It also served to provide Bruno Kreisky's socialist government with a usable and heroic past. See Helmut Konrad, "Geschichte der Arbeiterbewegung in Lehre und Forschung," in Karl R. Stadler, ed., *Rückblick und Ausschau: 10 Jahre Ludwig Boltzmann Institut für Geschichte der Arbeiterbewegung* (Vienna, 1978), and Helmut Gruber, "History of the Austrian Working Class: Unity of Scholarship and Practice," *International Labor and Working-Class History* 24 (Fall 1983), especially 61.

19. Marie Jahoda, Paul F. Lazersfeld, and Hans Zeisl, *Die Arbeitslosen von Marienthal: Ein soziologischer Versuch über die Wirkung langdauernder Arbeitslosigkeit* (Leipzig, 1933).

20. See Franz Kreuzer (in conversation with Marie Jahoda), *Des Menschen hohe Braut: Arbeit, Freizeit, Arbeitslosigkeit* (Vienna, 1983), 7–8.

21. No doubt this view will be disputed in Vienna even today, and the city's leading role in the interwar years defended by citing the presence of various great persons (Ludwig Wittgenstein, Richard Strauss, Robert Musil, Sigmund Freud, etc.). It is fruitless to argue on the terrain of great personalities (one can surely find a surprisingly large number in Budapest as well at the time). The cultural demotion of Vienna went hand in hand with its demotion from capital of an empire to that of a small, impoverished republic.

22. See Otto Bauer, *Bolschewismus oder Sozialdemokratie* (Vienna, 1920); Max Adler, *Politische oder soziale Demokratie* (Berlin, 1926); and Karl Renner, "Demokra-

tie und Rätesystem," *Der Kampf* 14 (1921). See also the analysis in Raimond Löw, *Otto Bauer und die russische Revolution* (Vienna, 1980), 42–55, and Marramao, "Zum Problem der Demokratie."

23. See Helmut Gruber, *Léon Blum, French Socialism, and the Popular Front: A Case of Internal Contradictions* (Ithaca, N.Y., 1986); idem, "The German Socialist Executive in Exile, 1933–1939: Democracy as Internal Contradiction," in Wolfgang Maderthaner and Helmut Gruber eds., *Chance und Illusion—Labor in Retreat* (Vienna, 1988).

24. See the brilliant exposition of this dilemma by George Orwell in *The Road to Wigan Pier* (New York, 1958), 135–36, 160–69.

25. This tendency of organic leaders to abandon their milieu for the values of the dominant culture prevailed in other interwar socialist parties. Fredrich Ebert and Otto Wels, successive heads of the German SPD, are clear examples.

26. See Quintin Hoare and Geoffrey N. Smith, eds., *Selections from the Prison Notebooks of Antonio Gramsci* (New York, 1971), and the perceptive analysis in Jerome Karabel, "Revolutionary Contradictions: Antonio Gramsci and the Problem of Intellectuals," *Politics & Society* 6:2 (1976). Ultimately, Gramsci questioned whether the proletariat could creat its own stratum of intellectuals before the conquest of state power. See Antonio Gramsci, *The Modern Prince and Other Writings* (London, 1957), 49–50.

27. See J. Robert Wegs, "Working-Class Respectability: The Viennese Experience," *Journal of Social History* 15:4 (Summer 1982).

28. The controversy about mass culture continues to rage today. For an interesting exposition, see the position paper by Michael Denning entitled "The End of Mass Culture" and the critical commentaries by Janice Radway, Luisa Passerini, William Taylor, and Adelheid von Saldern in *International Labor and Working-Class History* 37 (Spring 1990). See also Denning's response in the same journal, "The Ends of Ending Mass Culture," ibid., 38 (Fall 1990).

29. Unfortunately a history of the Catholic church during the First Republic has not yet been written. Considering the continued prominence of the church in public life, (Sunday Mass on the national radio station, for instance), there is little likelihood that someone will dare to puncture the gentle self-criticism the church has used to cover its true past.

30. William J. McGrath calls this substitution of culture for politics, whose roots lay in prewar Austria, "the politics of metaphor" or "politics of illusion." See his *Dionysian Art and Populist Politics in Austria* (New Haven, Conn., 1974). The concept originated with Carl E. Schorske in "Politics in a New Key: An Austrian Triptych," *The Journal of Modern History* 39 (1967).

31. For an excellent summary of economic conditions, see Hans Kernbauer and Fritz Weber, "Von der Inflation zur Depression: Österreichs Wirtschaft, 1918–1934," in E. Talos and W. Neugebauer, eds., *"Austrofaschismus": Beiträge über Politik, Ökonomie und Kultur, 1934–1938* (Vienna, 1984).

32. Most authorities list 557,000 unemployed, or 26 percent of the potential labor force, for 1933. See for instance Dieter Stiefel, *Arbeitslosigkeit: Soziale, politische und wirtschaftliche Auswirkungen am Beispiel Österreichs, 1918–1938* (Berlin, 1979), 28–29. But official statistics left out large groups such as the long-term unemployed and the young, who had been prevented from entering the labor market. An upward revision might increase the number of unemployed by 200,000, bringing the total to 38 percent. See Ernst Bruckmüller, *Sozialgeschichte Österreichs* (Vienna, 1985), 500.

33. The decline was from 896,763 in 1923 to 520,162 in 1932. See Fritz Klenner, *Die österreichischen Gewerkschaften* (Vienna, 1953), I: 657; II: 960.

Chapter 2

1. Carl E. Schorske, *Fin-de-siècle Vienna: Politics and Culture* (New York, 1980). Four of Schorske's seven chapters were published in essay form between 1961 and 1973 and exerted considerable influence on two works published before Schorske's magnum opus. See Allan Janik and Stephen Toulmin, *Wittgenstein's Vienna* (New York, 1973), and William McGrath, *Dionysian Art and Populist Politics in Austria* (New Haven, Conn., 1974). In all three works the tendency is very strong to project elite culture as all culture and to offer the former as an emblem for society as a whole. For a critique of this tendency, see Dieter Schrage, "Klimt—Ikone und Waschbottich: Zum Traum und Wirklichkeit um 1900," in Hubert Ehalt, et al., *Glücklich ist, wer vergisst . . . ?: Das andere Wien um 1900* (Vienna, 1986).
2. The others were London, Paris, and Berlin.
3. The film (with the English subtitle *The Joyless Street*) was based on a novel by the Austrian sexual reformer Hugo Bettauer, published in 1923 and serialized in *Neue Freie Presse*. Before public screening the film underwent considerable cutting to reduce the unrelenting realism. Even so, in England public showings were prohibited. See Siegfried Kracauer, *From Caligari to Hitler: A Psychological Study of the German Film* (New York, 1959), 167–70.
4. Walter Ruttmann's film *Berlin, die Symphonie einer Grossstadt* (1927) attempted to present the farrago of everyday life in the metropolis by using the technique of montage. Ibid., 182–88.
5. The best source for the birth pains of the new republic is still Charles A. Gulick, *Austria from Habsburg to Hitler* (Berkeley, Calif., 1948), I: 43–65.
6. See Klemens von Klemperer, "The Habsburg Heritage: Some Pointers for a Study of the First Austrian Republic," in Anson Rabinbach, ed., *The Austrian Socialist Experiment: Social Democracy and Austromarxism, 1918–1934* (Boulder, Colo., 1985), 13.
7. Karl's statement had been written by Ignaz Seipel, minister of social welfare in the last monarchical government, in such an ambiguous way as to leave open the possibility of a Habsburg restoration. The word "abdicate" was never used. See Robert Stöger, "Der christliche Führer und die 'wahre Demokratie': Zu den Demokratiekonzeptionen von Ignaz Seipel," in *Archiv: Jahrbuch des Vereins für Geschichte der Arbeiterbewegung* 2 (1986): 54–67, especially 55. Stöger places this and other instances within the context of Seipel's passionate authoritarianism.
8. See Hans Hautmann, *Die verlorene Räterepublik: Am Beispiel der Kommunistischen Partei Deutschösterreichs*, 2nd enlarged ed. (Vienna, 1971), 71–80.
9. John Bunzel, "Arbeiterbewegung, 'Judenfrage' und Antisemitismus am Beispiel des Weiner Bezirks Leopoldstadt," in Gerhard Botz et al., *Bewegung und Klasse: Studien zur österreichischen Arbeitergeschichte* (Vienna, 1978), 744; and Karl M. Brousek, *Wien und seine Tschechen: Integration und Assimilation einer Minderheit im 20. Jahrhundert* (Vienna, 1980), 31–35.
10. Felix Czeike, *Geschichte der Stadt Wien* (Vienna, 1981), ch. 8.
11. Renate Benik-Schweitzer and Gerhard Meissl, *Industriestadt Wien: Die Durchsetzung der industriellen Marktproduktion in der Habsburgerresidenz* (Vienna, 1983), 35, 38, 46–47, 135, 141.
12. Among those which remained were Tabakregie Ottakring & Favoriten with 1,000 workers; L. Röscher & Co. (Dunlop) with 1,800 workers and employees;

Ankerbrotfabrik with 2,000 workers and employees; Weineberger Ziegelfabrik-und Baugesellschaft with 2,500 workers; Hofherr-Schrantz-Clayton-Shuttelworth (agricultural machinery) with 2,500 workers; and Wiener Lokomotifabrik with 2,000 workers. Some of these, such as Ankerbrot, used the latest technology and means of production. See *Das Neue Wein,* II (Vienna, 1928).

13. During the war, women made up 50% of the labor force in the metals industry. Their expulsion from production in 1918–19 led to a huge increase in the number of unemployed (130,000) requiring some kind of relief. By 1920 and for the following years the number of unemployed decreased by two-thirds. See Berthold Unfried, "Arbeiterschaft und Arbeiterbewegung, 1917–1918," *Sozialdemokratie und Habsburgerstaat, 1867–1918* (Vienna, 1987), 334.

14. See Bunzel, "Arbeiterbewegung, Judenfrage," 745–46; Anton Staudinger, "Christlichsoziale Judenpolitik in der Grüdungsphase der österreichischen Republik," *Jahrbuch für Zeitgeschichte 1978* (Vienna, 1979), 28.

15. For the food deprivation of the Viennese working population during the war, see Hans Hautmann, "Hunger ist ein schlechter Koch: Die Ernährungslage der österreichischen Arbeiter im Ersten Weltkrieg," in Botz, *Bewegung und Klasse.*

16. See Jan Tabor, "An dieser Blume gehst du zugrunde: Bleich, purpurrot, weiss—Krankheit als Inspiration," Ehalt, *Glücklich ist.*

17. See for instance Otto Bauer, *Die österreichische Revolution* (Vienna, 1923), and Julius Deutsch, *Aus Österreichs Revolution* (Vienna, 1920). The assumption in those works and current to the present is that the establishment of the republic was itself revolutionary. Virtually all of Otto Bauer's writings referred to here and elsewhere are quite easily accessible in a reprint edition: Arbeitsgemeinschaft für die Geschichte der österreichischen Arbeiterbewegung, *Otto Bauer: Werkausgabe* (Vienna, 1975–80), 9 vols.

18. See Helmut Gruber, *International Communism in the Era of Lenin* (New York, 1967), 191–96; Hautmann, *Verlorene Räterepublik*, 214–18; and Ernst Bruckmüller, *Sozialgeschichte Österreichs* (Vienna, 1985), 456–67.

19. See Joseph Ehmer, "Wiener Arbeitswelten um 1900," in Ehalt, *Glücklich ist,* 196–97. I have extrapolated the figures for 1919 from Ehmer's data.

20. Joseph Ehmer and Heinz Fassmann, "Zur Sozialstruktur von Zuwanderern im 19. Jahrhundert," in *Immigration et société urbaine en Europe occidentale, XVIe–XXe siècle* (Paris, 1985), 42–44; Michael John, *Wohnverhältnisse sozialer Unterschichten im Wien Kaiser Franz Josephs* (Vienna, 1984), 197–209.

21. See Michael John, *Hausherrenmacht und Mietereland: Wohnverhältnisse und Wohnerfahrung der Unterschichten in Wien, 1890–1923* (Vienna, 1982), 108–30; Michael John, *Wohnverhältnisse,* 95–121.

22. On spontaneous versus instigated street violence, see Gerhard Botz, *Gewalt in der Politik: Attentate, Zusammenstösse, Putschversuche, Unruhen in Österreich 1918 bis 1938* (Munich, 1983), 22–72.

23. For the stress of postwar life on working-class children and youth, see Hans Safrian and Reinhard Sieder, "Gassenkinder, Strassenkämpfer: Zur politischen Sozialisation einer Arbeitergeneration in Wien 1900 bis 1938," in Lutz Niethammer and Alexander von Plato, eds., *Wir kriegen jetzt andere Zeiten* (Berlin, 1985), 120–25; and Reinhard Sieder, "Behind the Lines: Working-Class Families in Wartime Vienna," in Richard Wall and Jay Winter, eds., *The Upheaval of War: Family, Work and Welfare in Europe, 1914–1918* (Cambridge, 1988).

24. See Hans Hautmann and Rudolf Kropf, *Die österreichische Arbeiterbewegung vom Vormärz bis 1945,* 3rd rev. ed. (Vienna, 1974), 122–23.

25. Otto Bauer made it quite clear that the SDAP feared worker spontaniety and

sought to control and contain the mass movement. See *Österreichische Revolution*, 84ff. On the danger to democracy posed by soviet-styled experiments, see Otto Bauer, "Rätediktatur oder Demokratie?," *Die Arbeiter-Zeitung*, March 28, 1919. For the reaction of the Austrian workers' councils to the Hungarian Soviet Republic, see Julius Braunthal, *Die Arbeiterräte in Deutschösterreich* (Vienna, 1919).

26. See Rolf Reventlow, *Zwischen Alliierten und Bolschewiken: Arbeiterräte in Österreich, 1918–1923* (Vienna, 1969), 69ff, 124; Botz, *Gewalt Politik*, 72–80.

27. For attitudes toward proletarian dictatorship in the SDAP see Raimond Löw, *Otto Bauer und die russische Revolution* (Vienna, 1980), 42–55.

28. See Gruber, *International Communism Lenin*, 177–78.

29. An example of the rationalizing role of theory is Bauer's explanation of the elections to the Constituent Assembly in 1919. The socialists had received 40.76% of the vote and had the largest number of seats, but could not govern alone and had to share power with their sworn opponents, who shattered the coalition one year later. Characterizing the election results several years later, Bauer said that the socialists had captured "predominant power in the Republic." See Anson Rabinbach, *The Crisis of Austrian Socialism: From Red Vienna to Civil War, 1927–1934* (Chicago, 1983), 22.

30. Virtually every socialist party in the interwar years was in the hands of an oligarchical leadership which spoke in the name of the party but had little use for internal democracy, alternate views, factions, or grass-roots initiatives. See Helmut Gruber, *Léon Blum, French Socialism, and the Popular Front: A Case of Internal Contradictions* (Ithaca, N.Y., 1986), 1–3 and idem., "The German Socialist Executive in Exile, 1933–1939: Democracy as Internal Contradiction," in Wolfgang Maderthaner and Helmut Gruber, eds., *Chance und Illusion: Labor in Retreat* (Vienna, 1988), 185–89 and preface.

31. See Fritz Klenner, *Die Österreichischen Gewerkschaften* (Vienna, 1953), I: 520; Gulick, *Austria*, I: xi, 258–59. Membership by Viennese workers in the Catholic Workers' Association amounted to only 7% of the total socialist union membership. Ibid., 27–28, 266–67.

32. See Alfred G. Frei, *Rotes Wien: Austromarxismus und Arbeiterkultur* (Berlin, 1984), 58–59; Hans Hautmann and Rudolf Hautmann, *Die Wohnbauten der Gemeinde Wien* (Vienna, 1980), 31–32.

33. See Peter Kulemann, *Am Beispiel Austromarxismus: Sozialdemokratische Arbeiterbewegung in Österreich von Hainfeld bis zur Dollfuss-Diktatur* (Hamburg, 1979), 304–7.

34. By 1927 the socialist vote in Vienna reached 60.3%. In the same year 55.5% of the votes cast for the SDAP came from party members, as compared to 21.5% in 1919. See Frei, *Rotes Wien*, 60.

35. Gulick, *Austria*, I: 690.

36. A second coalition government was formed in October 1919. But conservative resistance to the socialists' reform demands led to the breakup of the coalition in June 1920. From then on the SDAP behaved very much like a social democratic oppositon in refusing to consider participation in coalition governments.

37. For the reform legislation, see Gulick, *Austria*, I: 175ff; Julius Braunthal, *Die Sozialpolitik der Republik* (Vienna, 1919); and Otto Bauer, *Der Weg zum Sozialismus* (Vienna, 1919).

38. It was actually a euphemism for the forty-eight-hour week usually involving five full days and a Saturday of half-day work. Women's weekly hours under the law were reduced to forty-four.

39. See Gerhard Meissl, "Minutenpolitik: Die Anfänge der 'Wissenschaftlichen Betriebsführung' am Beispiel der Wiener Elektroindustrie vor dem Ersten Weltkrieg," in Helmut Konrad and Wolfgang Maderthaner, eds., *Neuere Studien zur Arbeitergeschichte* (Vienna, 1984), I: 41–100.

40. From the beginning, the socialist trade unions won more than 80% of the seats on the Chambers of Workers and Employees.

41. On socialization and factory councils, see Hautmann and Kropf, *Österreichische Arbeiterbewegung*, 130–31, 136–37; Erwin Weissel, *Die Ohnmacht des Sieges: Arbeiterschaft und Sozialisierung nach dem Ersten Weltkrieg in Österreich* (Vienna, 1976), 144–49, 252–99; Klaus Klenner, "Die Ursachen des Versagens der gemeinwirtschaftlichen Anstalten in der Ersten Republik" (Ph.D diss., University of Vienna, 1959); Otto Bauer, *Die Sozialisierungsaktion im ersten Jahre der Republic* (Vienna, 1919); Edward März and Fritz Weber, "Otto Bauer und die Sozialisierung," in D. Albers et al., *Otto Bauer und der "dritte Weg"* (Frankfurt/Main, 1978), 74–82; and Robert Stöger, "Sozialisierrungs- und Verstaatlichungsdiskussion in der österreichischen Sozialdemokratie in der Ersten und Zweiten Republik: Ein Vergleich," *Archiv: Jahrbuch des Vereins für Geschichte der Arbeiterbewegung*, III (1987): 66–71.

42. The head of the Catholic trade unions, Leopold Kunschak, and Christian Social leader Ignaz Seipel, were also members and did their best to keep the Commission on the plane of discussion.

43. Here, as in other domains, the socialists explained their failure to accomplish their program as the result of pressures or anticipated actions by the Allied powers. In the case of socialization, it was suggested that foreign credit would be cut off in the event of expropriations.

44. See Maren Seliger, *Sozialdemokratie und Kommunalpolitik in Wien* (Vienna, 1980), 24. In the prewar period the restricted suffrage disenfranchised large segments of the working class and made Vienna a Christian Social stronghold.

45. Ibid, 65–82. This effective dual status became formally constitutional only in January 1922. In the Viennese provincial election of 1923 the SDAP attained 65% of the mandate. See Robert Danneberg, *Die sozialdemokratische Gemeindeverwaltung in Wien* (Vienna, 1928), 10. For the political struggle over the Vienna problem and the distinction between Vienna as capital as opposed to "greater Vienna," see Wilfried Posch, "Lebensraum Wien: Problem der Raumordnung und des Wohnungswesens, 1918–1978," *Austriaca: Cahiers universitaires d'information sur l'Autriche* 12 (May 1981): 139–45.

46. But a substantial part of the provincial/municipal budget was derived from a federal apportionment of national taxes. The latter was in the hands of the government coalition of Christian Socials and Pan-Germans and other conservative parties after 1920. It placed a powerful weapon in the hands of the socialists' opponents, who at first threatened to and after 1929 did make use of it.

47. For the forms and means of taxation, see Rainer Bauböck, *Wohnungspolitik im sozialdemokratischen Wien, 1919–1934* (Salzburg, 1979), 128–39; Frei, *Rotes Wien*, 84; and for the most detailed presentation, Gulick, *Austria;* I: 354–407.

48. For the following, see John, *Hausherrenmacht*, 3–26; Helmut Weihsmann, *Das Rote Wien: Sozialdemokratische Architektur und Kommunalpolitik, 1919–1934* (Vienna, 1985), 35–38; Seliger, *Sozialdemokratie*, 91–105; Bauböck, *Wohnungspolitik*, 26–38; and Gulick, *Austria*, I: 421–32.

49. Although these figures suggest a substantial increase in the real wages of workers in postwar Vienna, lowered rents only made a marginal difference. Real wages had risen substantially for the generation before the war, but the economic

dislocations after 1918 erased most of these gains. Thus the drastic rent reductions in the worker budget mainly compensated for other losses. See Michael John, "Wohnpolitische Ausseinandersetzungen in der Ersten Republik insbesondere ausserhalb des Parlaments," in Konrad and Maderthaner, eds., *Neuere Studien*, I: 247–54.

50. See *Österreichische Revolution*, introduction.

51. My list of titles in use during the First Republic is by no means complete (or even perfectly accurate), being derived from the memory of elderly Viennese acquaintances. Title mania is even more pronounced in the Second Republic, especially among socialists. One well-known former director of a principal Viennese archive had seven titles before his name, all of which had to be included in any correspondence with him if one expected a favorable response.

52. With one exception, Austrian socialist rituals and their relation to older cultural forms have not been studied. See the excellent Ph.D dissertation by Béla Rasky, "Arbeiterfesttage: Die Fest- und Feierkultur der sozialdemokratischen Bewegung in der Ersten Republik Österreich, 1918–1934," (University of Vienna, 1985), especially 386–95. For an interesting introduction to the symbolism of German worker rituals, see Gottfried Korff, "Rofe Fahnen und geballte Faust: Zur Symbolik der Arbeiterbewegung in der Weimarer Republic," in Dietman Petina, *Fahnen, Fäuste, Körper: Symbolic und Kultur der Arbeiterbewegung* (Essen, 1986).

53. See Reinhard Sieder, "Gassenkinder," *Aufrisse: Zeitschrift für politische Bildung* 5:4 (1984): 8–11.

54. See Bruckmüller, *Sozialgeschichte*, 504–5.

55. On the question of viability and *Anschluss*, see the excellent succinct article by Bruce F. Pauley, "The Social and Economic Background of Austria's 'Lebensunfähigkeit,'" in Rabinback, ed., *Austrian Socialist Experiment*, 21–37. See also Otto Bauer, *Acht Monate Auswärtige Politik* (Vienna, 1919); K. W. Rothschild, *Austria's Economic Development Between Two Wars* (London, 1947); Lajos Kerekes, "Wirtschaftliche und soziale Lage Österreichs nach dem Zerfall der Doppelmonarchie," in Rudolf Neck and Adam Wandruska, eds., *Beiträge zur Zeitgeschichte* (St. Pölten, 1976); and Gulick, *Austria*, I: 52–55.

56. Pauley, "Lebensunfähigkeit," 29–30.

57. Klemens von Klemperer, *Ignaz Seipel: Christian Statesman in a Time of Crisis* (Princeton, N.J., 1972), 177ff.

58. But in 1900 German was the language spoken at home by only 36% of the Viennese: 23% spoke Czech, 17% spoke Polish, and 13% spoke Ruthenian, while the rest were divided among Slovenian, Serbo-Croation, Italian, Rumanian, and Hungarian. See Eva Viethen, "Wiener Arbeiterinnen: Leben zwischen Familie, Lohnarbeit und politischem Engagement" (Ph.D diss., University of Vienna, 1984), 168.

59. C. A. Macartney, *The Social Revolution in Austria* (Cambridge, 1926), 98.

60. In 1923 there were 112,000 Czechoslovakian citizens resident in Vienna. Combined with the 100,000 to 120,000 Czechs with Austrian citizenship, the total community was a significant enclave within the Viennese population. See Albert Lichtblau, "Ceská Viden: Von der tschechischen Grossstadt zum tschechischen Dorf," *Archiv: Jahrbuch des Vereins für Geschichte der Arbeiterbewegung* 3 (1987): 34–41, 45, n. 18.

61. Ibid., 41–44.

62. For the racial anti-Semitism of Georg von Schönerer and Karl Leuger, see Schorske, *Fin-de-Siècle Vienna*, 126–32, 138–43.

63. See John W. Boyer, *Political Radicalism in Late Imperial Vienna: Origins of the Christian Social Movement* (Chicago, 1981), ch. 6.

64. Joseph Roth, "Juden auf Wanderschaft—Wien," in Ruth Beckermann, ed., *Die Mazzesinsel: Juden in der Wiener Leopoldstadt, 1918–1938* (Vienna, 1984), 35. Roth ends by observing: "But they are unemployed proletarians. A peddler is a proletarian."

65. See Staudinger, "Christlichsoziale Judenpolitik," 39. Staudinger points out that these concentration camps of 1920 cannot be confused with the brutal Nazi institutions. Yet, he argues, the notion of internment of Jews was not without influence on later Nazi anti-Semitism.

66. The most visible of these include the writers Jakob Wasserman, Richard Beer-Hofmann, Arthur Schnitzler, Stefan Zweig, Franz Werfel; the theater director Max Reinhardt; the film directors Michael Kertesz and Alexander Korda; the social scientists Emil Lederer and Paul Lazersfeld; the psychoanalysts Sigmund Freud, Otto Fenichel, and Wilhelm Reich; the scientists Rudolf Carnap and Moritz Schlick; four Nobel laureates in medicine; the businessmen Julius Meindl and Gustav Heller; and the banker Louis Nathanial Rothschild. This list merely skims the surface of Jews prominent and visible in the public life of Vienna.

67. The second echelon of SDAP functionaries who were Jews included David Joseph Bach, Paul Federn, Otto Leichter, Käthe Leichter, Otto Felix Kanitz, Benedikt Kautsky, Edgar Zilsel, Zoltan Ronai, Fritz Rosenfeld, Paul Speiser, Oskar Pollak, Marianne Pollak, Karl Kautsky, Jr., Leopold Thaller, and Margarete Hilferding. Lists like this one tend to be haphazard in the absence of a biographical dictionary of Austrian socialism. The nearest thing, though lacking in rigor of data and uniformity, is Georges Haupt, et al., *Dictionnaire biographique due movement ouvrier international: Autriche* (Paris, 1971).

68. *Die Reichspost*, Dec. 24, 1918.

69. Cited in Peter G. J. Pulzer, *The Rise of Political Anti-Semitism in Germany and Austria* (New York, 1964), 318.

70. Staudinger, "Christlichsoziale Judenpolitik," 19.

71. Ibid., 36–42. Seipel saw several versions of the proposed laws' text and considered a final version as "juridically and politically acceptable" but not timely. See also Anton Pelinka, *Stand oder Klasse? Die Christliche Arbeiterbewegung Österreichs* (Vienna, 1972), 297–300.

72. Like Karl Lueger, Seipel was prepared to admit that there were some decent Jews. See Alfred Pfoser, "Der Wiener 'Reigen' Skandal: Sexualangst als politisches Syndrom der Ersten Republik," in Konrad and Maderthaner, eds., *Neuere Studien*, III: 684–86.

73. The fear that social democracy would be identified with Jewry goes back to the SDAP's founder, Victor Adler, who at the time of the Dreyfus Affair distanced himself from the controversy. By contrast, Léon Blum became a leading Dreyfusard. See Jack Jacobs, "Austrian Social Democracy and the Jewish Question in the First Republic," in Rabinbach, ed., *Austrian Experiment*, 158.

74. See the pamphlets *Der Judenschwindel*, (Vienna, 1923) *Wenn Judenblut vom Messer spritzt*, (Vienna, [1932?]) *Der Jud ist schuld*, (Vienna, n.d.) and Danneberg's important tract *Die Schiebergeschäfte der Regierungsparteien: Der Antisemitismus im Lichte der Tatsachen* (Vienna, 1926). All these were published by the SDAP publishing house. The same tactic of fighting fire with fire, but also of using anti-Semitic stereotypes to show up the anti-Semites, was used in articles appearing in *Die Arbeiter-*

Zeitung and *Der Kampf*. For the above, see Ibid., 158–60, 165–66. See also Bunzel, "Arbeiterbewegung 'Judenfrage,'" 760–61, and Beckermann, *Mazzes-insel*, 20. For citations of anti-Semitic election posters of the SDAP and the use of a Yiddish dialect parody by a socialist in parliament, see George E. Berkley, *Vienna and Its Jews: The Tragedy of Success, 1880s–1980s* (Cambridge, Mass., 1988), 160–61, 166–67.

75. In Weimar Germany the Socialist party (SPD) regularly defended German Jews from the rising anti-Semitism. Jacobs, "Austrian Social Democracy," 163.

76. Cited in Joel Colton, *Léon Blum: Humanist in Politics* (Durham, N.C., 1987), 6.

77. See Pierre Birnbaum, *Un mythe politique: La "République juive" de Léon Blum à Pierre Mendès-France* (Paris, 1988), 61–85. It has been generally assumed that Austromarxism became the theoretical means of providing Austrian socialism with the ideals of the Enlightenment so largely absent from the intellectual environment of the old Dual Monarchy. The relationship between Enlightenment and German culture on the part of Austromarxism is one of the subjects of the second part of this chapter.

78. See *Wiener Diözesanblatt* (Vienna, 1919), 1–3, and *St. Pöltner Diözensanblatt* (Vienna, 1919), 5–7.

79. One afternoon a week was set aside for religious instruction in the schools. Since a majority of the pupils were Catholic, a priest came to the classroom, and children of other faiths had to take their instruction elsewhere. Marriage ceremonies had to be religious as well as of secular record. Only those who had legally established themselves as having no religion *(Konfessionslos)* were entitled to a strictly secular marriage.

80. It would be difficult to find another republic in Europe at that time where religious officials could hold any public office, least of all as head of government.

81. In France, by comparison, leading Catholic intellectuals such as François Mauriac and Jacques Maritain, the Catholic youth, and even Cardinal Verdier, the archbishop of Paris, considered collaboration on social issues with working-class organizations during the early 1930s. In Catholic publications such as *Esprit* and *Vendredi*, a strong anticapitalist current prevailed. See James Stell, "'La main tendue,' the French Communist Party and the Catholic Church, 1935–37," in Martin S. Alexander and Helen Graham, eds., *The French and Spanish Popular Fronts: Comparative Perspectives* (Cambridge, 1989), 97–100.

82. See "'Aus christlicher Verantwortung am Schicksal der sozialistischen Bewegung teilnehme': Gespräch mit Otto Bauer, dem Vorsitzenden der Religiösen Sozialisten, über die Entwicklung zum 12. Februar 1934," *Mitbestimmung* 13:5 (1984); 26–29.

83. See Wolfgang Maderthaner, "Kirche und Sozialdemokratie: Aspekte des Verhältnisses von politischen Klerikalismus und sozialistischer Arbeiterschaft bis zum Jahre 1938," in Konrad and Maderthander, eds., *Neuere Studien*, III: 538–41.

84. Although the Catholic majority in Vienna was 87%, only 10% of these Catholics attended Sunday church service. See Ernst Hanish, "Der politische Katholizismus als ideologischer Träger des 'Austrofaschismus,'" in E. Talos and W. Neugebauer, eds., *"Austrofaschismus: Beiträge über Politik, Ökonomie und Kultur, 1934–1938* (Vienna, 1984), 55–56.

85. See *Der Pionier: Mitteilungsblatt des Landesvereines Wien des "Freidenkerbund Österreichs"* 4:9 (Sept. 1929): 5–6. Resignations in 1927 reached 120,000, or every fiftieth Catholic. More than 80% of these were in Vienna, and 94% of those were workers. See Anton Burghardt, "Kirche und Arbeiterschaft," in Ferdinand Kloster-

mann et al., *Kirche in Österreich, 1918–1965* (Vienna, 1966), 271. By 1933 resignations reached a total of nearly 200,000. See *Pioneer* 8:1 (Jan.–Feb. 1933): 101.

86. See Eva Viethen, "Wiener Arbeiterinnen," 192–201.

87. The best explanation offered is that the SDAP was simply unprepared to shoulder the responsibility of creating a secular republic. Its much touted leftwing under Friedrich Adler and Otto Bauer was no more ready than anyone else for the collapse of the old regime and the vacuum of power which opened up unforeseen possibilities. See Berthold Unfried, "Positionen der 'Linken' Innerhalb der österreichischen Sozialdemokratie während des 1. Weltkrieges," in Konrad and Maderthaner, eds., *Neuere Studien*, II: 319–60.

88. See Gerhard Steger, *Rote Fahne Schwarzes Kreuz: Die Haltung der Sozialdemokratischen Arbeiterpartei Österreichs zu Religion, Christentum und Kirchen von Hainfeld bis 1934* (Vienna, 1987), 296–99. See also Paul M. Zulehner, *Kirche und Autromarxismus: Eine Studie zur Problematik Kirche-Staat-Gesellschaft* (Vienna, 1967).

89. Otto Bauer, *Sozialdemokratie, Religion, und Kirche: Ein Beitrag zur Erläuterung des Linzer Programms* (Vienna, 1927), 51–60.

90. A mere listing of these would fill a small volume. I have found the following books particularly suggestive. By far the analytically sharpest and speculatively most original treatment of Austromarxism is scattered throughout Rabinbach's *Crisis*, passim and chs. 1–2, and supports the growing opinion that this is the most important political study of Austrian socialism to appear since Gulick's work forty years ago. At the nontheoretical end of the spectrum, the encyclopedic volume by Ernst Glaser, *Im Umfeld des Austromarxismus: Ein Beitrag zur Geschichte des österrichischen Sozialismus* (Vienna, 1981), is an indispensable compendium of every intellectual and cultural figure in the German-speaking world who might in any way be associated with Austromarxism. Even such nonsocialist moths as Hermann Broch, Elias Canetti, Alfred Adler, and Karl Popper are shown to be drawn to the Austromarxist flame, giving weight to the latter's fashionable modernity and attractiveness based on the absence of theoretical clarity. Other useful works include: Kuhlmann, *Beispiel Austromarxismus*, 31–38, 250–52, 256–71, and 380–88; Tom Bottomore, "Introduction," in Tom Bottomore and Patrick Goode, eds., *Austro-Marxism* (Oxford, 1978); Alfred Pfoser, *Literatur und Austromarxismus* (Vienna, 1980), 9–33; Raimund Löw, Siegfried Mattl, Alfred Pfabigan, *Der Austromarxismus—eine Autopsie: Drei Studien* (Frankfurt/Main, 1986); and one of the earliest studies, now quite unfairly relegated to the dustbin of history, Norbert Leser's *Zwischen Reformismus und Bolschevismus: Der Austromarxismus als Theorie und Praxis* (Vienna, 1968), which, in the conclusion especially, analyzes the "myth of unity" in Austromarxism.

91. In the late 1970s a number of European socialist parties, while searching for programs that went beyond exhausted social democracy and would appeal to new audiences, reported to have found a historical forebear in Austromarxism and particularly in Otto Bauer's "third way." See, for instance, Giacomo Marramao, *Austromarxismo e socialismo di sinistra fra le due guerre* (Milan, 1978), and Jean-Pierre Chevenement, *Les Socialistes, les communistes et les autres* (Paris, 1977). The latter is echoed in the official French Socialist party program: *Projet socialiste pour la France des années 80* (Paris, 1980). There is no indication that any of the European socialists flirting with Austromarxism had a very clear idea of what attracted them: a vague association between the abortive Austrian factory councils and *autogestion*, or between the Viennese worker culture and *"le goût de vivre"*? The French socialists in power seem to have settled for a mixture of social-democratic/liberal reforms and have kept their distance from any theoretical justification, Austromarxist or otherwise.

92. Gulick, *Austria,* II: 1363–1400. After locating Austromarxism somewhat vaguely as a dynamic prewar non-Bernsteinian revisionism, Gulick congratulates it for being a pragmatic and nondogmatic theory adapted to the challenges after 1918. Later in his discussion Gulick is more sensible and specific in calling Austromarxism the theory of Otto Bauer on the "balance of class forces." That, from this writer's view, is a splendid beginning, but it leaves out the cultural potentiality opened up for socialism, which was so central to Bauer's formulation.

93. "Was ist Austromarxismus?," *Arbeiter-Zeitung,* Nov. 3, 1927, cited in Hans-Jörg Sandkühler and Rafael de la Vega, eds., *Austromarxismus: Texte zu Ideologie und Klassenkampf* (Vienna, 1970), 49–54. In November 1927 the party was still shaken by the bloody confrontation between workers and police of July 15, which was widely viewed as a setback. Bauer seems to have avoided any engagement of Austromarxism with the concrete divisions in the party.

94. *Marxism: An Historical and Critical Study* (New York, 1963), 302ff. The other groups associated with the political storms of 1905–1906 and identified as a generation by Lichtheim include the German-Polish group: K. Liebknecht, R. Luxemburg, L. Jogiches, A. L. Parvus, and K. Radek; the Mensheviks, including Trotsky; and the Bolsheviks around Lenin.

95. For fine points on biographical information I am indebted particularly to two psychohistorical studies: Peter Loewenberg, *Decoding the Past: The Psychohistorical Approach* (New York, 1983), and Mark E. Blum, *The Austro-Marxists, 1890–1918: A Psychobiographical Study* (Lexington, Ky., 1985). See also Julius Braunthal, *Victor und Friedrich Adler: Zwei Generationen Arbeiterbewegung* (Vienna, 1965), and Jacques Hannak, *Karl Renner und seine Zeit* (Vienna, 1965).

96. Leon Trotsky, who traveled in the circle of Austromarxists before the war, observed that behind the thinkers he saw "a phalanx of young Austrian politicians, who have joined the party in the firm conviction that an approximate familiarity with Roman law gives a man the inalienable right to direct the fate of the working class." Cited in Isaac Deutscher, *The Prophet Armed: Trotsky, 1879–1921* (Oxford, 1963), 185–86.

97. See Otto Bauer, *Die Nationalitätenfrage und die Sozialdemoratie,* Marx-Studien II (Vienna, 1907), and Rudolf Hilferding, *Das Finanzkapital: Eine Studie über die jüngste Entwicklung des Kapitalismus,* Marx-Studien III (Vienna, 1910).

98. Rudolf Springer [Karl Renner], *Der Kampf der österreichischen Nation um den Staat* (Leipzig and Vienna, 1902), and Bauer, *Nationalitätenfrage.*

99. Max Adler, *Kausalität und Teleologie im Streite um die Wissenschaft,* Marx-Studien I (Vienna, 1904), and Friedrich Adler, *Ernst Machs überwindung des mechanischen Materialismus* (Vienna, 1918).

100. According to Alfred Pfabigan: "The characteristics of cultural life in Vienna at the turn of the century inevitably became the characteristics of Austromarxism." "Die austromarxistische Denkweise," in Löw, Mattl, Pfabigan, *Autopsie,* 103. This schematic view overstates the case. It is the cultural and political tensions in Vienna which exerted a special influence on the Austromarxists.

101. For the Austrian socialists' awareness of and grappling with the nationality problem, see Hans Mommsen, *Die Sozialdemokratie und die Nationalitätenfrage im Habsburgischen Vielvölkerstaat* (Vienna, 1963), especially pp. 362–422 on the impact of the Russian Revolution of 1905 and the disintegrating effect of nationalism on the SDAP.

102. Bruno Bettelheim, *Freud's Vienna and Other Essays* (New York, 1990), 3–17.

103. Lichtheim further traces this radicalization among the Viennese Marxists

to "the inherited conviction that East Europe was about to enter a revolutionary era.... Vienna thus became a centre of political radicalism as well as theoretical Marxism." *Marxism*, 303.

104. "In the context of the '*fin de siècle*'," H. Stuart Hughes reminds us, "the thought of the eighteenth century seldom figured in its pure or original form: it appeared overlaid with the late nineteenth-century accretions that had deformed it—materialism, positivism, and the more vulgar forms of humanitarianism." *Consciousness and Society: The Reorientation of European Social Thought, 1890–1930* (New York, 1958), 97.

105. Bottomore, *Austro-Marxism*, 8–10.

106. For the influence of Mach on Austromarxism directly and indirectly through the Vienna Circle, See Friedrich Stadler, *Vom Positivismus zur "Wissenschaftlichen Weltauffassung": Am Beispiel der Wirkungsgeschichte von Ernst Mach in Österreich von 1895 bis 1934* (Vienna, 1982).

107. It is at the Café Central that Trotsky met the Austromarxists. The postwar Austromarxists, led by Bauer, made the Central their home away from home.

108. See Siegfried Mattl, "Einleitung," in Löw, Mattl, Pfabigan, *Autopsie*, 5.

109. On Friedrich Adler, see Braunthal, *V. and F. Adler*, chs. 17–19. On the councils of 1918–1919 and the revolutionary climate, see Gruber, *Communism Era of Lenin*, 175–81. For Adler's abortive attempt to unify the socialist parties, see Julius Braunthal, *History of the International* (New York), 1967), II: chs. 9–11. Though Adler was largely absent from the Austrian scene, he continued to play a role as a staunch supporter of Otto Bauer in his interpretation of Austromarxism as well as on purely tactical questions.

110. But Renner was already an outsider in relation to the new so-called left which gained prominence in the SDAP during the war. At that time, Friedrich Adler had attacked Renner for his patriotic position on the war, calling him the "Leuger of social democracy."

111. See Hanno Dreschler, *Die Sozialistische Arbeiterpartei Deutschlands (SADP): Ein Beitrag zur Geschichte der deutschen Arbeiterbewegung am Ende der Weimarer Republik* (Meisenheim am Glan, 1965), 21–23.

112. I apologize to those who enjoy a full presentation of complicated ideas for riding roughshod over philosophic elegances and even essences in my attempt to extract only those themes pertinent to the subject of this book.

113. See Pfabigan, *Adler*, ch. 7, which has a section entitled "Kant becomes a social democrat." This chapter gives the best critical reading of Adler's philosophical Marxist excursions.

114. The following works of Adler, in addition to *Kausalität*, are particularly useful here: "Die sozialistische Idee der Befreiung," *Der Kampf* 11 (1918); *Der Sozialismus und die Intellektuellen* (Vienna, 1910); *Der soziologische Sinn der Lehre von Karl Marx* (Leipzig, 1914); and *Kant und der Marxismus* (Berlin, 1925).

115. Rabinbach's characterization of the contradiction is telling: "a permanent tension between automatic and teleological laws of history and the practical efforts of the human subject." In other words, a contradiction between preparatory cultural strategy and "objective" reality. *Crisis*, 125–26.

116. See Hilferding, *Finanzkapital* (Berlin, 1955), 174–75; Lichtheim, *Marxism*, 310–14; Bottomore, *Austro-Marxism*, 33–34.

117. See Pfabigan, *Max Adler*, 314–23.

118. Mattl suggests that the Austromarxist founders were intellectually old-fashioned in clinging to the already outmoded nineteenth-century philosophic ideas, and

that they were unfamiliar with the newest work in positivism, econometrics, and psychoanalysis, all of which were were very current in the Vienna of their time. See *"Einleitung,"* in Löw, Mattl, Pfabigan, *Autopsie,* 7.

119. No doubt German culture was more developed than the others in the Dual Monarchy. But the issue went beyond the question of the quality of ideas. In the Vienna of 1914, with a 2.1 million population of which non-Germans represented more than a third, 1,475 German-language newspapers and periodicals were published as compared to 60 in all other languages. *Statistiches Jahrbuch der Stadt Wien* (Vienna, 1918), 32 (1914): 482–83. By the outbreak of World War I, Renner's philo-Germanism had turned into chauvinism. See Blum, *Austro-Marxists,* 172–74.

120. *Crisis,* 7.

121. Even as late as 1924, when Max Adler tried to provide the theoretical justification for the SDAP *Bildung* program already put into practice, he insisted that education for the proletariat must come from "the book" and not from experience. See *Neue Menschen: Gedanken über sozialisistische Erziehung* (Berlin, 1924), 109–10. Part of the working-class culture being developed in Vienna, involving the largest number of participants, had little or nothing to do with book learning (sports and festivals, for instance).

122. See Kulmann, *Austromarxismus,* 34–38; Bottomore, *Austro-Marxism,* 15–22; and Peter Heintel, *System und Ideologie: Der Austromarxismus im Spiegel der Philosophie Max Adlers* (Vienna, 1967), 15ff.

123. Hughes, *Consciousness,* 107–8.

124. For the following discussion of this key concept underlying the socialist cultural experiment, see especially the sophisticated analysis in Rabinbach, *Crisis,* 39–45, 60–63, 119–20. See also Bauer, *Österreichische Revolution,* 132–40, 175–84, 228, 259, 258–60; Bauer, "Das Gleichgewicht der Klassenkräfte," *Der Kampf* 17 (Jan. 1924): 57–67; as well as Kulemann, *Beispiel,* 238–42, and Pfoser, *Literatur,* 18–21.

125. Particularly a balance of classes leading to a pause in history. See "Ruhepausen der Geschichte," *Der Kampf* 3 (Sept. 1910), and "Volksvermehrung und soziale Entwicklung," ibid., 7 (April 1914).

126. At virtually the same time Trotsky wrote about the contradictory relationship between culture and revolution. In Western Europe, he argued, the richer the history of the working class—the more education, tradition, and accomplishments—the more difficult it would be to gather it into a revolutionary unity, because the privileges of bourgeois democracy and freedoms would tie them to the bourgeois order. Leon Trotsky, *Fragen des Alltaglebens: Die Epoche der "Kulturarbeit" und ihre Aufgaben* (Hamburg, 1923), 24.

127. The "inheritor party role," to which the SDAP subscribed, hewed to the orthodox Marxist evolutionary position that in some dim future when the bourgeoisie would be unable to rule, social democracy would assume power with a minimum of resistance. See Peter Nettl, "The German Social Democratic Party as a Political Model, 1890–1914," *Past and Present* 30 (1965): 67.

128. Lowenberg calls Bauer's ambivalence and doubt leading to the avoidance of decisions and actions "obsessing." See *Decoding,* 181. Bauer's cultural optimism and political immobility were sustained in the SDAP by conceptions of "loyalty" which, by reducing criticism to a ritual, foreclosed inner-party democracy. See Leser, *Zwischen,* 350–51.

129. Dieter Groh, *Negative Integration und revolutionärer Attentismus: Die deutsche Sozialdemokratie am Vorabend des Ersten Weltkrieges* (Frankfurt/Main, 1973). Groh's characterization of German prewar social democracy is well suited to the postwar SDAP, especially its substitution of *Bildung* for political action.

130. See Adler, *Neue Menschen*, 22–24, 92, 109, and Pfabigan, *Adler*, 198–213.

131. Hans Kelsen, "Dr. Otto Bauer's politische Theorie," *Der Kampf* 17 (Feb. 1924): 50–56. Kelsen put his finger on a major weakness in the Bauerian Austromarxist formulation adopted by the SDAP: its lack of influence in the workplace, where the trade unions were largely ineffective.

132. Otto Leichter, "Zum Problem der sozialen Gleichgewichtzustände," *Der Kampf* 17 (May 1924): 184.

133. See "Das 'Linzer Programm' der Sozialdemokratischen Arbeiterpartei Österreichs, 1926," in Albert Kadan and Anton Pelinka, eds., *Die Grundsatzprogramme der österreichischen Parteien: Dokumentation und Analyse* (Vienna, 1979), 81–88, for the social and cultural program. See also Gulick, *Austria*, II: 1389–1400.

134. It was the most important right-wing paramilitary organization, initially organized in 1918–19 to defend the as-yet-undetermined Austrian borders. It became the paramount provincial armed force poised against Vienna during the republic. It was supported by big business and finance, large landowners such as Ernst Rüdiger Stahrenberg, Catholic political leaders like Seipel, Dollfuss, and Schuschnigg, Catholic priests, petty bourgeois, and peasants. It was strongly monarchist and used force and terror to oppose socialism and any idea or act contrary to Catholic teaching. As the principal exponent of Austrofascism before 1934, it was overtaken by the Nazi SA after 1930. See Bruce F. Pauley, *Hahnenschwanz und Hackenkreuz: Steirischer Heimatschutz und österreichischer Nationalsozialismus* (Vienna, 1972), and C. Earl Edmondson, *The Heimwehr and Austrian Politics, 1918–1936* (Athens, Ga., 1978).

135. Point III of the program entitled "The Struggle for Control of the State." "*Das Linzer Programm,*" 78–81.

136. Adler had been a member of the program committee. See Dreschler, *Sozialistische Arbeiterpartei*, 30–32. For Adler's position as well as his intervention in the debates, see *Protokoll des sozialdemokratischen Parteitages, 1926* (Vienna, 1926), 199–200, 286, 292, and 310–14, and Leser, *Zwischen*, 382–98, for the debates in general. The congress accepted the program unanimously.

137. *Protokoll, 1926,* 272.

138. In reading "Das Programm der Christlichsozialen Partei, 1926" one gets the impression that open confrontation was not so unexpected in the opposition camp. Kadan and Pelinka, eds., *Grundsatzprogramme*, 115–16. The program takes as its guidelines the principles of Christianity, that is, the ethics and morals of the Catholic church. It decisively rejects every attempt to create the dictatorship of a class, demands the cultivation of "German behavior," and "combats the predominance of the demoralizing Jewish influence on intellectual and economic life."

139. For the events surrounding July 15 and its implications, see Gerhard Botz, *Die Ereignisse des 15. Juli 1927: Protokoll des Symposiums in Wien am 15 Juni 1977* (Vienna, 1979), 17–59; Botz, *Gewalt*, 141–60; Rabinbach, *Crisis*, 32–34, 48–50; Otto Leichter, *Glanz und Elend der Ersten Republik: Wie es zum österreichischen Bürgerkrieg kam* (Vienna, 1964), 45–68; Leser, *Zwischen*, 199–428; Gulick, *Austria*, I: 717–71; Pfoser, *Literatur*, 24–26.

140. Botz, *Gewalt*, 144. Aside from the inflammatory article by Friedrich Austerlitz, editor-in-chief of *Die Arbeiter-Zeitung,* no socialist leader or trade union official or Schutzbund functionary gave the workers any guidance or leadership during the crucial hours before the outbreak of violence. Only when the Palace of Justice was already on fire and the workers prevented the fire trucks from getting through, did Mayor Seitz intervene to secure a passage for the vehicles.

141. For a one-sided defense of Seipel's actions, see Klemperer, *Seipel*, 262–69.

142. See Frei, *Rotes Wien*, 60, and Gulick, *Austria*, I: 712–13.
143. See *Protokoll des sozialdemokratischen Parteitages, 1927* (Vienna, 1927), 138ff.
144. See Alfred Pfabigan, "Revolutionärer Geist: Max Adler (1873–1937) und der Austromarxismus," *Archiv: Jahrbuch des Vereins für Geschichte der Arbeiterbewegung* 3 (1987): 61–62. Adler's attack on the party caused a scandal. Henceforth he was isolated in the party, was no longer delegated to congresses, and had difficulty being published in Austria. As a result, his activity shifted more to Berlin.
145. Cited by Pfoser, *Literatur*, 25.
146. Gulick, *Austria*, I: 717–71. He concludes this chapter with the thought that Austrian labor and democracy had begun the slow descent to defeat.
147. That such absolute loyalty to the idol of unity could no longer be relied on is exposed in Rabinbach's *Crisis*, where the ultimately ineffectual opposition to Bauer from the party's left is the major theme.
148. That the SDAP succeeded in receiving 59 percent of the vote in Viennese municipal elections to the very end of the republic suggests that the voter confidence of the workers continued to be strong despite the setback of July 15.

Chapter 3

1. When the socialist mayor Jakob Reumann assumed office in May 1919, he promised no less than a quantitive and qualitative improvement in the social net and asserted that the well-to-do would have to shoulder a large part of the cost. *Arbeiter-Zeitung*, May 23, 1919, cited in Franz Patzer, *Streiflichter auf die Wiener Kommunalpolitik, 1919–1934* (Vienna, 1978), 11–12.
2. See Julius Bunzel, "Der Wohnungsmarkt und die Wohnungspolitik," in Julius Bunzel, *Beiträge zur stadtischen Wohn- und Siedelwirtschaft: III. Wohnungsfragen in Österreich* (Leipzig, 1930), 107; Felix Czeike, *Wirtschafts- und Sozialpolitik der Geimeinde Wien, 1919–1934* (Vienna, 1959), 16–17. The reader is alerted to the fact that the numerical figures and related statistics on virtually all aspects of municipal socialism vary, at times considerably. This is due in large part to the use of different statistical yearbooks with varying systems of notation to provide comparative and long-range information. I will indicate those instances where the differences become significant.
3. The soundest analysis of socialist communal policies is still to be found in Rainer Bauböck, *Wohnungspolitik im sozialdemokratischen Wien, 1919–1934* (Salzburg, 1979), and Maren Seliger, *Sozialdemokratie und Kommunalpolitik in Wien: Zu einigen Aspekten sozialdemokratischer Politik in der Vor- und Zwischenkiriegszeit* (Vienna, 1980).
4. Czeike, *Wirtschafts/Sozialpolitik*, 55–56.
5. For the origins and development of the concept of "ordentliche Arbeiterfamilie," see Joseph Ehmer, "Familie und Klasse: Zur Entstehung der Arbeiterfamilie in Wien," in Michael Mitterauer and Reinhard Sieder, eds., *Historische Familienforschung* (Frankfurt/Main, 1982).
6. See Karl Sablik, *Julius Tandler: Mediziner und Sozialreformer* (Vienna, 1983), 70–74. Environmentalism also led the socialists to embrace the ego psychology of Alfred Adler, with its promise of personality restructuring, in preference to the more pessimistic psychoanalysis of Siegmund Freud.
7. That includes some 3,604 dwellings created with mortgage funds between 1919 and 1923 (before the first innovative housing program based on a new system of taxation), as well as those still under construction in 1934. See Charles A. Gulick, *Austria: From Habsburg to Hitler* (Berkeley, Calif., 1948), I: 434, 450. Bauböck, *Wohn-*

ungspolitik, 152, gives a total of 63,071 domiciles built, and other sources fluctuate by as much as 3,000 for reasons not discernable.

8. The new municipal housing accounted for 10.4% of all Viennese housing in 1934. Density of domicile occupation had declined from 4.14 in 1910 to 3.03 in 1934. Bauböck, *Wohnungspolitik*, 152.

9. See Karl Honay, "Aufbauarbeit in Krisenzeiten. Der Wiener Stadthaushalt im Jahre 1932," *Der Sozialdemokrat* 1 (1932): 6–8. For similar self-congratulation, see Robert Danneberg, *Zehn Jahre Neues Wien* (Vienna, 1929), 50–56. Both were members of the city council.

10. Bauböck, *Wohnungspolitik*, 154, and Käthe Leichter, *So leben wir . . . 1320 Industrie-arbeiterinnen berichten über ihr Leben* (Vienna, 1932), 84.

11. For examples, see Klaus Novy, "Der Wiener Gemeindewohnungsbau: 'Sozialisierung von unten,'" *ARCH: Zeitschrift für Architekten, Sozialarbeiter und kommunalpolitische Gruppen* 45 (July 1979); Hans Hautmann and Rudolf Hautmann, "Hubert Gessner und das Konzept des Volkswohnungspalasts," *Austriaca: Cahiers universitaires d'information sur l'Autriche* 12 (May 1981); Wolfgang Speiser, *Paul Speiser und das rote Wien* (Munich, 1979), 50–52; and Hans Hautmann and Rudolf Kropf, *Die österreichische Arbeiterbewegung vom Vormärz bis 1945* (Vienna, 1974), 146–48. A refreshing contrast to the tendency to heroize and celebrate a mythic socialist Vienna is Helmut Weihsmann's *Das Rote Wien: Sozialdemokratische Architektur und Kommunalpolitik, 1919–1934* (Vienna, 1985).

12. Bunzel, "Wohnungsmarkt/Wohnungspolitik," lists 9,720 such accommodations still existing in 1923 (109–10). He also indicates that in 1919 16.9% of all apartments still harbored subtenants and bed renters (107).

13. See J. Robert Wegs, *Growing Up Working Class: Continuity and Change Among Viennese Youth, 1890–1938* (Pennsylvania, 1989), 39–42; Peter Feldbauer, *Stadtwachstum und Wohnungsnot: Determinanten unzureichender Wohnungsversorgung in Wien, 1848–1914* (Vienna, 1977), 202–4.

14. For previous tenant insecurity, the power of landlords, and tenant nomadism, see Michael John, *Hausherrenmacht und Mieterelend: Wohnverhältnisse und Wohnerfahrung der Unterschichten in Wien, 1890–1923* (Vienna, 1982), 29–67.

15. Czeike, *Wirtschafts/Sozialpolitik*, 16–17.

16. See Wilfred Posch, *Die Wiener Gartenstadt Bewegung: Reformversuch zwischen Erster und Zweiter Gründerzeit* (Vienna, 1981), and Klaus Novy, "Selbsthilfe als Reformbewegung: Der Kampf der Wiener Siedler nach dem 1. Weltkrieg," *ARCH* 55 (March 1981).

17. Czeike, *Wirtschafts/Sozialpolitik*, 30–31.

18. See Bauböck, *Wohnungspolitik*, 91–99; Seliger, *Sozialdemokratie und Kommunalpolitik*, 99–102; Czeike, *Wirtschafts/Sozialpolitik*, 78–83, 89–90.

19. See Alfred Georg Frei, *Rotes Wien: Austromarxismus und Arbeiterkultur—Sozialdemokratische Wohnungs- und Kommunalpolitik, 1919–1934* (Berlin, 1984), 83–84.

20. Czeike, *Wirtschafts/Sozialpolitik*, 39–41; Bauböck, *Wohnungspolitik*, 123–38; Seliger, *Sozialdemokratie/Kommunalpolitik*, 115–36; Frei, *Rotes Wien*, 83–84; Gulick, *Austria*, I: 449–59.

21. The right-wing press went into a frenzy over this tax, calling it pure expropriation. Breitner was smeared with anti-Semitic slurs, to which the SDAP failed to respond forcefully. Ironically, even the wealthiest tenants were the beneficiaries of the rent-control law and, despite the allegedly unbearable tax, paid no more than 37% of prewar rents. Czeike, *Wirtschafts/Sozialpolitik*, 40.

22. At the time the socialist press made much of the champagne and caviar, racehorses and sports cars of the rich being taxed for the benefit of the "Viennese little man." They neglected to mention that even the little man's entertainment, especially admissions to cinema, *Varieté*, circus, and football matches were somehow also classified as "luxuries."

23. See Adelheid von Saldern, "Sozialdemokratie und kommunale Wohnungsbaupolitik in den 20er Jahren—am Beispiel von Hamburg und Wien," *Archiv für Sozialgeschichte* 25 (1985): 195–97. Von Saldern uses 1925 as a sample year to indicate that the housing tax brought in 23% and luxury taxes 20% of the budget for the building fund. I would increase this income by 10–15% for the years through 1929. Czeike points out that part of the compromise by which the rent-control law was weakened in 1929 included specific federal subvention of Viennese housing projects under construction. In all likelihood this assistance was marginal. See Czeike, *Wirtschafts/Sozialpolitik*, 44, 91.

24. The Christian Socials applied pressure in parliament against rent control in 1925 in order to defeat the renewal of the housing requisitioning law as part of a compromise. For this and other attacks on rent control and the housing program, see the detailed presentation in Gulick, *Austria*, I: 466–503.

25. Seliger, *Sozialdemokratie und Kommunalpolitik*, 137–39.

26. Bauböck, *Wohnungspolitik*, 108–14.

27. Peter Feldbauer and Wolfgang Hösl, "Die Wohnungsverhaltnisse der Wiener Unterschichten und die Anfänge des genossenschaftlichen Wohn- und Siedlungswesens," in Gerhard Botz, Hans Hautmann, Helmut Konrad, Joseph Weidenholzer, eds., *Bewegung und Klasse: Studien zur österreichischen Arbeitergeschichte* (Vienna, 1978), 690–91. This plan of E. H. Aigde was rejected by conservative industrialists who feared that such concentrated housing would lead to worker solidarity and radicalism.

28. Feldbauer and Hösl, "Wohnungsverhältnisse," 698–99.

29. See the excellent textual and photographic presentation of the historical precedents of communal housing in Vienna in Weihsmann, *Rote Wien*, 67–99.

30. See Otto Bauer, *Der Weg zum Sozialismus* (Vienna, 1919), 116–21, for the following.

31. See the photo essay in the exhibition catalog *Mit uns zieht die neue Zeit: Arbeiterkultur in Österreich, 1918–1934* (Vienna, 1981), 70–72. The myth, created by SDAP spokesmen in the 1920s, that the cost of the building program was borne by the rich paying the housing tax, is repeated there (68).

32. See Michael John, "The Importance of Neighborhood Relationships and Grass Roots Movements in Red Vienna, 1919–1934," 8, 13. Unpublished paper available at the Verein für Geschichte der Arbeiterbewegung, Vienna. John collected some forty histories; Reinhard Sieder more than sixty; and Gottfried Pirhofer and Peter Feldbauer considerable additional ones. Most are available on tape and typed transcription at the Institut für Wirtschafts- und Sozialgeschichte of the University of Vienna.

33. See Franz Patzer, "Zeittafel sämtlicher Sitzungen des Wiener Gemeinderates von 1918 bis 1934 mit den wichtigsten Verhandlungspunkten, wie Kundgebungen, Wahlen, Beschlüsse, Anfragen, Anträgen, usw.," *Streiflichter auf die Wiener Kommunalpolitik, 1919–1934* (Vienna, 1978), 61–123.

34. The argument, for instance, that the ulterior motive of the socialist building program was to prepare strategically for civil war, was repeated ad nauseam. Ibid., 46.

35. See *Informationsblatt des Bezirksmuseums Rudolfsheim-Fünfhaus,* and Gottfried Pirhofer, "Gemeinschaftshaus und Massenwohnungsbau," *Transparent* 3/4 (1977): 38–42.

36. *Stenographischer Bericht über die Sitzung des Gemeinderates,* March 9, 1923. Wiener Stadt- und Landesarchiv. I am grateful to Joseph Ehmer for having drawn my attention to these protocols.

37. Ibid., April 24, 1925, 100–108. Archiv der Stadt Wien.

38. The German *Handwörterbuch des Wohnungswesens* of 1930 designated Heimhof as a model *Einküchenhaus.* See Pirhofer, "Einküchenhaus," 40. The Bezirksmuseum Rudolfsheim-Fünfhaus, 1150 Vienna, Rosinagasse 4, has a model of Heimhof on permanent exhibit as well as taped reminiscences of former tenants and a publicity film of 1922. The urban archaeology, done by the group of academics and teachers who run this district museum as volunteers, is first-rate.

39. "Einküchenhaus," *Arbeiter-Zeitung,* June 2, 1923.

40. See Leichter, *So leben wir,* 85–87.

41. For Käthe Leichter's study of homeworkers and female factory workers based on survey techniques, see ch. 6 on the family. The most original and famous social science analysis, done at the request of Otto Bauer by Marie Jahoda, Paul Lazersfeld, and Hans Zeisel, *Die Arbeitslosen von Marienthal: Ein soziographischer Versuch* (Leipzig, 1933), showed how inventive Austrian social science could be in studying worker life-styles and needs.

42. See Helmut Gruber, "Preface," in Wolfgang Maderthaner and Helmut Gruber, eds., *Chance und Illusion: Labor in Retreat* (Vienna, 1988), 25–27.

43. For this very useful schema of SDAP organization, see Mark E. Blum, *The Austro-Marxists, 1890–1918: A Psychobiographical Study* (Lexington, 1985), 11–14. "Austrian social democracy," Blum suggests, "provided [the intellectual] with an audience that ineluctably fed his ego while maintaining his isolation from the working class" (13).

44. How powerful they were in turning back or neutralizing the feeble left opposition which developed in the party after 1927 is amply demonstrated in Anson Rabinbach, *The Crisis of Austrian Socialism: From Red Vienna to Civil War, 1927–1934* (Chicago, 1983), especially ch. 6.

45. For an enlightening discussion on socialist party democracy, see Peter Kulemann, *Am Beispiel des Austromarxismus: Sozialdemokratische Arbeiterbewegung in Österreich von Hainfeld bis zur Dollfuss-Diktatur* (Hamburg, 1979), 309–18.

46. Ibid., 309, 315. Kulemann's figures are for 1929 and 1932 respectively. I have adjusted both to the 1932 date.

47. Ibid., 316–18. Kulemann gives figures for 1928–32 and indicates that the spread of salary differences grew larger with time, a problem the party secretary Danneberg was aware of but could do little to alter significantly.

48. No substantial and significant archives have been discovered for these three or for other principal leaders. It is doubtful, as one is always told, that such valuable records were completely destroyed during the war. There is no biography extant of Seitz. Otto Leichter's *Otto Bauer: Tragödie oder Triumph?* (Vienna, 1970), as well as Leon Kane's *Robert Danneberg: Ein pragmatischer Realist* (Vienna, 1980), are quite general and superficial.

49. There was a small amount of fierce criticism from communist tenants in the housing projects and working-class neighborhoods, as well as from tiny cells of anarchists. I will refer to these hectographed newsletters and broadsides in discussing social adaptation to the new housing.

50. In view of the 68,858 persons seeking apartments in 1924, the first building program did little to alleviate the crisis. See Czeike, *Wirtschafts/Sozialpolitik*, 17. Not all of the apartment hunters needed a new or different apartment. In many instances upgrading of existing habitations would have been acceptable.

51. Such exaggerations are prevalent even among some younger Austrian historians. See for instance Michael John, *Hausherrenmacht und Mieterelend, 1890-1923* (Vienna, 1982). One leafs with fascination through the photographs in this book, which illustrate only the most decrepit habitations and surroundings.

52. See Wegs, *Growing Up Working Class*, 25-27.

53. For brief periods I have lived in two such former tenements: in the 20th district in 1938-39 and in the 5th district in 1981. Both had a living room/kitchen, bedroom, and a half room; both (originally) did not have a toilet, running water, or gas. I would judge the total space in both to be $50m^2$ ($540 ft^2$). The ceilings were 3m (10 ft) high and therefore gave a larger aspect to the rooms than their size would suggest. In both apartments the kitchens faced a long interior building corridor but had a window as well. In that corridor a water tap *(Basena)* and a toilet had served all the apartments. The stairwells and corridors were wide and commodious. These examples are not meant to ennoble the far-from-ideal quality of the tenements but simply to indicate that the quality of such housing conformed to a wide spectrum. I would judge that 25% of such tenements were barely habitable; that 50% were just adequate, considering the quality of cheaper housing in Vienna in general; and that 25% were livable and even desirable as interior domicile and exterior structure.

54. See for example Anton Weber, *Die Wohnungsprobleme und die Gemeinde Wien* (Vienna, 1927), and *Die wohnungspolitik der Gemeinde Wien* (Vienna, 1929). In addition, numerous pamphlets rolled off the SDAP printing presses on the accomplishments of the socialist municipality, with housing in the forefront. The formula of presentation was quite simple: before/after, decaying/blooming, ugly/beautiful, suffering/rejuvenated, etc. Robert Danneberg, the party secretary, usually signed these achievement reports.

55. In the hopeless postwar real estate market, housing stock as well as land was a glut on the market. Landlords were happy to be rid of their unprofitable houses at one-quarter of their prewar prices. See Fritz C. Wulz, *Stadt in Veränderung: Eine architektur-politische Studie von Wien in den Jahren 1848-1934* (Stockholm, 1978), II: 439. It would have been possible for the municipality to acquire such tenements at very low cost and to improve the apartments by introducing electricity, water, and gas. Whether a part of the building funds invested in such renovation would have improved the housing of a larger number than could be satisfied by only new construction was never examined by the SDAP.

56. By 1928 the municipality owned 25% of the total Viennese land surface, and by 1933 over 30%. Ibid., II: 438; Bauböck, *Wohnungspolitik*, 14-43; Weihsmann, *Rote Wien*, 63, n. 46; Gulick, *Austria*, I: 457.

57. See Bunzel, "Wohnungsmarkt," 128.

58. Seliger, *Sozialdemokratie und Kommunalpolitik*, 137-38.

59. See Leo Adler, ed., *Neuzeitliche Miethäuser und Siedlungen* (Leipzig, 1931), with detailed photographs and drawings of postwar architecture in Berlin and Hamburg as well as in Holland and Sweden. Von Saldern points out "that between 1924 and 1930 Germany became the international center of the new architecture style." It was well received by the Socialist party, which sponsored it in public buildings and housing in Hamburg, Frankfurt, Berlin, Stuttgart, and other cities. See "Sozialdemokratie kommunale Wohnungsbaupolitik," 208-9. For the positive response of

the French Popular Front government to functionalist architecture, see Jean-Louis Cohen, "Architectures du Front populaire," *Le Mouvement Social* 46 (Jan.–March 1989): 49–59.

60. The publication of an exhibit, *Rotes Wien, 1919–1934: Kommunaler Wohnbau in der Zwischenkiregszeit* (Vienna, 1980–84), is accompanied by an excellent slide program (available at all Austrian cultural institutes) which highlights the traditional adherence of the municipal housing of red Vienna to the basic Viennese courtyard construction of both public and private buildings, as well as illustrating the hodgepodge of styles in structures and exterior decorations.

61. See Adelheid von Saldern, "Die Neubausiedlungen der Zwanziger Jahre," in Ulfert Herlyn, Adelheid von Saldern, Wulf Tessin, eds., *Neubausiedlungen der 20er und 60er Jahre* (Frankfurt/Main, 1987), 39–41.

62. Novy, "Wiener Gemeindewohnungsbau," 17–18; Bauböck, 145–48. Unlike more sophisticated crafts, it was argued, bricklaying could be learned by the inexperienced with relative ease.

63. See Barbara M. Lane, *Architecture and Politics in Germany, 1918–1945* (Cambridge, Mass., 1968), 83–107.

64. See Ferdinand and Lore Kramer, "Sozialer Wohnbau der Stadt Frankfurt am Main in den 20er Jahren," *Ausstellung Kommunaler Wohnbau in Wien* (Vienna, 1978). Despite the use of rationalized materials and methods in Frankfurt and other German cities, the expected savings in building costs and low rents were not realized, because mortgage rates were so high. The Viennese public housing was financed without mortgages.

65. In Frankfurt it took five years to build new housing for 11% of the population; in Vienna it took ten years to accommodate somewhat fewer people. See ibid.

66. See von Saldern, "Sozialdemokratie kommunale Wohnungsbaupolitik," 198.

67. Lane, *Architecture*, 103.

68. The ten largest projects, which I have inspected, contained 11,570 apartments: Sandleitn Hof, 1,587; Engelshof, 1,467; Karl-Marx-Hof, 1,325; Karl-Seitz-Hof, 1,273; Mithlingerhof, 1,136; Rabenhof, 1,109; George-Washington-Hof, 1,084; Siedlung Freihof, 1,014; Am Laaer Berg, 846; Wildganshof, 829. These were located in districts with existing concentrations of workers: the 3rd, 10th, 12th, 16th, 19th, 20th, and 21st.

69. For the particulars of municipal buildings as well as individual habitations and communal facilities, see Czeike, *Wirtschaft/Sozialpolitik*, 59–78; Weihsmann, *Rote Wien*, 92–369 (a detailed walking tour through the city, stopping at each municipal project for a detailed discussion); and Hans Hautmann and Rudolf Hautmann, *Die Gemeindebauten des Roten Wien, 1919–1934* (Vienna, 1980), passim.

70. See for instance Richard Wagner, *Der Klassenkampf um den Menschen* (Berlin, 1927); Robert Danneberg, *Das neue Wien* (Vienna, 1930); Karl Honay, "Sozialistische Arbeit in der kapitalistischen Gesellschaft," *Der Kampf* 5 (1929). See also the latter-day sympathizers: Gulick, *Austria*, I: 503–4; *Austellung Kommunaler Wohnbau in Wien* (Vienna, 1977).

71. See Reinhard Sieder, "Housing Policy, Social Welfare and Family Life in 'Red Vienna,' 1919–1934," *Oral History: Journal of the Oral History Society* 13:2 (1985): 35–48.

72. See Lane, *Architecture*, 103–14; von Saldern, "Sozialdemokratie und Wohnungsbaupolitik," 222–30; *Das Wohnungswesen der Stadt Frankfurt A.M.* (Frankfurt/Main, 1930). But with the onset of the depression, the size was reduced to 30 to 50

square meters. Even so, rents in German public housing were beyond the means of the average worker family, because 60–70% of it was due to high mortgages. Such apartments were within the means of skilled workers, employees, and functionaries. See von Saldern, "Neubausiedlungen," 33–37, 53.

73. Between 1926 and 1930 ten thousand new apartments in Frankfurt were equipped with these kitchens. See Margarete Schütte-Lihotzky, "Vienne–Francfort: Construction de logements et rationalisation des travaux domestiques," *Austriaca: Cahiers universitaires d'information sur d'Autriche* 12 (May 1981), 129–38.

74. These are best enumerated and illustrated in Hautmann and Hautmann, "Hubert Gessner," 118–19, and the photographs in idem., *Gemeindebauten*.

75. Hautmann and Hautmann, "Hubert Gessner," 118, list 33 central laundries with a total of 830 workplaces. That allows for 302,950 washdays for 63,000 tenants or 4.8 washdays per tenant per year. Sieder, "Housing Policy," 10–11, cites 5,032,847 baths taken in municipal bathhouses (other than the three in the projects) as a sign of increased cleanliness. But even a conservative estimate of use works out to one bath in two weeks per person.

76. For the negative reaction of tenants to the communal facilities, see Dieter Langewiesche, "Politische Orientierung und soziales Verhalten: Familienleben und Wohnverhältnisse von Arbeitern im 'roten' Wien der Ersten Republik," in Lutz Niethammer, ed., *Wohnen im Wandel: Beiträge zur Geschichte des Alltags in der bürgerlichen Gesellschaft* (Wuppertal, 1979), 183–85.

77. Weihsmann, *Rote Wien*, 63, n. 61.

78. It was often young couples with one or two children who were able to leave the overcrowded quarters of parents or in-laws. In general they had waited from four to eight years before being selected by the housing office. See Sieder, "Housing Policy," 6–9.

79. See Dr. Ph. Vass, *Die Wiener Wohnungswirtschaft von 1917 bis 1927* (Jena, 1928), 38–39.

80. *Das neue Wien: Städtewerk* (Vienna, 1926), I: 235. Those residing in Vienna since birth received four points. Single persons and couples married less than a year could not get into the first group of needy, even if their points totaled ten.

81. The Austrian Communist party (KPÖ) criticized the SDAP and the whole municipal reform program for having abandoned the weakest portion of the working class: the unemployed, those on relief, the homeless and evicted. See for instance KPÖ, *"Waldhotel": Die Geschichte einer Delogierung im sozialdemokratischen Wien* (Vienna, 1931). The attack was too vehement for the points made, but eviction had increasingly become a problem after 1929, when the Christian Socials, using implicit threats of violence, forced the SDAP to agree to a certain weakening of the rent-control law.

82. Nearby shops and *Gasthäuser* were also important sites of spontaneous socialization. See John, "Importance of Neighborhood Relationships," 3–6.

83. See John, Hausherrenmacht, 108–16, and "Anhang: Interview mit Herrn Merinsky Ottokar, 18., Hildebrandgasse 21, am 3.7. 1981"; Gottfried Pirhofer and Reinhard Sieder, "Zur Konstitution der Arbeiterfamilie im Roten Wien: Familienpolitik, Kulturreform, Alltag und Ästhetik," in Michael Mitterauer and Reinhard Sieder, ed., *Familienforschung* (Vienna, 1982), 351–57; Wegs, *Growning Up Working Class*, 47–51.

84. Reinhard Sieder, "Working-Class Family Life in Wartime Vienna," in Richard Wall and Jay Winter, eds., *The Upheaval of War: Family, Work and Welfare in Europe, 1914–1918* (Cambridge, 1988), 126.

85. See Gottfried Pirhofer, "Ansichten zum Wiener kommunalen Wohnbau der zwanziger und frühen dreissiger Jahre," in Helmut Fielhauer and Olaf Bockhorn, eds., *Die andere Kultur: Volkskunde, Sozialwissenschaften und Arbeiterkultur: Ein Tagungsbericht* (Vienna, 1982), 237–38.

86. Pirhofer and Sieder, "Konstitution der Arbeiterfamilie," 354.

87. As we shall see in the discussion of the family in Ch. 6, every effort was made by the administration of the municipal houses to wean the workers away from their former habits and life-styles: women were told to become active in the party culture outside the home; local party meetings were removed from the neighborhood *Gasthaus* to the project meeting rooms; children were discouraged from both free play on the project grounds and street play; and various pressures were exerted to produce new norms of cleanliness, public behavior, and housekeeping.

88. Gottfried Pirhofer and Reinhard Sieder, "Familie und Wohnen im Roten Wien," in International Conference of Labour Historians, *ITH-Tagungsbericht 16* (Vienna, 1981), 190–91.

89. Pirhofer and Sieder, "Arbeiterfamilie," 351–57. The feeling of pressure from the building management, complaints about the strict regimentation, and the desire to escape from the controls of municipal housing are a constant refrain in virtually all oral history accounts.

90. There were complaints in the anarchist press that tenants who protested about the housing management were threatened with eviction. See "Das 'rote Paradies' in Wien: Ein offener Brief über sozialdemokratische Auslandslügen," *Erkenntnis und Befreiung: Organ des Herrschaftslosen Sozialismus* 11:49 (1929): 3.

91. A rare find in the SDAP archives is an exchange of letters between a loyal SDAP and trade union member and the party executive. The member, Aegidius Finker, complained that the district party representative had overturned the election of tenants' representatives twice, because the official candidate had been rejected by the community tenants. Some twenty members of the Schutzbund present but out of uniform (and not from the district) threatened to break up the meeting. When Finker and others present expressed outrage at the behavior of the SDAP representatives, they were threatened with loss of their jobs and informed that they were permanently unqualified for apartments in the new municipal housing. Finker minced no words in charging the city council with donating 60–70 percent of all municipal housing to their favorites. The SDAP's reply dismissed Finker's charges by instructing him to make his complaint to the appropriate party organ. It further informed him that the party was engaged in removing corrupt tenants' representatives, a task in which it would not be hindered. See Aegidius Finker, "An den Parteivorstand der S.D.A.P., Wien," March 14, 1928, and "Herrn Aegidius Finker," March 22, 1928, Allgemeines Verwaltungsarchiv Wien (AVA), SD-Parteistellen, Karton 93.

92. The most common complaint was about the arbitrary behavior of SDAP cadres in acting like a police force in the housing projects, and about their tendency to monopolize and dominate tenants' meetings, preventing complaints from being aired. See *Prolet im Gemeindebau: Selbsthilfsorgan der Mieter des Schummeierhofes und Umgebung*, Dec. 1930 and Jan. 1931; *Das Alsergrunder Arbeiterblatt*, Jan. 1929; *Proletarierviertel: Häuserzeitung der Mieter von Hernals oberhalb der Wattgasse*, Oct. 1932; *Der Rote Sandleitner Prolet*, Jan. 1933.

93. There is nothing vaguely socialist about calling for experts to improve the quality and management of families and child rearing. Liberal reformers and statist interventionists have made such calls for over a century, and these demands have by now been largely satisfied. See Christopher Lasch, *Haven in a Heartless World: The*

Family Besieged (New York, 1976). In the SDAP, the first order of experts were the functionaries themselves. See Kuhlemann, *Beispiel Marxismus,* 313–24.

94. A celebratory parade initiated by the tenants on the eve of the official opening of the Karl-Marx-Hof was not permitted to take place. See Frei, *Rotes Wien,* 110–16.

95. See Langewiesche, "Politische Orientierung," 175, where he reports the triumphal speech of a leading Austrian socialist at an SPD meeting in Würzburg in 1933. The speaker clamined that the "world-renowned" municipal housing had been paid for with the building tax alone.

96. See von Saldern, "Sozialdemokratie kommunale Wohnungspolitik," 194–99; F. and L. Kramer, "Sozialer Wohnbau Frankfurt"; Weihsmann, *Rote Wien,* 159–65; Marc Bonneville, *Villeurbanne: naissance et métamorphose d'une banlieue ouvrière* (Lyon, 1978); Jean-Paul Flammand, *Loger le peuple: essai sur l'histoire du logement social en France* (Paris, 1989), ch. 3; Annie Fourcaut, *Bobigny: banlieue rouge* (Paris, 1986), ch. 4; John Burnett, *A Social History of Housing, 1815–1985* (London, 1986), 234–47; and Sean Glynn and John Oxborrow, *Interwar Britain: A Social and Economic History* (London, 1976), ch. 8. The point made here is that these attempts to provide public housing were not free of problems of various sorts or necessarily better than what the SDAP accomplished, but that the Viennese housing program was only one example of such social democratic reform efforts.

97. That these massive structures were indeed fortresses created by the socialists to protect their worker denizens in case of civil war was a popular charge among Christian Social critics of the municipal housing program. The fragility of these brick-and-mortar buildings, demonstrated by the destruction inflicted on them even by the World War I artillery of Dollfuss in 1934, should have put these allegations to rest, but they linger on even in the work of otherwise sound historians. For the refutation of the "red fortress" theory, see Gerhard Kapner, "Der Wiener kommunale Wohnbau: Urteilen der Zwischen- und Nachkriegszeit," in Franz Kadrnoska, ed., *Aufbruch und Untergang: Österreichische Kultur zwischen 1918 und 1938* (Vienna, 1981), passim, but especially 149–59.

98. It has been suggested that the municipal housing projects were enclaves on the city's periphery, leaving the urban power center untouched; that instead of creating a new "ring" or proletarian belt of housing near the city center, the municipal socialists opted for a defensive position from the outset. See Gottfried Pirhofer and Michael Tripes, *Am Schöpfwerk neu Bewohnt: Ungewohntes vom Wiener Gemeindebau* (Vienna, 1981), 22–25, 35–36. For the danger of the SDAP's confusing cultural with political power, see Anton Pelinka, "Kommunalpolitik als Gegenmacht: Das 'rote Wien' als Beispiel gesellschaftsverändernder Reformpolitik," in K.-H. Nassmacher, *Kommunalpolitik und Sozialdemokratie: Der Beitrag des demokratischen Sozialismus zur kommunalen Selbstverwaltung* (Bonn, 1977), 63–77.

99. I am greatful to Peter Marcuse for having drawn my attention to this gestaltist perspective. See his article "The Housing Policy of Social Democracy: Determinants and Consequences," in Rabinbach, ed., *Austrian Experiment,* 212–13.

100. See Gottfried Pirhofer, "Wirtschaftspolitik," and "Politik am Körper," *Ausstellungskatalog Zwischenkriegszeit—Wiener Kommunalpolitik, 1918–1938* (Vienna, 1980), 21, 65–67.

101. Unfortunately the one detailed biography—Karl Sablik, *Julius Tandler: Mediziner und Sozialreformer: Eine Biography* (Vienna, 1983)—although generally informative, lacks any real insight into Tandler's social Darwinist and eugenic orientation on the "population question." Sablik also fails to appreciate the consider-

able courage with which Tandler faced the virulent and rampant anti-Semitism in the medical faculty and also among his Christian Social colleagues on the municipal council, where he stood his ground against infamous slander without concerted or decisive support from his own party.

102. See Czeike, *Wirtschafts/Sozialpolitik,* 159–65. This remains the best source on the detailed aspects of health and welfare programs, their organization, extent, cost, and accomplishments (153–211). It should be used in conjunction with Patzer, *Streiflichter,* which provides both a chronological and subject index of issues before the municipal council, for which protocols can be found. Also informative on the specifics of health and welfare programs, but not very analytical, is Gulick, *Austria,* I: 505–43.

103. Julius Tandler, "Gemeinde und Gesundheitswesen," *Die Gemeinde: Halbmonatschrift für sozialdemokratische Kommunalpolitik* 8 (1920): 165–69.

104. See Julius Tandler, *Wohltätigkeit oder Fürsorge?* (Vienna, 1925). An excerpt of this pamphlet is reprinted in Junius, *Sozialismus und persönliche Lebensgestaltung: Texte aus der Zwischenkriegszeit* (Vienna, 1981), 123–25, with a discussion of socialist *caritas,* which was not charity but spontaneous welfare actions not sponsored by the municipality but by workers for workers at the grass roots and by initiatives from below (122).

105. See Franz Karner, *Aufbau der Wohlfahrtspflege der Stadt Wien* (Vienna, 1926).

106. *Das neue Wien: Städtewerk,* I: 602–5. The official report credits the education of mothers to breast feeding and rational infant care for the sharp decline in infants deaths.

107. Of some 10,000 first-grade schoolchildren tested for tuberculosis in 1925–26, 39 percent of the boys and 31 percent of the girls were positive. See Wegs, *Growing Up Working Class,* 19.

108. For the following, see Hermann Hartmann, *Die Wohlfahrtspflege Wiens* (Jena, 1929), 98–100; Speiser, *Rote Wien,* 49–50; and Felix Czeike, *Liberale, Christlichsoziale und Sozialdemokratische Kommunalpolitik (1861–1934): Dargestellt am Beispiel der Gemeinde Wien* (Vienna, 1962), 99–101, 107–10.

109. See Philipp Frankowski and Dr. Karl Gottlieb, *Die Kindergärten der Gemeinde Wien* (Vienna, 1927), 9, 11, 46–48.

110. See David Crew, "German Socialism, the State and Family Policy, 1918–1933," *Continuity and Change* 1:2 (1986).

111. It was a major subject of discussion at the socialist womens' conference preceding the important party congress of 1926, at which the main guidelines of socialist policy, including the role of women and the family, were promulgated. See *Frauenarbeit und Bevölkerungspolitik: Verhandlungen der sozialdemokratischen Frauenreichskonferenz, Oktober 29–30, 1926 in Linz* (Vienna, 1926), and Dr. Margarete Hilferding, *Geburtenregelung* (Vienna, 1926), on the danger of reproducing the eugenically unfit. The socialists were not alone in their concern about the decline of the birth rate and quality of future generations. These matters preoccupied most national governments, racists, imperialists, as well as social reformers of every stripe. See Michael Teitelbaum and Jay Winter, *Fear of Population Decline* (Orlando, Fla., 1985). In France, during the interwar period such concerns led to first steps toward family allowances based on the number of children. See Cicely Watson, "Population Policy in France: Family Allowances and Other Benefits: I," *Population Studies* 7 (1953–54): 263–86.

112. See Doris Beyer, "Sexualität—Macht—Wohlfahrt: Zeitgemässe Erinnerungen an das 'Rote Wien,'" *Zeit Geschichte* 14:11/12 (Aug./Sept. 1987): 453. I am

grateful to Beyer for drawing my attention to a revealing essay by Julius Tandler, "Die Gefahren der Minderwertigkeit," in Franz Breunlich, *Das Wiener Jugendhilfswerk* (Vienna, 1928), 3–6.

113. Julius Tandler, *Ehe und Bevölkerungspolitik* (Vienna, 1924), 20–22.

114. See Sieder, "Housing Policy, Social Welfare," 13.

115. *Stenographischer Bericht über die Sitzung des Gemeinderates*, December 18, 1928.

116. See Dr. Karl Kautsky, Jr., "Die Eheberatung im Dienste der Wohlfahrtspflege," *Blätter für das Wohlfahrtswesen der Stadt Wien* 24 (1925): 26–28. Next to Tandler, Kautsky was one of the most important medical authorities in the SDAP, particularly involved in questions of sexuality, contraception, and abortion. See ch. 6.

117. Ibid., p. 26.

118. Sablik, *Tandler*, 278–80.

119. Beyer, "Sexualität—Macht—Wohlfahrt," 454–55.

120. The figure is for 1932. See Gulick, *Austria*, I: 510.

121. See *Das neue Wien: Städtewerk* (Vienna, 1928), III: 214.

122. See Pirhofer and Sieder, "Konstitution der Arbeiterfamilie," 332: Beyer, "Sexualität—Macht—Wohlfahrt," 457; Sablik, *Tandler*, 283.

123. Ibid., 284–85; Pirhofer and Sieder, "Konstitution der Arbeiterfamilie," 332.

124. See *Das neue Wien*, III: 215–18.

125. Pirhofer and Sieder, "Konstitution der Arbeiterfamilie," 332.

126. By 1927 almost 20,000 children were wards of the municipality. Sieder, "Housing Policy, Social Welfare," n. 50.

127. See Czeike, *Wirtschafts/Sozialpolitik*, 165–71.

128. By 1927 Vienna already had more than 6,000 social workers. See Sablik, *Tandler*, 290.

129. Kamilla Heidenreich, a social worker in a youth clinic, wrote: "Her work is the emanation of the power of love in the world; it is the extension of motherhood and, thus, the domain of woman." Quoted in Sieder, "Housing Policy, Social Welfare," n. 28.

130. See Beyer, "Sexualität—Macht—Wohlfahrt," 457; Sablik, *Tandler*, 224–25.

131. Sieder, "Housing Policy, Social Welfare," 15, offers figures for 1925–27.

132. See *Der Kaisermühlner Prolet* 5 (1932) and 6 (1933), and *Döblinger Echo* 1 (1929).

133. See *Rund um die Friedrich-Kaiser-Gasse* 1 (Feb. 1933), and *Der rote Beobachter von Mödling* 1 (Oct. 1932) and 2 (April 1933).

134. See Pirhofer and Sieder, "Konstitution der Arbeiterfamilie," 332–34, and Sieder, "Housing Policy, Social Welfare," 16–20. I am enormously indebted to Sieder and Pirhofer for their exemplary scholarship in the social history of the First Republic. Their imaginative use of oral history and other sources and their subtle and imaginative interpretations meet the highest international standards.

135. Municipal intervention on health and welfare grounds was directed against working-class women in particular: as homemakers (cleanliness and neatness), as rearers of children, as setters of the moral tone and, ultimately (see ch. 6) as the emotional mainstay of her husband and children.

136. See Wegs, *Growing Up Working Class*, 63–64.

137. There was a close relationship between the Public Welfare Office and the

municipal police, which acted as a source of information and enforcement. See Beyer, "Sexualität—Macht—Wohlfahrt," 459.

138. See Wulz, *Stadt in Veränderung*, 410–11, and Sablik, *Tandler*, 269–74. Earlier that year the Municipal Health Department had taken over the cemeteries of the city as communal property.

139. The number of cremations rose from 835 in 1923 to about 4,000 yearly by 1933.

140. It was a victory nonetheless. Pirhofer calls the crematorium "a symbol of red Vienna. . . . It signified the final liberation of the body from the power of the Roman Catholic Church." See "Politik am Körper," 69.

141. The Viennese socialists were not alone in seeking to transform the worker family. In Düsseldorf, for instance, an aim of the welfare policies was to make the worker family orderly, stable, and disciplined. But nowhere more than in Vienna was the goal of socialists so comprehensively linked to a new humanity envisioned in the emerging working-class culture. See Crew, "German Socialism, Family Policy," 7, 9.

142. See Oskar Achs and Eva Tesar, "Aspekte sozialistischer Schulpolitik am Beispiel Täublers und Furtmüllers," in Helmut Konrad and Wolfgang Maderthaner, eds., *Neuere Studien zur Arbeitergeschichte* (Vienna, 1984), III: 567.

143. Stefan Zweig, *Die Welt von Gestern: Erinnerungen eines Europäers* (1942; Frankfurt/Main, 1970), 35–36.

144. See Erik Adam, "Austromarxismus und Schulreform," *Die Schul- und Bildungspolitik der österreichischen Sozialdemokratie in der Ersten Republik* (Vienna, 1983), 301. This is an exceptionally well documented and analytical study of educational reform efforts.

145. Ibid., 278.

146. See Ernst Glaser, *Im Umfeld des Austromarxismus: Ein Beitrag zur Geistesgeschichte des österreichischen Sozialismus* (Vienna, 1981), 309–16. Glaser points out (305) that this program was immediately attacked by the church and the Christian Social party. Their pamphlet *Freimauschelei: Die Ziele der Gründer und Protektoren des Vereins Freie Schule*, was an anti-Semitic diatribe linking Jews and Freemasons allied against Christianity.

147. See Czeike, *Wirtschafts/Sozialpolitik*, 271–72.

148. Otto Glöcket publicized the program in *Das Tor der Zukunft* (Vienna, 1917).

149. See Irene Wondratsch, "Schulreform und Volksbildung in Wien der Zwischenkiregszeit," *Ausstellungskatalog Kommunalpolitik*, 86.

150. See Horst Pfeiffle, "Otto Glöckels gescheiterte Schulreform: Ein Überblick," *Schulheft* 37 (1985): 4. Perhaps the most important of these smaller achievements for working-class youth was moving vocational continuing education for apprentices from evenings and Sunday mornings to workdays.

151. See Oskar Achs, "Otto Glöckel—Leben und Werk," in Oskar Achs, ed., *Otto Glöckel: Ausgewählte Schriften und Reden* (Vienna, 1985), 10–13.

152. See Otto Glöckel, *Drillschule—Lernschule—Arbeitsschule* (Vienna, 1928). The integrated health, welfare, and education approach also included hot lunches for the needy in the schools and kindergartens, and the use of school buildings as after-school centers where homework, remedial studies, shop work, and play were offered in the neighborhoods.

153. This was accomplished by creating the publishing house Deutscher Verlag für Jugend und Volk, in which the municipality was the principal stockholder. See Czeike, *Wirschafts/Sozialpolitik*, 282.

154. See Wolfgang Maderthaner, "Die Schule der Freiheit—Otto Glöckel und

die Wiener Schulreform," *Archiv: Mitteilungsblatt des Vereins für Geschichte der Arbeiterbewegung* 24:3 (July/Sept. 1984): 6.

155. For a clear statement of the original proposal, see Otto Glöckel, *Die österreichische Schulreform: Einige Feststellungen im Kampfe gegen die Schulverderber* (Vienna, 1923), 11.

156. Glöckel, *Drillschule*, passim.

157. Achs, "Glöckel," 12–13.

158. Maderthaner, "Schule der Freiheit," 9.

159. Ibid., 7–8, and especially Glaser, *Umfeld*, 309–16.

160. See Glöckel, *Österreichische Schulreform*, 18. Glöckel kept on the surplus teachers and used them to reduce average class size to thirty pupils until 1929, and thereafter to thirty-three. See Czeike, *Wirtschafts/Sozialpolitik*, 281–82.

161. See Viktor Belohoubek, *Die ersten Zehn Jahre der österreichischen Bundeserziehungsanstalten* (Vienna, 1931).

162. See Felix F. Strauss, "Schule und Heim in der Bundeserziehungsanstalt Wiener Neustadt im Rahmen der österreichischen Reformpolitik," *Brenner Schriften* (Innsbruck, 1990).

163. See Henrietta Kotlan-Werner, *Kunst und Volk: David Joseph Bach, 1874–1947* (Vienna, 1977), 174.

164. See Joseph T. Simon, *Augenzeuge: Erinnerungen eines österreichischen Sozialisten—Eine sehr persönliche Zeitgeschichte* (Vienna, 1979), 31–33.

165. Glöckel, *Drillschule*, 3. This publication of 1928 was preceded five years earlier by a pamphlet in which he already listed with pride the many foreigners who came to see the Viennese attempts at educational reform. But the interwar period was fertile in educational experiments everywhere and such interest or exchanges were the rule rather than exceptional. Glöckel, *Österreichische Schulreform*, 51. Everyone of Danneberg's reports on the accomplishments of the municipality, updated every year, contained paeans to Vienna's educational trailblazing. The socialist press abounded in similar tributes, extolling Vienna as the Mecca of innovation and reform.

166. See for instance Speiser, *Rote Wien*, 75; Achs, "Glöckel," 15; Glaser, *Umfeld*, 301–2; Ralph Grossmann and Rudolk Wimmer, *Schule und politische Bildung I: Die historische Entwicklung der politischen Bildung in Österreich* (Klagenfurt, 1979), 56–123.

167. See R. H. Samuel and R. H. Thomas, *Education and Society in Modern Germany* (London, 1949), chs. 2–3 and 7, and Peter Lundgreen, *Sozialgeschichte der deutschen Schule im Überblick, Teil II: 1918–1980* (Göttingen, 1981).

168. See John Stevenson, *British Society, 1914–1945* (Harmondsworth, 1984), 250–57, and Walford Johnson, John Whyman, George Wykes, *A Short Economic and Social History of Twentieth Century Britain* (New York, 1967), 168–76. These experiments in "progressive" education were private undertakings aimed at the middle class.

169. See Lawrence A. Cremin, *The Transformation of the School: Progressivism in American Education* (New York, 1969), ch. 6, and Rush Welter, *Popular Education and Democratic Thought in America* (New York, 1962), ch. 18.

170. See "Das 'Linzer Programm' der Sozialdemokratischen Arbeiterpartei Österreichs," in Albert Kadan and Anton Pelinka, eds., *Die Grundsatzprogramme der österreichischen Parteien: Dokumentation und Analyse* (Vienna, 1979), 86.

171. Wilhelm Weinhäupl, *Pädagogik vom Kinde aus: Viktor Fadrus—Ein Leben für die Schulreform* (Vienna, 1981), 110.

172. For the following, see Wondratsch, "Schulreform," 93–94; Pfeiffle,

"Glöckel gescheiterter Schulreform," 4; Michael Sertl, "Vom Mythos zur Wirklichkeit: Zwei Annäherungen an die Realität der zwanziger Jahre," *Schulheft,* 26–27.

173. Rumors have circulated ever since that the powerlessness revealed by the SDAP on July 15, 1927, and especially the demonstrated inability of the Schutzbund to protect either the proletariat or the republic, was interpreted in Seipel's coterie as "die Blutangst der jüdischen Sozialisten." This notion that the Jewish leaders in the SDAP were afraid of spilling blood convinced them, so the theory goes, that street politics with the use of the Heimwehr would be an effective way of checking, reducing, and ultimately removing the socialist competition.

174. For the following, see Franz Ronzal, *Schule und Kirche* (Vienna, 1926), 21–45.

175. The Christian Social party's program for 1926 called for the creation of confessional schools throughout the country that would segregate the children on the basis of religion. Until such time, the program demanded the reintroduction of religious practices (prayers, Mass, confession, etc.) in all schools. Bishops' pastoral letters were used to coerce parents into supporting the church's position on education, and Catholic action groups were organized to demonstrate opposition to secular reforms.

176. State secondary education at the time was neither free nor compulsory, but it was secular. Religious education was left to the parents entirely. One day a week was set aside for children to partake of such instruction or not. See D. W. Brogan, *The Development of Modern France, 1870–1939* (New York, 1966), 154–55. In the interwar period that "special" day was frequently used for Scout outings, sports, music lessons, domestic chores, and visits to the dentist.

177. In Weimar Germany the churches (Catholic and Protestant) succeeded in holding on to their religious education position in the schools and in the late 1920s began a rollback of earlier secular reforms. Here too the opportunity for a complete separation of church and school existed only during the first formative years of the republic. See Geoffrey Field, "Religion in the German Volksschule, 1890–1928," *Leo Baeck Institute Year Book,* 1980, 41–72.

178. See Sertl, "Mythos zur Wirklichkeit," 34–35.

179. Apparently Glöckel, who invoked Max Adler's "Neue Menschen" as the basis for his position, had completely misconstrued Adler's somewhat convoluted distinction between practical politics (to be banned from the schools) and the politics inherent in a socialist education.

180. The criticism, by the prominent socialist psychoanalyst Siegfried Bernfeld, of the reformers' adherence to a tepid liberal faith in pedagogical neutrality, in place of a proper class analysis, was particularly incisive. See Adam, "Austromarxismus und Schulreform," 283.

181. See Wegs, *Growing Up Working Class,* 95–97.

182. In a study of working-class schoolgirls done by a new-style psychologically trained teacher, nearly half the girls were kept at home regularly to do household or child-rearing chores. See Margarete Rada, *Das reifende Proletariermädchen* (Vienna, 1931), 78.

183. See Friedrich Scheu, *Ein Band der Freundschaft: Schwarzwald-Kreis und Entstehung der Vereinigung Sozialistischer Mittelschüler* (Vienna, 1985), 178.

Chapter 4

1. In recent years German historians in particular have been fixated on the theoretical distinctions among working-class party culture, worker everyday culture,

elite or dominant culture, working-class subculture, and worker culture per se. Although their often brilliant demonstrations of the manipulation of abstractions have the quality of a tour de force, they do not appear to have greatly influenced empirical work. For the best of these studies related to our subject, see Adelheid von Saldern, "Arbeiterkulturbewegung in Deutschland in der Zwischenkriegszeit," in Friedhelm Boll, ed., *Arbeiterkulturen zwischen Alltag und Politik: Beiträge zum europäischen Vergleich in der Zwischenkriegszeit* (Vienna, 1986); Dieter Langewiesche, "Politik—Gessellschaft—Kultur: Zur Problematik von Arbeiterkultur und kulturellen Arbeiterorganisationen in Deutschland nach dem 1. Weltkrieg," *Archiv für Sozialgeschichte* 22 (1982); idem., "Arbeiterkultur in Österreich: Aspekte, Tendenzen und Thesen," in Gerhard A. Ritter, ed., *Arbeiterkultur* (Königstein, 1979); Helene Maimann, "Zum Stellenwert der Arbeiterkultur in Österreich, 1918–1934," in Internationale Tagung der Historiker der Arbeiterbewegung, *Arbeiterkultur in Österreich, 1918–1934* (Vienna, 1981); and Gerd Strom, Michael Scholing, and Armin Frohmann, "Arbeiterkultur zwischen Gegenkultur und Integration: Ein Literaturbericht," *Internationale wissenschaftliche Korrespondenz zur Geschichte der deutschen Arbeiterbewegung* 22:3 (Sept. 1986).

The concept "Socialist Party culture" used here comprises the cultural activities sponsored and directed by the SDAP on behalf of the workers. Such activities involved the workers' private sphere and excluded both direct political activity and life at the workplace. A more comprehensive view of "worker culture" would include these as well as various worker subcultures. The latter, originating largely in the preindustrial period, were strongly marked by artisanal forms of production and related social structures and, more distantly, by an agrarian, Catholic-dominated milieu. The notion that a distinct worker culture existed outside the dominant bourgeois cultural mainstream is clearly rejected.

2. See Dieter Langewiesche, *Zur Freizeit des Arbeiters: Bildungsbestrebungen und Freizeitgestaltung österreichischer Arbeiter im Kaiserreich und in der Ersten Republik* (Stuttgart, 1979), 388–89, and Joseph Weidenholzer, *Auf dem Weg zum "Neuen Menschen": Bildungs- und Kulturarbeit der österreichischen Sozialdemokratie in der Ersten Republik* (Vienna, 1981), 90–91. Both the aggregate and individual organization memberships lack precision, because the *Jahrbuch der österreichischen Arbeiterbewegung* (Vienna, 1926–31) on which they are based provides sketchy and sometimes dubious data.

3. These are my adjustments from national figures which were 650,000, 520,000, and 260,000 respectively. Weidenholzer, *Auf dem Weg*, 91, and Langewiesche, *Freizeit*, 388.

4. For criticism of such bureaucratic tendencies, see Walter Fischer, "Der historische Materialismus als historische Methode," and Anni Farchy, "Zum Problem des Parteiapparates," *Der Kampf* 21 (1928): 18–24, 170–75.

5. See Ferdinand Flossmann, "Gegen die Zersplitterung der Kräfte," *Der Vertrauensmann* 2:3/4 (1926): 6.

6. Using capitalist production and the culture industry as examples, Joseph Luitpold Stern suggested similar rationalization for the party's cultural enterprises. "Rationalisierung der Arbeiterbildung," *Bildungsarbeit* 15–10 (Oct. 1928): 189–92.

7. Its director from 1918 to 1922 and 1932 to 1934 was the powerful party cultural ideologue Joseph Luitpold Stern. From 1922 to 1932 the more pragmatic party functionary Leopold Thaller was in charge. See Weidenholzer, *Auf dem Weg*, 100–101.

8. Its director (more appropriately boss) was David Joseph Bach. See Henrietta Kotlan-Werner, *Kunst und Volk: David Joseph Bach, 1874–1947* (Vienna, 1977), 68–69.

9. See Oskar Negt and Alexander Kluge, *Öffentlichkeit und Erfahrung: Zur Organisationsanalyse von bürgerlicher und proletarischer Öffentlichkeit* (Frankfurt/Main, 1972), 375–76.

10. See Viktor Adler, *Aufsätze, Reden, Briefe* (Vienna, 1902), XI: 21–23.

11. Englebert Pernerstorfer, "Die Kunst und die Arbeiter," *Der Kampf* 1 (1907): 38.

12. The earliest worker cultural organizations were virtually dominated by bourgeois liberalism. That influence extended into the republican period, with liberal teachers exercising influence over worker cultural organizations. Worker choirs often under their direction, for instance, tended to keep performance of working-class songs to a minimum. See Helmut Konrad, "Die Rezeption bürgerlicher Kultur in der österreichischen Arbeiterbewegung," in Helmut Fielhauer and Olaf Bockhorn, eds., *Die andere Kultur: Volkskunde, Sozialwissenschaft und Arbeiterkultur* (Vienna, 1982), 51–60.

13. See Kurt W. Rothschild, "Bildung, Bildungspolitik und Arbeiterbewegung," in Gerhard Botz, Hans Hautmann, Helmut Konrad, eds., *Geschichte und Gesellchaft* (Vienna, 1973), 337–43.

14. Otto Bauer's own thoughts on the subject are dispersed marginally throughout his work, mainly as the Austromarxist abstraction about raising worker consciousness and sensibility (a "revolution of souls") as part of the revolutionary process. See ch. 2.

15. For the following, see Max Adler, *Neue Menschen: Gedanken über sozialistische Erziehung* (Berlin, 1924), 29, 50–53, 66–68, 71–73, 109–10.

16. Adler attempted to clarify his ideas in a lecture delivered in Dresden in 1926, but all that emerged was a restatement: worker culture must be revolutionary and aim at a reform of consciousness. See Ernst Glaser, *Im Umfeld des Austromarxismus: Ein Beitrag zur Geistesgeschichte des österreichischen Sozialismus* (Vienna, 1981), 343–44.

17. Max Adler, *Kulturbedeutung des Sozialismus* (Vienna, 1924), 2–3.

18. Otto Neurath, "M. Adler, Neue Menschen," *Der Kampf* 18 (1925): 118–19.

19. See Klaus-Dieter Mulley, "Demokratisierung durch Visualisierung: Zur Geschichte des Gesellschafts- und Wirtschaftsmuseums in Wien," in Helmut Konrad and Wolfgang Maderthaner, eds., *Neuere Studien zur Arbeitergeschichte* (Vienna, 1984), III; Alfred Pfoser, "Das Glück in der Vernunft: Überlegungen zur Otto Neuraths Kulturtheorie und zur austromarxistischen Lebensform," in Friedrich Stadler, ed., *Arbeiterbildung in der Zwischenkriegszeit: Otto Neurath und Gerd Arntz* (Vienna, 1982), 168–72; and Glaser, *Umfeld*, 58–59.

20. Otto Neurath, *Lebensgestaltung und Klassenkampf*, "Schriftenreihe Neue Menschen," ed. Max Adler (Berlin, 1928), 5–8. His applauding of the increasing rationalization and concentration of capitalism, as advantageous to a future socialist society, resembled Kautskyan determinism on the one hand, and on the other approached the Comintern's position on fascism and capitalist concentration after 1928. See Gert Schäfer, *Die Kommunistische Internationale und der Faschismus* (Offenbach, 1973), and Theo Pirker, *Komintern und Faschismus: Dokumente zur Geschichte und Theorie des Faschismus* (Stuttgart, 1965).

21. Otto Neurath, "Bauformen und Klassenkampf," *Bildungsarbeit*, 13:4 (April 1926): 61–64.
22. Otto Neurath, "Proletarische Lebensgestaltung," *Der Kampf* 21:7 (July 1928): 318–20.
23. Ibid., 26–27.
24. See the excellent exposé of how Otto Bauer, Friedrich Adler, and other prominent leaders handled a youthful left opposition to the party's *attentisme* during the last years of the republic, forcing it to pull back from a real challenge to party policy, in Anson Rabinbach, *The Crisis of Austrian Socialism: From Red Vienna to Civil War, 1927–1934* (Chicago, 1983), chs. 5–7.
25. His report of cultural work done in the unions stresses how little time and money was spent by them for such purposes. He points with pride to fifty-three trade union newspapers with an edition of 860,000, as if these were actually read by workers who received them free as a part of union membership. See "Richard Wagner, "Unsere freigewerkschaftliche Bildungsarbeit," *Bildungsarbeit* 13:7/8 (July–Aug. 1926): 113–14.
26. Richard Wagner, "Grundfragen proletarischer Bildungsarbeit," Ibid., 10:1 (Jan. 1923): 1–2.
27. Richard Wagner, "Arbeiterbildung und Demokratie," Ibid., 18:2 (Feb. 1931): 9–11.
28. See Joseph Luitpold Stern, "Leben die Bücher bald?," *Der Sozialdemokrat: Monatschrift der Organisation Wien* 8 (1925): 56.
29. The ranting against trash and kitsch in the SDAP's publications was exposed to a devastating critique by Karl Kraus, who accused the party of hypocrisy in running advertisements for kitsch in *Die Arbeiter-Zeitung* while attacking it in its editorial columns. See Alfred Pfabigan, *Karl Kraus und der Sozialismus: Eine politische Biographie* (Vienna, 1976), 320–21.
30. David Joseph Bach, "Programm für das Jahr 1927/28," *Kunst und Gewerkschaft* 6 (1927): 1–8. See also Viktor Stein, "Betriebsräte, es geht um eine gute Sache," *Kunst und Volk* 3 (1928): 8, and Hans Tietze, "Demokratie und Kunstförderung," *Der Kampf* 26 (1933): 303–5.
31. See Joseph Luitpold Stern, *Klassenkampf und Massenbildung* (Vienna, 1924), 13–14, 29–34.
32. See Alfred Pfoser, *Literatur und Austromarxismus* (Vienna, 1980), 142.
33. See Alfred Pfoser, "Joseph Luitpold Stern und die Arbeiterbüchereien," in Institut für Wissenschaft und Kunst, *Joseph Luitpold Stern* (Vienna, 1988), 15–20. For the idealist-nationalist ideology of Langbehn, see Fritz Stern, *The Politics of Cultural Despair: A Study of the Germanic Ideology* (New York, 1965), part II.
34. David Joseph Bach, *Kunst und Volk: Eine Festgabe der Kunststelle* (Vienna, 1923), 116–17.
35. See Walter Fischer, "Aesthetische Dogmatik und proletarische Kunstpolitik," *Der Kampf* 19 (1926): 339–44.
36. See Richard Wagner, "Sozialistischer Kulturbund," *Bildungsarbeit* 13:3 (March 1926): 41–43.
37. See *Der Vertrauensmann* 8:7 (July 1932): 3.
38. After the violence of July 15, 1927, which put a peaceful democratic evolution in question, a commonly expressed opinion in circles of the Christian Social party was that the socialists were prisoners of a "Jüdische Blutangst" (Jewish fear of spilling blood). Stripped of its vicious anti-Semitism, this judgment contained an

important element of truth. Until 1934 the SDAP clung to the legal protections of the republic, which its opponents sought to dismantle by every means.

39. See the very suggestive analysis in Pfoser, *Literatur*, 113–14.
40. *Arbeiter-Zeitung*, Dec. 5, 1930, 8.
41. See Pfoser, "Joseph Luitpold Stern," 17.
42. See Alois Jalkotzky, "Die Parteipresse," *Der Kampf* 23:10 (Oct. 1930): 406.
43. *Freizeit des Arbeiters*, 121.
44. Jalkotzky, "Parteipresse," 407–10.
45. See Peter Kulemann, *Am Beispiel des Austromarxismus: Sozialdemokratische Arbeiterbewegung in Österreich von Hainfeld bis zur Dollfuss-Diktatur* (Hamburg, 1979), 22.
46. Stephan Schreder, "Der Zeitungsleser: Eine soziologische Studie mit besonderer Berücksichtigung der Zeitungsleser Wiens" (Ph.D. diss., University of Basel, 1936), cited in Langewiesche, *Freizeit*, 123–24.
47. Kulemann, *Beispiel*, 23.
48. See Alexander Potyka, "*Das kleine Blatt* (1927–1934): Ideologie und Tagesgeschehen für den kleinen Mann" (Ph.D. diss., University of Vienna, 1983), 10–11.
49. Although this was a publication for women by women, the SDAP made Max Winter the editor-in-chief to supervise the female staff.
50. See Christina Kronaus, "Zwischen Avantgarde und Gartenlaube: Literatur und Politik am Beispiel des Fortsetzungsromans in den sozialdemokratischen Medien, 1920–1934" (Ph.D. diss., University of Vienna, 1985), 190–95.
51. See Käthe Leichter, *So leben wir . . . 1320 Industriearbeiterinnen berichten über ihr Leben* (Vienna, 1932), 114.
52. Langewiesche, *Freizeit*, 121.
53. Helene Maimann, ed., *Dei ersten 100 Jahre: Österreichische Sozialdemokratie, 1888–1988* (Vienna, 1988), 351.
54. Leichter, *So leben wir*, 116.
55. "Das geistige Leben der Arbeiterjugend," *Bildungsarbeit* 20:9 (Sept. 1933): 172–73. Some respondents obviously listed more than one publication as the total is more than 100%.
56. For the following, see the Schreder study cited in Langewiesche, *Freizeit*, 124–26.
57. In a broad critique of the party's cultural efforts, Oskar Pollak complained that editorials and leaders in the party press were too difficult for popular consumption. See "Warum haben wir keine Kunstpolitik?, *Der Kampf* 22:2 (Feb. 1929): 86.
58. Even these figures would have to be reduced by 20%, if only the Vienna market was considered.
59. Kulemann, *Beispiel*, 23–24.
60. Middle-class book clubs, Pfoser observes, probably reached ten times as many worker readers as socialist ones. The same may be said about serialized fiction in the large-circulation middle-class tabloids. Pfoser, *Literatur*, 82–83.
61. See for instance *Der Metalarbeiter*, *Der Galanteriearbeiter*, and *Der Textilarbeiter* for tedious content and style.
62. Langewiesche, *Freizeit*, 122.
63. See Joseph Zech, "Zur Frage des geistigen Leben in unser Partei," *Der Kampf* 18:12 (Dec. 1925): 47.
64. For the obstacle of Viennese dialect to various kinds of worker education,

see Robert Wegs, *Growing Up Working Class: Continuity and Change Among Viennese Youth, 1890–1938* (University Park, Pa., 1989), 90–91.

65. See Weidenholzer, *Auf dem Weg,* 108–17, and Langewiesche, *Freizeit,* 282–85.

66. See Franz Senghofer, "Vom Einzelvortrag zur Vortragsreihe," *Bildungsarbeit* 12:10 (Oct. 1926): 191.

67. See Max Adler, "Wandlung der Arbeiterklasse," *Der Kampf* 26 (1933): 379–80.

68. See Weidenholzer, *Auf dem Weg,* 113–17, and Langewiesche, *Freizeit,* 285.

69. For the following, see Weidenholzer, *Auf dem Weg,* 127–53.

70. Worker proportion of the SDAP membership was 47.49%, but in the *Parteischulen* it was only 24.7–40.3%; employees were 11.8% of membership and accounted for 30.4–47.5% of the students; and civil servants were 12.3% of membership and accounted for 21–24.7 of the students. The preponderance of employees and civil servants in party schools became greater in the late 1920s and early 1930s. It revealed, among other things, the close relationship between party leadership and the employees of the socialist municipality. Ibid., 145–46.

71. On fallacious assumptions regarding the relationship between reading and informing, reception and perception of print and other media, see Michel de Certeau, *The Practice of Everyday Life,* trans. Steven F. Randall (Berkeley, 1984), ch. 12.

72. Joseph Luitpold Stern, *Handbuch für Arbeiterbibliothekare* (Vienna, 1914).

73. The list of undesirable writers included the ever-popular masters of light fiction: Karl May, Hedwig Courths-Mahler, Edgar Wallace, and Ludwig Ganghofer. But even such authors as Joseph Ferch and Hugo Bettauer, close to the socialist camp, were considered unworthy.

74. For similar campaigns against trash and kitsch and for ennoblement in Weimar Germany, see Adelheid von Saldern, "The Political Striving for 'Good Taste' and 'Good Morals' in the Weimar Republic," paper presented at the International Colloquium on Mass Culture and the Working Class, Paris, 1988.

75. Stern, "Rationalisierung," 189–90.

76. Stern, *Handbuch,* 19.

77. Ibid., 9.

78. For the following, see Pfoser, *Literatur,* 88–90, 92, 96, 107, and 109.

79. These were unpaid party cadres who took short training courses offered by the party to prepare them for their work and assure their understanding of the Bildungszentrale's aims.

80. See Langewiesche, *Freizeit,* 145–49.

81. Ibid., 160–61, 166.

82. Four-fifths of Viennese SDAP members in 1929 were workers (blue-collar workers, employees, housewives, pensioneers). One-fifth or 80,000 members were from the middle class. Only 58.6% of those who voted for the SDAP in 1932 were party members. This meant that 282,885 voters for the SDAP were not affiliated with it. See Kulemann, *Austromarxismus,* 302–3, and Alfred Georg Frei, *Rotes Wien* (Berlin, 1984), 59–60. This extension of the SDAP from its working-class base into the middle class paralleled developments in other socialist parties. For Germany, see Sigmund Neumann, *Die Parteien der Weimarer Republik* (Stuttgart, 1965), 33–36; for France, see Georges Lefranc, *Les Gauches en France* (Paris, 1965), and Eugen Weber, "Un demi-siècle de glissement à droite," *International Review of Social History* 5 (1960).

83. These 32,000 worker subscribers amounted to 8% of SDAP membership.

84. Langewiesche. *Freizeit*, 175–77.
85. For the following, see Pfoser, *Literatur*, 97–100.
86. In 1913 this last category amounted to 13%. See Langewiesche, *Freizeit*, 150, 174.
87. See Wolfgang Bass, "Die Wissenschaft in den Arbeiterbüchereinen," *Bildungsarbeit* 16 (1929).
88. See "Meistentlehnte Schriftsteller und Werke der Gruppe Dichtung in Wien, 1932," *Bildungsarbeit* 20:7 (July 1933): 144. For a very similar choice by female industrial workers, see Leichter, *So leben wir*, 117.
89. See Pfoser, *Literatur*, 138–40.
90. See Sophie Lazersfeld, "Vom Fusel zum Drama," *Kunst und Volk* 3:1 (Jan. 1928): 6–7.
91. See Otto Spranger, "Zur Intensivisierung der Bildungsarbeit," *Der Kampf* 25 (1932): 176–77.
92. See Joseph Luitpold Stern, "In Wien gibt es Bildung," *Arbeiter-Zeitung*, March 27, 1932, 7.
93. See Kotlan-Werner, *Kunst und Volk*, 69.
94. See Pfoser, *Literatur*, 60, 63.
95. See David Joseph Bach, "Warum haben wir keine sozialdemokratische Kunstpolitik?," *Der Kampf* 22:3 (March 1929): 140, and idem., "Program für das Jahr 1927/28," *Kunst und Volk* 2:6 (June 1927): 1–3.
96. See Charles A. Gulick, *Austria: From Habsburg to Hitler* (Berkeley, Calif., 1948), I: 668.
97. See David Joseph Bach, "Die Kunststelle der Arbeiterschaft," *Arbeiter-Zeitung*, Oct. 30, 1921, 7, and idem., "Zum Beginn," *Kunst und Volk* 1:1 (Feb. 1926): 1.
98. Pfoser, *Literatur*, 61–62, makes that claim, mentioning Brecht and Weil's *Dreigroschenoper*, Ernst Toller's *Hoppla, wir leben*, Sergei Tretyakov's *Brüll China*, and the works of Georg Kaiser among such experiments sponsored by the Kunststelle. His claim is repeated in the catalog *Mit uns zieht die neue Zeit* (Vienna, 1983), 138, for the exhibition of working-class culture of the First Republic in Vienna.
99. See Internationale Tagung der Historiker der Arbeiterbewegung, *Arbeiterkultur in Österreich, 1918–1934* (Vienna, 1981), 141.
100. Cited in Pfoser, *Literatur*, 63.
101. For the following, see Otto Leichter, "Warum haben wir keine sozialdemokratische Kulturpolitik?," *Der Kampf* 22:2 (Feb. 1929): 83–86.
102. See David Joseph Bach, "Der Arbeiter und die Kunst," *Der Kampf* 7:10 (Oct. 1913): 46.
103. Bach, "Warum haben wir," 139–48.
104. See Werner Jank, "Zur Arbeitermusikbewegung in der Ersten Republik," *Arbeiterkultur in Österreich*, 132.
105. See Reinhard Kannonier, *Zwischen Beethoven and Eisler: Zur Arbeitermusikbewegung in Österreich* (Vienna, 1981), 64–65.
106. Bach, "Arbeiter und Kunst," 41–42.
107. See Langewiesche, "Arbeiterkultur in Österreich," 45.
108. These programs did not differ from those of the famous Vienna Philharmonic, whose favorites were "Beethoven, Brahms, Bruckner, supplemented by Mahler, Mendelssohn, Mozart, Schubert, Schuman, Strauss, Tschaikovsky, and Wagner. Singly, Schönberg's 'Verklärte Nacht' was on the program twice in 1930." See Manfred Wagner, "Zwischen Aufbruch und Schatten," in Franz Kadrnoska, ed.,

Aufbruch und Untergang: Österreichische Kultur zwischen 1918 und 1938 (Vienna, 1981), 388–89.

109. The decided preference for German classics extended to the modernists, with Mahler, Schönberg, and von Webern represented but Stravinsky, Debussy, Prokoviev, Poulenc, Janaček, Bartók, and other foreign composers rarely performed.

110. See Kotlan-Werner, *Kunst und Volk*, 46.

111. See August Foster, "Die Transportarbeiter im ersten Arbeiter-Symphoniekonzert," *Kunst und Volk* 3:2 (Oct. 1928): 4–5.

112. See Kannonier, *Zwischen Beethoven*, 3.

113. Spontaneous music making in the tenements—singing, dance music, street musicians, instrumentals on holidays—is reported in many oral histories of workers. See Michael John, *Wohnungsverhältnisse sozialer Unterschichten im Wien Kaiser Franz Josephs* (Vienna, 1984), 199–200, and Wegs, *Growing Up Working Class*, 49–50.

114. See Henrietta Kotlan-Werner, *Otto Felix Kanitz und der Schönbrunner Kreis: Die Arbeitsgemeinschaft sozialistischer Erzieher, 1923–1934* (Vienna, 1982), 302–3.

115. See Kannonier, *Zwischen Beethoven*, 87–88.

116. See Reinhard Kannonier, "Einige Gedanken zum Begriff 'Arbeitermusikkultur,'" in Fielhauer and Bockhorn, eds., *Andere Kultur*, 46.

117. See Jank, "Arbeitermusikbewegung," 128.

118. For the following, see ibid., 129–31.

119. See Kannonier, *Zwischen Beethoven*, 133.

120. Richard Wagner, "Der Klassenkampf im Proletarierheim," *Bildungsarbeit* 13:7/8 (July–Aug. 1926): 113–15.

121. See Anna Bloch, "Zur Rundfrage—das Bild: Bilder im proletarischen Heim," ibid., 15:3 (March 1928): 50–51.

122. See Lorenz Popp, "Kunst und Proletariat: Bemerkungen über Erziehung zum Kunstgenuss in der bildenden Kunst," ibid., 13:7/8 (July–Aug. 1926): 129–30.

123. For the following, see the excellent Ph.D. dissertation of Bela Rasky, "Die Fest- und Feierkultur der sozialdemokratischen Bewegung in der Ersten Republik Österreich 1918–1934" (University of Vienna, 1985), 234–35.

124. See Friedrich Scheuer, *Humor als Waffe: Politisches Kabarett in der Ersten Republik* (Vienna, 1977).

125. For a very similar development in Weimar Germany, see von Saldern, "Political and Cultural Striving," passim.

126. See Kannonier, "Einige Gedanken," 47.

127. See Wolgang Maderthaner, "Sport für das Volk," in *Die ersten hundert Jahre*, 174.

128. See for instance Robet F. Wheeler, "Organisierter Sport und organisierte Arbeit: Die Arbeitersportbewegung," in Ritter, *Arbeiterkultur*, 59.

129. See *Die ersten Hundert Jahre*, 351, for the Viennese figures. The national total was 240,000 putting Vienna in an atypical minority position regarding organizational strength, as compared to the provinces. For all sports statistics on the national level, see Reinhard Krammer, *Arbeitersport in Österreich: Ein Beitrag zur Geschichte der Arbeiterkultur in Österreich* (Vienna, 1981), passim and especially 267–68. It should be kept in mind that the worker sports movement was developing rapidly in the interwar years, especially in Central Europe. German worker sports registered 1.5 million members (socialist and communist), and the Czechoslovakian (socialist) membership stood at 200,000 for the period under consideration. See Wheeler, "Organisierter Sport," 63–64.

130. Krammer, *Arbeitersport*, 128–30.

131. See Jacques Hannak, "Die Vertrustung der Leibesertüchtigung," *Der Kampf* 13:4 (April 1920): 159.

132. See Maderthaner, "Sport," 175, and Krammer, *Arbeitersport*, 118–19.

133. Ibid., 82–89. The soccer players joined ASKÖ only in 1926, after Vienna's championship workers' team Rapid-Hütteldorf had become professional and left the fold. See Maderthaner, "Sport," 175.

134. See Wheeler, "Organisierter Sport," 69.

135. See Krammer, *Arbeitersport*, 73.

136. See Julius Deutsch, *Sport und Politik: Im Auftrage der sozialistischen Arbeiter-Sports-Internationale* (Berlin, 1928), 23–25.

137. See Julius Deutsch, *Unter roten Fahnen: Vom Rekord zum Massensport* (Vienna, 1931), 12.

138. See Jacques Hannak, "Die neue Grossmacht—Sport: Zum Arbeiter- Turn- und Sportfest in Wien," *Der Kampf* 19:7 (July 1926): 273–80.

139. See Hans Gastgeb, "Bildung und Sport der Arbeiterklasse," *Bildungsarbeit* 14:7/8 (July–Aug. 1927): 121–22. The pseudoreligious admonitions here and elsewhere—as in the case of Bach, who directed librarians to become "moral confessors" of their readers—is a dimension of SDAP leadership mentality easily overlooked.

140. See Jacques Hannak, "Sport und Kunst," *Kunst und Volk* 2:7 (Oct. 1927): 2–3.

141. Cited in Langewiesche, *Freizeit*, 382.

142. See Deutsch, "Unter roten Fahnen," 6.

143. See "Panem et Circenses," *Der Kampf* 26:1 (Jan. 1926): 38.

144. With the increasing sports reportage in *Das kleine Blatt* and *Kuckuck*, the emphasis on winning and breaking records and the number of spectators—otherwise condemned as demeaning concerns of bourgeois sports—acquired major importance. But even the more staid *Arbeiter-Zeitung* proudly announced on August 27, 1931, that Austria had been the victor in the International Workers' Olympics in July.

145. See Krammer, *Arbeitersport*, 159, 177.

146. See Wheeler, "Organisierter Sport," 68.

147. An important aspect of the emerging discipline structure was the use of a military-derived vocabulary for exercise commands and for describing the role and activities of worker athletes. For the parallel development in Weimar Germany, see Peter Friedemann, "Die Krise der Arbeitersportbewegung am Ende der Weimarer Republik," in Friedhelm Boll, ed., *Arbeiterkulturen zwischen Alltag und Politik: Beiträge zum europäischen Vergleich in der Zwischenkriegszeit* (Vienna, 1986), 232–40.

148. For the following, see Krammer, *Arbeitersport*, 122, 125–26, 192–200, 214. Deutsch had championed military preparedness as early as 1924.

149. The depoliticizing of ASKÖ followed on the heels of identical measures in the Schutzbund. Julius Deutsch was the linking figure between the two. See Ilona Duczynska, *Workers in Arms: The Austrian Schutzbund and the Civil War of 1934* (New York, 1978).

150. See Maderthaner, "Sport," 177.

151. Illustrated publications of the party such as *Kuckuck* and *Das kleine Blatt*, as well as the poster art used from the late 1920s on, prominently featured the body of the worker athlete.

152. See the suggestive analysis by Gerhard Hauk, "'Armeekorps auf dem Weg zur Sonne'—Einige Bemerkungen zur kulturellen Selbstdarstellung der Arbeiter-

bewegung," in Dietmar Petzina, ed., *Fahnen, Fäuste, Körper: Symbolik und Kultur der Arbeiterbewegung* (Essen, 1986), 79–86.

153. For the political paralysis of the SDAP, see Rabinbach, *Crisis*, ch. 4.

154. See Krammer, *Arbeitersport*, 222–25, and also *Festschrift zur 2. Arbeiterinternationale* (Vienna, 1931).

155. In response to the request for such a stadium by Deutsch and Tandler, Mayor Seitz and Finance Councillor Breitner pushed the necessary appropriation through the municipal council. No doubt the customary socialist reservations about spectator events were overcome in anticipation of the symbolic importance of the event. See Hans Gastgeb, *Vom Wirtshaus zum Stadium: 60 Jahre Arbeitersport in Österreich* (Vienna, 1952), 68–70.

156. See *Das kleine Blatt,* July 27, 1931, 1–2.

157. Quoted in *Mit uns zieht die neue Zeit,* 96.

158. See Alfred Pfoser, "Massenästhetik, Massenromantik, Massenspiel: Am Beispiel Österreichs—Richard Wagner und die Folgen," *Das Pult* 66 (1982): 64.

159. For this and the following I am much indebted to Rasky, *Arbeiterfesttage,* 121–22, 131–32.

160. Thus the Sunday feast took the place of Sunday mass, a spring festival supplanted Corpus Christi, and winter solstice replaced Christmas. A consecration of youth (puberty rite) was established in place of first communion and confirmation. See "Wir müssen die Kirchenfeste überwinden," *Bildungsarbeit* 13:1 (Jan. 1926): 32. To these were added special labor holidays such as Mayday, a March holiday to commemorate the revolution of 1848, the founding of the republic on November 12, and later the commemoration of the victims of July 15, 1927. There were similar efforts to substitute secular for religious celebrations in France. In the anticlerical Parisian worker suburb of Bobigny, "red baptisms" were celebrated publicly to counteract attempts by the Catholic church to rechristianize the community. See Tyler Stovall, "French Communism and Suburban Development: The Rise of the Paris Red Belt," *Journal of Contemporary History* 24:3 (July 1989): 447.

161. Rasky, *Arbeiterfesttage,* 162.

162. See Pfoser, "Massenästhetik," *Pult,* 60.

163. For the origins of politics as an aesthetic-emotional experience in late-nineteenth-century Austria, see Carl E. Schorske, *Fin-de-Siècle Vienna: Politics and Culture* (New York, 1981), and William J. McGrath, *Dionysian Art and Populist Politics in Austria* (New Haven, Conn., 1974). These origins in Central Europe are traced back even further by George L. Mosse, *Nationalization of the Masses: Political Symbolism and Mass Movements in Germany from the Napoleonic Wars Through the Third Reich* (New York, 1975).

164. See for instance Joseph Luitpold Stern, "Functionärschulen für Feiern," *Kunst und Volk* 17 (1930): 8–9.

165. See Pfoser, "Massenästhetik," *Pult.* 67–70. In Hermann Bahr's novel *Österreich in Ewigkeit* (1929) the suggestion is made that "Bolshevik Vienna" be abandoned altogether and that "only a Vendée can save Austria." Ibid., 70.

166. See for instance Wilhelm Ellenbogen, "Richard Wagner und das Proletariat," *Der Kampf* 7:6 (May 1913): 41–43, and David Joseph Bach, "Der Arbeiter und die Kunst," ibid., 7:10 (Oct. 1913): 41–46.

167. See Richard Lorenz, ed., *Proletarische Kulturrevolution in Sowjetrussland* (Munich, 1969), 12–13, 163–71.

168. See Alfred Pfoser, "Massenästhetik, Massenromantik und Massenspiel," *Arbeiterkultur in Österreich,* 126.

169. See Mosse, *Nationalization*, ch. 7.
170. See Fritz Rosenfeld, "Gedanken zum Sprechchor," *Der Kampf* 19:2 (Feb. 1926): 85–86.
171. See Elisa Karau, "Was ist Sprechchor?," *Bildungsarbeit* 13:1 (Jan. 1926): 11.
172. See Wolfgang Schumann, "Die Sprechchorbewegung," *Kunst und Volk* 2:6 (Sept. 1927): 9–11.
173. See Rasky, *Arbeiterfeste*, 179–82, and "Saure Wochen, Frohe Feste," in *Berichte zur Kultur- und Zeitgeschichte* (Vienna, 1931/1932), 70–76.
174. This pageant closely resembled one offered at the International Worker Olympics at Frankfurt in 1924. Devised by Alfred Auerbach and entitled "Struggle for the World," it traced "the struggle through history between the masses and their rulers." See James Wickham, "Working-Class Movement and Working-Class Life: Frankfurt am Main during the Weimar Republic," *Social History* 8:3 (Fall 1983): 338–39.
175. *Das kleine Blatt* and *Der Kuckuck* offered huge photo essays; *Die Arbeiter-Zeitung* (July 19, 1931, 8), rhapsodized about the festival's power to enchant the senses while inflaming the convictions, worldview, and soul of the audience. But the stadium, as a municipal institution, served other purposes, such as almost weekly football matches. In September 1933, Catholic Day was celebrated there with a mass festival in which 8,000 Catholic youth performed a religious play before an audience of 60,000 which collectively intoned the Te Deum for the closing. See Pfoser, "Massenästhetik," *Arbeiterkultur*, 126.
176. See the interpretation and powerful critique of SDAP festival culture by Roberto Cazzola, "Die proletarischen Feste: Zwischen revolutionärer Propädeutik und ästhetischem Ritualismus," *Wiener Tagebuch* 4 (April 1981): 18–20. "While the workers were toppling the idol," he observes, "reactionaries were working to topple the republic."
177. Pfrimer had announced that he was taking control of Austria in the name of the Heimwehr. But the fact that the latter did not support his band of adventurers and their foolish theatrics led many SDAP leaders to underestimate the seriousness of the incident. See Gerhard Botz, *Gewalt in der Politik: Attentate, Zusammenstösse, Unruhen in Österreich 1918 bis 1934* (Munich, 1976), 181–86.
178. Cazzola, "Proletarische Feste," 20. In his brilliant and prescient essay "The Mass Ornament," Siegfried Kracauer had warned in 1927 that "production and consumption of the ornamental patterns divert from the necessity to change the current order. Reason is impeded when the masses into which it should penetrate yield to emotions provided by the godless mythological cult. Its social meaning is much like that of the Roman *circus games* sponsored by tyrants." See *New German Critique* 5 (Spring 1975): 75.
179. See Fini Samek-Klupp, "Festspiele für grössere Jugendfeier," *Sozialistische Erziehung* 10 (1930): 94.
180. For the following, see Rasky, *Arbeiterfesttage*, 183, 225–26, 384.
181. See Anson Rabinbach, "Ernst Fischer and the Left Opposition in Austrian Social Democracy: The Crisis of Austrian Socialism, 1927–1934" (Ph.D. diss., University of Wisconsin, 1973), ch. 3. See also Rasky, *Arbeiterfesttage*, 227–29, and Maderthaner, "Sport," 177.
182. Like many other aspects of indigenous worker culture, the traditional worker feast continued to be practiced at the local level. There, even political celebrations were forced to include features of popular entertainment. Such feasts continued to resist being turned into demonstrations. Locally, summer feasts and bath-

ing expeditions were particularly popular among youth organizations. See Rasky, *Arbeiterfesttage*, 382–83.

183. Pfoser calls it a play ritual of aesthetic arrangements. See *Literatur*, 75.

184. It is interesting to note that the three socialist symbols were imported from Weimar Germany: the salute from the communist Rot-Frontkämpfer-Bund and the emblem and greeting from the socialist Eiserne Front. See Gottfried Korff, "Rote Fahnen und geballte Faust: Zur symbolik der Arbeiterbewegung in der Weimarer Republik," in Petzina, ed., *Fahnen*, 34–47.

185. Helmut Konrad suggests that these young people were probably closer to the ideal of "neue Menschen" than the athletes of ASKÖ. See "Foreword," Krammer, *Arbeitersport*, viii.

186. In Weimar Germany there was a similar popular resistance to an all-encompassing socialist party culture. "Beneath the formalization of the labor movement culture," Geoff Ely observes, "was a popular culture that remained relatively impermeable to the former's attractions and rationalizing methods.... The distance between the formal and quotidien cultures was also reproduced *inside* the labor movement itself, because the former neglected whole dimensions of experience— 'a broad spectrum of expectations, anxieties, and hopes,' or the contradictory fullness of the working-class 'lifeworld'—even of its own card-carrying members." See "Labor History, Social History, *Alltagsgeschichte:* Experience, Culture, and Politics of the Everyday—A New Direction for German Social History?," *The Journal of Modern History* 61:2 (June 1989): 311–12.

187. This was a familiar problem for Socialist parties in the interwar years. The mercurial rise in membership of the French SFIO following the strike wave of 1936 could not be accommodated by the existing party structure, and leaders of the individual federations were not inclined to find or support new ways of reaching the majority of uninitiated workers. See Helmut Gruber, *Léon Blum, French Socialism, and the Popular Front: A Case of Internal Contradictions* (Ithaca, N.Y., 1986), 5–7, 14–16, 52.

188. Max Adler accused party managers and bureaucrats of having developed a mentality of ownership over their cultural preserves. See "Wandlung der Arbeiterklasse," 372–80. Käthe Leichter warned that the party elite reached by the cultural program was in danger of becoming a social elite as well, and complained that the party had completely neglected the large number of unemployed. See "Bildungsarbeit für Arbeitslose," *Bildungsarbeit* 19 (1932): 220.

189. See for instance Felix Kanitz, "Individualpsychologie in der Arbeiterbewegung," *Bildungsarbeit* 14:10 (Oct. 1927); Erwin Wexberg, "Alfred Adler's Individualpsychologie und die sozialistische Erziehung," *Die sozialistische Erziehung* 4:12 (Dec. 1924); Paul Lazersfeld, "Marxismus und Individualpsychologie," ibid., 7:5 (May 1927); "Individualpsychologie und Sozialismus," ibid., 7:11 (Nov. 1927); Kotlan-Werner, *Kanitz*, 189–90; Glaser, *Umfeld*, 273–87.

190. The SDAP never felt comfortable with the work of Sigmund Freud, praising it faintly in its publications while keeping its distance from it. In all likelihood this was a response to his pessimism about the ability of social or economic transformation to ameliorate the painful transition from the "pleasure to the reality principle" or to make humans more happy or fulfilled. The human costs in the avoidance of pain were made all too clear in Freud's *Das Unbehagen in der Natur*, published in 1930, whose first edition of 12,000 was sold out that year. See Johannes Reichmayr and Elisabeth Wiesbauer, "Das Verhältnis von Sozialdemokratie und Psychoanalyse in Österreich zwischen 1900 und 1938," in Wolfgang Huber, ed., *Beiträge zur Ge-*

schichte der Psychoanalyse in Österreich (Vienna, 1978), and Peter Gay, *Freud: A Life for Our Time* (New York, 1989), 543–53. For the SDAP's attitude toward the work of Wilhelm Reich, see ch. 6.

191. It has been suggested that this dismal image of the worker reflected the latent fear of the proletariat by intellectuals. See Hauk, "Armeekorps," 79.

192. See *Des Menschen hohe Braut: Arbeit, Freizeit, Arbeitslosigkeit*, Franz Kreuzer in conversation with Marie Jahoda fifty years after the inquiry *Die Arbeitslosen von Marienthal* (Vienna, 1983), 7–8.

Chapter 5

1. It is important to remember that the party was unable to affect the workplace, the one area crucially important in establishing both possibilities and limits in the workers' public and private sphere. The trade unions' virtual impotence in the face of continuous depressed economic conditions and intransigent owners protected by the Christian Social national government gave an artificial cast to the entire Viennese socialist experiment. See Gerhard Botz, "Streik in Österreich 1918 bis 1975: Probleme und Ergenbnisse einer quantitativen Analyse," in Gerhard Botz et al., *Bewegung und Klasse: Studien zur österreichischen Arbeitergeschichte* (Vienna, 1978), 807–20, and Dieter Stiefel, *Arbeitslosigkeit: Soziale, politische und wirtschaftliche Auswirkung— am Beispiel Österreichs, 1918–1938* (Berlin, 1979).

2. For a unique and brilliant analysis of the emblematic qualities of mass culture (first published in the *Frankfurter Zeitung* in 1927), and the extent to which its surface manifestations reveal the deep qualities of an epoch, see Siegfried Kracauer, "The Mass as Ornament," trans. Barbara Correll and Jack Zipes, *New German Critique* 5 (Spring 1975): 67–76.

3. For a sophisticated discussion of the place and power of mass culture in modern society, see Michael Denning, "The End of Mass Culture," and the related critiques of his position in *International Labor and Working-Class History* 37 (Spring 1990), as well as Denning's reply in the same journal, 38 (Fall 1990). For the role of commercial culture in the creation of worker traditions, see Eric Hobsbawm, "Mass-Producing Traditions: Europe, 1870–1914," in Eric Hobsbawm and Terence Ranger, eds., *The Invention of Tradition* (Cambridge, 1983).

4. For worker reaction to commercial culture in the late nineteenth century, see Peter Bailey, *Leisure and Class in Victorian England, 1830–1855* (London, 1978), and Gareth Stedman Jones, "Working Class Culture and Working Class Politics in London, 1870–1890," *Journal of Social History* 9 (Summer 1974). For the relationship of popular culture to the above, see Gareth Jowett, "Towards a History of Popular Culture," *Journal of Popular Culture* 9 (1975): 2.

5. On the relationship of working-class party cultures, worker subcultures, and elite culture and the role of socialist parties in the culture realm, see Stuart Hall, "Notes on Deconstructing 'the Popular':" in Ralph Samuel, ed., *People's History and Socialist Theory* (London, 1981), and Brigitte Emig, *Die Veredelung des Arbeiters: Sozialdemokratie als Kulturbewegung* (Frankfurt/New York, 1980). For a more theoretical consideration of the above, see Umberto Eco, *Apokalyptiker und Integrierte: Zur kirtischen Kirtik der Massenkultur* (Frankfurt, 1984).

6. Leading traditional critics from the right and left are united in their overestimation of mass culture's rise to dominance. See Jose Ortega y Gasset, "The Coming of the Masses," and Dwight Macdonald, "A Theory of Mass Culture," in Bernard

Rosenberg and David M. White, eds., *Mass Culture: The Popular Arts in America* (New York, 1957). In their desire to condemn mass culture, they overlook the struggle it had to wage against older commercial and noncommercial forms of leisure-time activities, and the length of time during which the latter were able to adapt and defend themselves.

7. Even the socialist municipal administration reduced the workweek in public utilities from 52–54 to 48 hours only in 1927. See Peter Kulemann, *Am Beispiel des Austromarxismus* (Hamburg, 1979), 346. But the average reduction of the workweek from over 60 hours before 1914 to under 50 hours meant a gain of 10 hours of free time a week for the worker.

8. See the superior Ph.D. dissertation by Ulrike Weber, "Wirtschaftspolitische Strategien der Freien Gewerkschaften in der Ersten Republik: Der Kampf gegen die Arbeitslosigkeit," (University of Vienna, 1986), 178–90.

9. See Hans Safrian, "'Wir ham die Zeit der Orbeitslosigkeit schon richtig genossen auch': Ein Versuch zur (Über-)Lebensweise von Arbeitslosen in Wien zur Zeit der Weltwirtschaftskrise um 1930," in Gerhard Botz and Josef Weidenholzer, eds., *Materialien zur Historischen Sozialwissenschaft: Mündliche Geschichte und Arbeiterbewegung* (Vienna, 1984).

10. The following account of commercial culture and the Viennese working class is based on three ground-breaking articles by Joseph Ehmer: "Vienna, anni settanta: osterie, struttura della classe operaia e cultura politica del movimento," *Movimento Operaio et Socialista* 8:1 (1985); "Vaterlandslose Gesellen und respektable Familienväter: Entwicklungsformen der Arbeiterfamilie im internationalen Vergleich, 1850–1930," in Helmut Konrad, ed., *Die deutsche und die österreichische Arbeiterbewegung sur Zeit der Zweiten Internationale* (Vienna, 1982); and "Rote Fahnen—blauer Montag: Soziale Bedingungen von Aktions- und Organisationsformen der frühen Wiener Arbeiterbewegung," D. Puls and E. P. Thompson, eds., *Wahrnemungsformen und Protestverhalten* (Frankfurt, 1979).

11. Whereas men ate more substantial meals in the *Gasthaus* and remained there for the duration of their lunch hour, making it their social domain, women consumed a generally meager lunch of leftovers at the workplace, which also served as the setting of their social networks. See Käthe Leichter, *So leben wir . . . 1320 Industriearbeiterinnen berichten über ihr Leben* (Vienna, 1932), 80.

12. In its multiple functions the *Gasthaus* was analogous to the English pub and French café, as a central locale of working-class sociability.

13. See Alfred Frei, *Rotes Vienna: Austromarxismus und Arbeiterkultur* (Berlin, 1984), 108–15.

14. "Vienna, anni settanta," 21–22.

15. *Handbuch der Gemeinde Wien* (Vienna, 1935), 122.

16. See for instance Emanuel Häussler, "Heurigendämmerung?," *Neues Wiener Tageblatt*, Sept. 25, 1932.

17. For the following, see Hans Pemmer and Nini Lackner, *Der Prater: Von den Anfängen bis zur Gegenwrt* (Vienna, 1974), and Bertrand M. Buchmann, *Der Prater: Die Geschichte des unteren Werd* (Vienna, 1979).

18. See Benedikt Kautsky, *Die Haushaltstatistik der Wiener Arbeiterkammer, 1925–1934*, supplement of *International Review of Social History* 2 (1935): 245–46, and Fritz Klenner, *Die österreichischen Gewerkschaften* (Vienna, 1953), II: 892–94.

19. None of the various official statistical yearbooks yields information about the unit cost of commercial culture products. Nor do the sparse records of trade unions or proprietor associations in the food or alcohol trades shed any light on the sub-

Notes to Pages 118–121   229

ject.One is thrown back on the vague memories of the aged for an approximation of prices of common consumer articles and services: 10 cigarettes cost 30 Groschen; a glass of beer, 10–20 Groschen; a sausage, 20 Groschen; an ice cream cone, 30 Groschen; circus and variety admission, 50 Groschen to 5 Schillings. The price structure between 1925 and 1933 remained fairly stable, although family wages declined after 1930 owing to massive unemployment and wage cuts.

20. For the following, see Berthold Lang, "Zirkus und Kabarett," in Franz Kadrnoska, ed., *Aufbruch und Untergang: Österreichische Kultur zwischen 1918 und 1938* (Vienna, 1981); Österreichisches Circus- und Clown-Museum, *Circus und Varieté in Wien 1918 bis 1938* (Vienna, 1980); idem., *Der österreichische Circus* (Vienna, 1978); idem., *Unterhaltungskunst in Wien um 1900* (Vienna, 1979); Rudolf Weys, *Cabaret und Kabarett in Wien* (Vienna, 1970); Ernst Günther, *Geschichte des Varietés* (Berlin, 1978); and Felix Czeike, *Das grosse Groner Wien Lexikon* (Vienna, 1974). I am also indebted to Mr. Berthold Lang, director of the Österreichisches Circus- und Clown-Museum, for allowing me to sample the museum's vast collection of circus and Varieté posters and memorabilia, and for information not available in print.

21. Lang, "Zirkus," 305, 307–8.

22. *Der Kuckuck* 2:20 (May 18, 1930): 8; *Der österreichische Circus*, 14.

23. Pemmer and Lackner, *Prater*, 95.

24. *Circus und Varieté in Wien*, 5–6.

25. Circuses were generally limited to five warm-weather months and performed on weekends, when four or five performances were given.

26. Some of the most famous permanent prewar circus structures, such as the Busch, were converted into movie theaters as early as 1920. See Lang, "Zirkus," 309; *Circus Gestern und Heute: Mitteilungsblatt der Gesellschaft der Freunde des Österreichischen Circusmuseums* 3 (March/April 1982): 9.

27. Günthers, *Geschichte Varietés*, introduction.

28. Virtually all the great waltz conductors, including the scions of the Strauss family, made a point of taking the baton of *Varieté* orchestras. The same was true for composers of operettas, such as Robert Stoltz and Ralph Benatzky.

29. *Circus und Varieté in Wien*, 7. *Varieté* made a great effort to advertise. Borrowing the techniques of the American circus impresario P. T. Barnum, it developed the sensational poster to a fine art and displayed it widely throughout the city.

30. Lang, "Zirkus," 308; *Cirkus und Varieté in Wien*, 9.

31. *Circus und Varieté in Wien*, 9–10; Lang, "Zirkus," 308–9.

32. *Circus und Varieté in Wien*, 7–10; *Unterhaltungskunst Wien*, 5, 7.

33. These included Varieté Westend, Margaretner Orpheum, Brigittinauer Orpheum, Favoritner Kolosseum, Tivoli Varieté, Steiner's Künstlerspiele, Rosensäle, and Metropol-Varieté. See Lang, 309; *Unterhaltungskunst Wien*, 18.

34. Lang, "Zirkus," 303.

35. See Fritz Klingenbeck, *Unsterblicher Walzer* (Vienna, 1943), and Robert Wegs, *Growing Up Working Class: Continuity and Change Among Viennese Youth, 1890–1938* (University Park, Pa., 1989), 119.

36. See Hans Safrian and Reinhard Sieder, "Gassenkinder, Strassenkämpfer: Zur politischen Sozialisation einer Arbeitergeneration in Wien 1900 bis 1938," in Lutz Niethammer and Alexander von Plato, eds., *Wir kriegen jetzt andere Zeiten* (Berlin, 1985); Reinhard Sieder, "'Vater, derf i aufstehn?': Kindheitserfahrungen in Wiener Arbeiterfamilien um 1900," in Hubert Ch. Ehalt, Gernot Heiss, and Hannes Stakl, eds., *Glücklich ist wer vergisst . . . das andere Wien um 1900* (Vienna, 1986); and Wegs, *Growing Up Working Class*, 68–73, 119–20.

37. Such play among boys sometimes became rough, almost ganglike struggles over territory. See Karl Klein, "Mit der Hollergasse gegen die Anschützgasse," in Heinz Blaumeiser et al., *Ottakringer Lesebuch: Lebensgeschichten* (Vienna, 1988), 44–45. Working-class children in Weimar Germany also regarded the street as their "home"—a territory that was liberating and gave free reign to a variety of uses and self-expression. See Detlev J. K. Peukert, *Jugend zwischen Krieg und Krise: Lebenswelten von Arbeiterjungen in der Weimarer Republik* (Cologne, 1987), 77–82.

38. Although the worker weekend began at noon on Saturday, married women workers were engaged in catch-up household chores until Sunday afternoon, when they too were at leisure. Single female workers more often had more time for hiking and other recreational activities. But for both, a trip into nature was a preferred form of release from the strains of the workweek. See Leichter, *So leben wir,* conclusion and appendix of case histories. A similar near reverence for nature (comparing it to liberation from the prison of factory life) can be found in a study of German female textile workers. See Deutscher Textilarbeiterverband, *Mein Arbeitstag—Mein Wochenende: 150 Berichte von Textilarbeiterinned* (Berlin, 1930), passim.

39. The noncommercial recreation of hiking and rambling was never quite out of touch with commercial culture. The satisfied but weary wanderer found ample opportunity to refresh himself at various *Gasthäuser* and *Heurigen* which lay not far from his chosen path. For an analysis of the significance of nature for workers in Weimar Germany, see Kaspar Maase, *Leben einzeln und frei wie ein Baum und brüderlich wie ein Wald: Wandel der Arbeiterkultur und Zukunft der Lebensweise* (Cologne, 1987), 55–56.

40. The municipal Kongressbad, opened in 1928, drew 448,555 paying bathers in the 1930 season. See Hans Hovorka, *Republik "Konge": Ein Schwimmbad erzählt seine Geschichte* (Vienna, 1988), 71.

41. Nude swimming, however, was forbidden by a law which was sometimes enforced. It took place, nevertheless, on several small islands in the Danube not easily accessible to the police. Lobau bathing received much coverage in the popular socialist press, particularly the illustrated *Kuckuck,* which had a penchant for nudity or seminudity. See the full-page spread on August 21, 1932, for instance.

42. See Fritz Keller, ed., *Lobau—die Nackerten von Wien* (Vienna, 1985), and Safrian, "'Wir ham die Zeit der Orbeitslosigkeit."

43. See Reinhard Krammer, *Arbeitersport in Österreich* (Vienna, 1981), vii–viii. Although Krammer's work is devoted to recounting the history of the SDAP's sport organization (ASKÖ), both in the introduction and conclusion the continued importance of unorganized and spontaneous sports is emphasized.

44. For a parallel situation in Weimar Germany, see Peukert, *Jugend,* 232–33.

45. *Handbuch der Gemeinde Wien* (Vienna, 1932). Official statistics of such "legal" garden plots did not include thousands of others created on waste land proximate to the city limits which were unauthorized but nonetheless tolerated. In 1918 there had been about 150,000 of these suburban plots of 100–300 square meters. See Hans Hautmann, "Hunger ist ein schlechter Koch: Die Ernährungslage der österreichischen Arbeiter im Ersten Weltkrieg," Botz, ed., *Bewegung und Klasse,* 670. For a photo essay of the garden plots as well as the laws governing their rental, see *Nach der Arbeit: Bilder und Texte zur Freizeit, 1870–1950* (Vienna, 1987), 7–11, 42–43.

46. Ironically, the galloping unemployment of the depression years created an even larger amount of worker "leisure," which the mass culture industries were able to exploit despite widespread impoverishment.

47. See the irony bordering on contempt of the trade union educator Richard

Wagner, "Klassenkampf im Proletarierheim," *Bildungsarbeit: Blätter für sozialistisches Bildungswesen* 13: 7–8 (1926).

48. See Edmund Reismann, "Bürgerliche und proletarische Vergnügungen," *Sozialistische Erziehung (Die Praxis)* 8:10 (Oct 1929): 227–28.

49. M. Feldman,"Bürgerliche und proletarische Vergnügungen," ibid., 8:11 (Nov. 1929): 247–48.

50. Gerta Morberger, "Bürgerliche und proletarische Vergnügungen," ibid., 8:12 (Dec. 1929): 274.

51. Trötzmüller, "Wir und das Tanzen," ibid., 9:1 (Jan. 1930): 22–23.

52. "Moderner Tanz und Massenorganisation," ibid., 9:2 (Feb. 1930): 42.

53. Karl Czernetz, "Tanzen?," ibid., 9:3 (March 1930): 68–69.

54. What the SDAP leaders failed to recognize was that their repeatedly invoked need for abstinence from pleasure kept working-class youth from joining the SAJ. Its Viennese membership in 1932 was 10,490, or 2.6 percent of the SDAP total. But the cohort of 18- to 20-year-olds in the population alone constituted 3.07 percent, whereas the SAJ recruited from 14- to 21-year-olds. See Käthe Leichter, "Die Struktur der Wiener Sozialdemokratie," *Der Kampf* 25:6 (June 1931): 262–67, and Wolfgang Neugebauer, *Bauvolk der kommenden Welt: Geschichte der sozialistischen Jugendbewegung in Österreich* (Vienna, 1975), 138–40.

55. For the SDAP's Adlerianism, see Ernst Glaser, *Im Umfeld des Austromarxismus: Ein Beitrag zur Geistesgeschichte des österreichischen Sozialismus* (Vienna, 1981), 273–87. For the relationship between psychoanalysis and the SDAP, see Elizabeth Wiesbauer and Johannes Reichmayr, "Das Verhältnis der Psychoanalyse zu der Sozialdemokratie," in Wolfgang Huber and Erika Weinzierl, eds., *Beiträge zur Geschichte der Psychoanalyse in Österreich* (Vienna, 1978); Michael Pollak, "Intellektuelle Aussenseiterstellung und Arbeiterbewegung: Das Verhältnis der Psychoanalyse zur Sozialdemokratie zu Beginn des Jahrhunderts," in Botz, ed., *Bewegung und Klasse;* and Karl Fallend, "Psychoanalyse und Politik im Wien der zwanziger Jahre: Wilhelm Reich—Dozent der Psychoanalyse, Sexualberater und rebellischer Parteigenosse" (Ph.D. diss., University of Salzburg 1987).

56. For the history of Austrian film, see Walter Fritz, *Kino in Österreich, 1896–1930* (Vienna, 1981); idem., "Glanz und Elend des Spielfilms in der Ersten Republik," *Der österreichische Film in der Ersten Republik* (Vienna, 1968); Franz Grafl, "Hinein in die Kinos!: Ein Beitrag zur Aufarbeitung der österreichischen Arbeiterfilmbewegung, 1918–1934," in Kadrnoska, ed., *Aufbruch;* and Ernst Glaser, "Die Stummfilmkritik unter besonderer Berücksichtigung des Sowjetfilms," typescript, Allgemeines Verwaltungsarchiv des österreichischen Staatsarchiv. For the most analytic treatment, see the unpublished manuscript of Theodor Venus, "'Hinein in das Kino!': Sozialdemokratische Film- und Kinopolitik von 1918 bis 1934," In Österreichische Gesellschaft für Kulturpolitik, "Arbeiterkultur in Österreich, 1918–1934: Endbericht" (available at the Verein für die Geschichte der Arbeiterbewegung in Vienna).

57. Fritz Rosenfeld, "Der Wiener Film von Gestern und Morgen," *Arbeiter-Zeitung*, Feb. 11, 1932.

58. Stabilization of the currency put an end to the capitalization of film production based on the repayment of loans with a constantly inflated currency.

59. A total of 59 Austrian sound films were produced between 1929 and 1934. See Walter Fritz, *Die österreichischen Spielfilme der Tonfilmzeit (1929–1938)* (Vienna, 1968), 1–59. Of the 484 foreign films imported in 1928, for instance, 240 were produced in the United States, 210 in Germany, 14 in France, and a few in England, the

U.S.S.R., and other countries. See "Österreichs Filmproduktion," *Neues Wiener Tageblatt*, November 10, 1928, and Glaser, "Stummfilmkritik," 3–4. The introduction of sound film increased the number and ratio of American imports. For the profound impact of American film on European cinema, see Victoria deGrazia, "The American Challenge to European Cinemas, 1920–1960," *Journal of Modern History* 61 (March 1989).

60. After the restructuring of the German and British industries, huge conglomerates such as Universal Film (UFA) and the Korda (London Films) and Balcon (Gaumont British) organizations became dominant. In France the collapse of Pathé and Gaumont left a vacuum filled by dozens of small companies, often producing only one or two films, centered around particular internationally acclaimed directors. For French and German film, see Paul Monaco, *Cinema and Society: France and Germany during the Twenties* (New York, 1973); Museum of Modern Art, *Fifty Years of French Film* (New York, 1985); Paul Leglise, *Histoire de la politique du cinéma français: le cinéma et la IIIe République* (Paris, 1970); Siegfried Kracauer, *From Caligari to Hitler: A Psychological Study of the German Film* (New York, 1956); Thomas G. Plummer et al., *Film and Politics in the Weimar Republic* (New York, 1982); and Anton Kaes, Kino-Debatte (Munich, 1978). For Britain, see Charles Barr, ed., *All Our Yesterdays: 90 Years of British Cinema* (London, 1986), and James Curran and Vincent Porter, eds., *British Cinema History* (Totowa, N.J., 1983).

61. In France neither the SFIO nor the Confédération Générale du Travail (CGT) had a comprehensive cultural program. Yet it was the CGT which in 1938 commissioned Jean Renoir and the cinema collective Groupe Octobre to make "La Marseillaise," a magnificent tableau calling for the defense of the republic against its internal and external enemies. More than 60% of the financing came from 2-franc subscriptions by the trade union rank and file. See Pascal Ory, "De Ciné-Liberté à La Marseillaise, espoirs et limites d'un cinéma libéré (1936–1938)," *Le Mouvement Social* 91 (April 1975), and Jonathan Buchsbaum, *Cinema Engagé: Film in the Popular Front* (Urbana, Ill., 1988), 250, 261–62, 269–70.

62. In Weimar Germany, where independent film production was at the mercy of UFA, the Socialist party produced *Die Schmiede* (1924), *Freies Volk* (1926), *Brüder* (1929), and *Lohnbuchhalter Kremke* (1930). The Communist party produced *Mutter Krausens fahrt ins Glück* (1929) and *Kuhle Wampe oder Wem gehört die Welt?* (1932). See Helmut Korte, ed., *Film und Realität in der Weimarer Republik* (Frankfurt/Main, 1980), and Bruce Murray, *Film and the German Left in the Weimar Republic: From Caligari to Kuhle Wampe* (Austin, Tex., 1990). But even the commercially well-organized communist Promethius-Film company went into receivership during the filming of *Kuhle Wampe*. All these films had a good box office in Vienna and received very favorable reviews ("the kind of films the working class needs") in the socialist press. Yet the reviewers rarely asked why such films were not being produced at home.

63. Glaser, "Stummfilmkritik,' 2, mentions 150 theaters in 1915. The absence of any official statistics makes that number doubtful, but even so, given the small seating capacity of most theaters, film had not yet become a mass medium.

64. See "Zahlen um Film und Kino," *Bildungsarbeit* 15:1 (Jan. 1928): 35; "Statistik der Kinos in Österreich," *Österreichische Film-Zeitung*, January 15, 1927; *Filmkunst: Zeitschrift für Filmkultur und Filmwissenschaft* 107 (Oct. 1985): 8–10.

65. See *Österreichische Film-Zeitung*, January 15, 1927, 14, and October 28, 1928, 2.

66. In 1928 there were 423,383 trade union members in Vienna. See *Statistisches Handbuch für die Republik Östereich, 1933* (Vienna, 1933). In the municipal election

of 1927 the SDAP received 694,557 votes, or 60.3% of the total. SDAP membership in the following year was 417,347. See Alfred Frei, *Rotes Wien*, 58–59. According to the SDAP Executive, the social composition of the party included 54.51% workers, 19.76% employees and managers, and 15.96% housewives. See Kulemann, *Beispiel*, 301–3. If we consider that in the second group at least 50% were white-collar workers and that a majority of the housewives were working class, then three-quarters of the SDAP membership were workers. In view of the potential number of worker filmgoers, the above estimates are on the low side.

67. Among workers in Chicago, moviegoing had become a habit by the 1920s. See Lizabeth Cohen, *Making a New Deal: Industrial Workers in Chicago, 1919–1939* (New York, 1990), 120–20. For a similar treatment of German workers, see W. L. Guttsman, *Workers' Culture in Weimar Germany: Between Tradition and Commitment* (New York, 1990), 263–74.

68. For the following, see Rudolf Lassner, "Theater- und Kinobesuch: Eine psychologische Analyse" (Ph.D. diss., University of Vienna, 1936), especially v, 1–2, 49, 56–62, 92–116. The study includes 356 persons of both sexes, and all ages beginning at 15. 270 responded to a questionnaire; 66 were interviewed. 25% of the sample was from the working class and was categorized as such in the analysis. For similar findings among youthful cinema audiences in Weimar Germany, see Alois Funk, *Film und Jugend: Eine Untersuchung über die psychischen Wirkungen des Films im Leben der Jugendlichen* (Munich, 1934), and Peukert, *Jugend*, 218–20.

69. See Ludwig Gesek, "Wann, wie oft und unter welchen Bedingungen geht die Jugend in Österreich ins Kino: Teilbericht über die Erhebung 'Jugend und Film'" (unpublished manuscript, 1933, in possession of Dr. Ludwig Gesek, Vienna), 1–61. Of 13,603 questionnaires sent out to schools and youth organizations throughout Austria, 10,054 were returned and analyzed. 25% of the respondents were from Vienna; the largest group (38.3%) were children of workers. Two further parts of the study, on children's perceptions of film content and the role of the film in the intellectual life of children, were never carried out.

70. Nor was it for their parents. Given their cold dwellings in wintertime, it is small wonder that the unemployed found refuge in movie theaters where, for a small price, hours of warmth could be enjoyed. This situation also prevailed in England and Weimar Germany. See Peukert, *Jugend*, 184–88; George Orwell, *The Road to Wigan Pier* (New York, 1958), 80–81; and John Stevenson, *British Society, 1914–45* (London, 1984), 396.

71. The argument was most cogently made by Hortense Powdermaker, *Hollywood, the Dream Factory: An Anthropologist Looks at the Movie-Makers* (New York, 1951), introduction. When the lights go out, she observes, critical faculties go out as well. The effect of the darkened theater is no doubt powerful, but are not other faculties stimulated by it as well: empathy, projection, assimilation? Ilya Ehrenburg's critique was an exercise in vulgar economic determinism to demonstrate that cinema was the powerful tool of leading industrialists. See *Die Traumfabrik* (Berlin, 1931). An even earlier version by the Austrian culture critic Richard Guttmann, *Die Kinomenschheit: Versuch einer prinzipiellen Analyse* (Vienna, 1916), charged that film educates the eye to see imprecisely.

72. The distinguished film critic and theorist Béla Balazs was the only one on the Viennese cultural scene to appreciate the unique visual power of film. In its expressive use of gestures he saw the emergence of the first international language and the beginnings of a new visual culture. See *Der sichtbare Mensch oder die Kultur des Films* (Vienna, 1924). Balazs wrote film reviews for *Der Tag* from 1922 to 1925, when he

left for Berlin as part of the cineast migration at the time. No film critic of comparable quality, with an understanding of film as a medium of mass culture, appeared in Vienna in the next decade. For his film criticism in Vienna, see Joseph Zsuffa, *Béla Balazs: The Man and the Artist* (Berkeley, Calif., 1987), 129–36.

73. The following is based on an analysis of *Paimanns Filmlisten: Wochenschrift für Lichtbild-Kritik,* 1924–32, an independent weekly listing current films with summaries; the trade publications *Der Filmbote: Zeitschrift für alle Zweige der Kinematographie,* 1925–27, *Österreichische Film-Zeitung,* 1927–29, and *Wiener Kino,* 1924; and the popular socialist daily *Das kleine Blatt,* 1929–33.

74. The roster of major directors included (to name but a few): G. W. Pabst, Joseph Sternberg, René Clair, Jean Renoir, Julien Duvivier, Marcel Pagnol, William Wellman, G. W. Griffith, Charles Chaplin, Michael Curtiz, Sergei Eisenstein, V. I. Pudovkin, and Jacques Feyder. The complaint in other countries about the lack of good films and the prevalence of kitsch seems also to have been the same. See Kracauer, *Caligari,* passim; Geneviève Guillaume-Grimaud, *Le Cinéma du Front Populaire* (Paris, 1986), 197; Tony Algate, "Comedy, Class and Containment: The British Domestic Cinema in the 1930s," in Curran and Porter, eds., *British Cinema,* 259ff; and Roger Dooley, *From Scarface to Scarlett: American Films in the 1930s* (New York, 1979), index of films.

75. See Johann Hirsch, "Kino und Massenbewusstsein: Die erfolgreichsten Filme, 1930/31," *Bildungsarbeit* 18:7–8 (July–Aug. 1931): 82–84.

76. See John Willett, *Art and Politics in the Weimar Period: The New Sobriety, 1917–1933* (New York, 1978), chs. 11, 15.

77. See Fritz, *Kino,* 138; Fritz, "Glanz," 3.

78. See "Allerweltsverdummungstrust" and "Vorstadtkino," *Arbeiter-Zeitung,* November 26, 1919, and January 18, 1920; David J. Bach, "Das Kino des Proletariats," ibid., October 1, 1922.

79. For the following, see "Die Filmreform," *Arbeiter-Zeitung,* May 17, 1924. Film producers boycotted the meeting.

80. See Bach, "Kino," and Ernst Weizmann, "Der Film und die Arbeiterschaft," ibid., May 1, 1924.

81. See for instance "Die Welt des Films," ibid., September 12, 1926, and August 14, 1932. Film exhibitors defended themselves against the charge of serving up a heavy diet of trash with the less than candid view that "the public decides what films are shown"—a stock answer in the international film world. See "Der Defraudant," *Der Filmbote* 43:9 (Oct. 23, 1926): 5–6.

82. See Glaser, *Umfeld Austromarxismus,* 496–97.

83. "Der blaue Engel," *Das kleine Blatt,* April 4, 1930.

84. In the exhortatory tone in which all of his writing was couched, Rosenberg urged youth to lead the way in "the struggle against and for the cinema." See "Wir und das Kino," *Der Jugendliche Arbeiter* 27:2 (Feb. 1928): 2–3. Rosenfeld repeated this argument on the radio. "Der Arbeiter und der Film: Vortrag, gehalten am 30. Jänner 1929 im Wiener Radio (Arbeiterkammerstunde)," *Bildungsarbeit* 16:2 (Feb. 1929): 17–20. He also attacked the alleged neutrality of the *Kulturfilm* or documentary as a bourgeois deception, for the selection of subjects alone reflected class bias. What the working class needed, he argued, was not hypocritical neutrality but honest proletarian films reflecting the class struggle. "Der 'neutrale' Kulturfilm," *Bildungsarbeit* 16:9 (Sept. 1919): 105–8. Compulsively, Rosenfeld repeated his critique of the capitalist film even in exile, long after the possibility of socialist influence on film

in Austria had passed. See "Film und Proletariat: Versuch einer Soziologie des Kinos,"*Arbeiterjahrbuch* (Karlsbad, 1934).

85. Hirsch, "Kino," 85.

86. "Sozialdemokratische Kinopolitik," *Der Kampf* 22:4 (April 1929): 192–97.

87. Gerhard Dreier, "Film und Partei," *Bildungsarbeit* 17:1–2 (Jan.–Feb. 1930): 18–20.

88. But even as late as 1932 the SDAP's theoretical organ featured an attack on the film as the repository of all the kitsch which had been driven out of literature and the fine arts. See Ernst Leonhard, "Der Film als ästhetische, wirtschaftliche und politische Erscheinung," *Der Kampf* 25:8–9 (Aug.–Sept. 1932).

89. For the following, see Venus, "'Hinein in das Kino,'" 210–21.

90. "Das Kino der Zehntausend," *Arbeiter-Zeitung*, April 3, 1927.

91. The bank was founded in 1922 as an economic enterprise by the SDAP, the trade unions, and the cooperative societies. In 1932, long past its best years, the Arbeiterbank had a capital stock of 4 million Schillings, deposits of 54 million, and profits of 713,000. See Kulemann, *Beispiel*, 319.

92. For the following, see Venus, "'Hinein in das Kino,'" 210–21.

93. In defending the new licensing procedures against its critics in the industry, the SDAP claimed that the previous control by the police was subject to political influence, whereas now city hall could help to improve the quality of films exhibited. The new law also contained a veiled form of censorship "to protect youth" under sixteen. To be exhibited as *Jugendfrei* (general admission), all films had to be screened by a municipal film committee. At the box office, age restrictions were difficult to enforce. See "Die Bundesregierung will das Wiener Kinogesetz verhindern," *Arbeiter-Zeitung*, August 22, 1926.

94. It appears that the Hambers' excellent connections at city hall—particularly with Breitner and Seitz—smoothed the way for their collaboration with Kiba.

95. The growth of Kiba increased the number of complaints in film industry publications about unfair competition from an enterprise that enjoyed the support of the *Arbeiterbank* and the municipal government. Ironically, it was charged that the socialists were politicizing the cinema. See for instance "Politisierung der Kinos," *Österreichische-Film-Zeitung*, February 19, 1927.

96. From 1925 to 1933 the major film trade publications fomented against the luxury tax on cinema, pointing out the favoritism shown to legitimate theater, which paid markedly lower taxes, and to the Kiba movie houses, which were given other economic privileges.

97. This included Kiba's twelve plus another fifteen commercial theaters. But Kiba's indirect control was even greater, because its purchasing and renting power made it possible to influence other distributors and to determine their choice of films. Venus, "'Hinein in das Kino,'" 218.

98. "Sozialdemokratische Kinopolitik," 195–96.

99. The SDAP's claim to be the champion of the quality film with social content was challenged in the controversy surrounding the exhibition of the American film "All Quiet on the Western Front" in 1931. Disturbances during the first few screenings and further threats of violence by right-wing groups made it possible for the national government to intervene on grounds of "threats to public order and safety." As in the case of Schnitzler's *Reigen* performance in 1921 (see ch. 6), the socialist mayor and city councillors initially took a firm stand on not giving in to the "film criticism of the street." But they capitulated only a few days later, and the film

was banned. Ironically, the SDAP organized bus tours to Bratislava, where the film was being shown (leading to the popular quip: "'Im Westen nichts neues'—im Osten gesehen"). See Alfred Pfoser, *Literatur und Austromarxismus* (Vienna, 1980), 199–201, and Grafl, "Hinein," 84–85.

100. Venus, "'Hinein in das Kino,'" 224. When Kiba remodeled the old Apollo-Theater into a luxurous movie palace at great expense, David Joseph Bach, Rosenfeld's superior at *Die Arbeiter-Zeitung,* took on the job of praising this creation and writing the subsequent film reviews. See Rosenfeld's letter of 1976, quoted in Henrietta Kotlan-Werner, *Kunst und Volk: David Joseph Bach, 1874–1947* (Vienna, 1977), 97–98.

101. The SDAP's failure to use the cinema law of 1926 to its advantage in gaining control over more theaters is similar to its feeble use of the *Wohnungsbeförderungsgesetz* in the early 1920s to gain municipal control of vacant or unused dwellings.

102. For a glimpse of just how ruthless the industry could be in the international center of the film world, see Robert Sklar, *Movie-Made America: A Cultural History of American Movies* (New York, 1979), ch. 14, and Gorham Kindem, ed., *The American Movie Industry: The Business of Motion Pictures* (Carbondale, Ill., 1982).

103. For a similar ambiguity regarding film among socialists in Weimar Germany, see Adelheid von Saldern, "Ennobling Mass Culture: The Political and Cultural Striving for 'Good Taste' and 'Good Morals' in the Weimar Republic," paper presented at the Colloquium on Mass Culture and the Working Class, 1914–70, Paris, 1988.

104. The denigration of film as a cheap form of amusement lacking the "noble and serious" qualities of elite culture was prevalent in all social strata. Frequent movie attendance was often admitted with a certain embarrassment, as being somehow culturally unworthy. See Lassner, "Kinobesuch," 5–7.

105. For the church's general position, see *Wiener Diözesanblatt,* July 10, 1926.

106. State ownerhip and/or control was usual in Europe and throughout the rest of the world except for the United States, where commercial advertisements formed a principal part of the revenue of privately owned stations. In Austria radio listeners paid modest users' fees.

107. See "Was ist die Ravag?," *Die Börse,* January 1, 1925; Theodor Venus, "Verhandlungen und Gründung der Österreichischen Radio-Verkehrs A. G." (Ph.D. diss. University of Vienna, 1982), part 3; idem., "Vom Funk zum Rundfunk—Ein Kulturfaktor entsteht," in *Geistiges Leben im Österreich der Ersten Republik* (Vienna, 1986).

108. SDAP proprietorship in Ravag (through the municipality) was 20%. The advisory council had representatives from the Chambers of Labor, Industry, and Agriculture, from trade associations of the radio industry and retailers, and from associations of listeners depending on membership. See Theodor Venus, "'Der Sender sei die Kanzel des Volkes': zur sozialdemokratischen Rundfunkpolitik der 1. Republik," in Österreichische Gesellschaft für Kulturpolitik, *Arbeiterkultur in Österreich,* 226–33.

109. Rintelin wrote in his memoirs that his intention from the beginning had been to thwart the Marxists in Vienna. He also participated in the Nazi putsch against the Dollfuss government in July 1934. See Ernst Glaser, "Die Kulturleistung des Hörfunks in der Ersten Republik," *Geistiges Leben,* 25–26. The conflict between business and culture orientations was made clear at the festive opening of Ravag, where Rintelin and Viennese mayor Seitz represented the two sides.

110. See Venus, "Sender Kanzel des Volkes," 239–40.

111. See Venus, "Verhandlungen und Gründung," 1329, and idem., "Sender Kanzel des Volkes," 244.

112. See Venus, "Verhandlungen und Gründung," 1331.

113. See "Rundfunkstatistik," *Rundfunkarchiv* 11:6 (June 1938): 252–56. One should add about 10% to the official statistics of illegal, non-fee-paying listeners (information from Theodor Venus).

114. See "Wer ist am Rundfunk interessiert?," *Bildungsarbeit* 17:4 (April 1930): 70–71. Radio listening had become habitual among American and German workers by the end of the 1920s. See Cohen, *Making a New Deal*, 129–39, and Guttsman, *Workers' Culture in Weimar Germany*, 256–62. For Britain, see D. L. LeMahieu, *A Culture for Democracy* (London, 1988, 141–54.

115. A simple tube set cost 212 Schillings; a seven-tube one, 400. Both were available with eighteen monthly payments, with 10 percent down. Thus, in addition to a down payment of 22 or 40 Schillings, monthly installments came to 10.50 or 20 Schillings. Radio receiver prices can be found in advertisements of virtually every issue of *Radio Welt: Illustrierte Wochenschrift für Jedermann*. Many workers, particularly those belonging to the Workers' Radio Club, also constructed their own tube radios at a fraction of the retail cost.

116. See for instance *Arbeiter-Zeitung*, January 6 and April 25, 1925; October 6 and 17, 1926. During this period the programming apportionment (in percentages) was music, 60–69; literature, 11–16; lectures, 11–17; and news, 7–9. See Venus, "Verhandlungen und Gründung," 1325.

117. *Radio-Welt* 2 (1925), 1–2. With all the criticism of Ravag not doing justice to the cultural singularity of Vienna, the municipal government made no attempt to establish a Vienna station independent of Ravag.

118. See Gerhard Botz, "Die 'Juli-Demonstranten', ihre Motive und die quantifizierbaren Ursachen des '15. Juli 1927," *Die Ereignisse des 15. July 1927: Protokoll des Symposiums in Wien am 15. July 1977* (Vienna, 1979), and idem., *Gewalt in der Politik: Attentate, Zusammenstösse, Putschversuche, Unruhen in Österreich, 1918 bis 1938* (Munich, 1983), 141–60, for the best accounts of what amounted to a miniature civil war and the turning point in the politics of the First Republic.

119. See the official Christian Social position in "Die Ravag im Dienste der sozialdemokratischen Partei," *Reichspost*, July 21, 1927.

120. The strategy of the German SPD on radio was very similar to that of the SDAP. It too sought pluralism and the aired contest of political ideas, and its efforts also foundered on the intransigence of its political opponents. Ultimately, German radio also became an organ of the government. See Horst D. Iske, *Die Film und Rundfunkpolitik der SPD in der Weimarer Republik: Leitfaden und Dokumente* (Berlin, 1985).

121. For the following, see Venus, "Sender Kanzel des Volkes," 233–34, 258, 264.

122. *Jahrbuch der österreichischen Arbeiterbewegung, 1931* (Vienna, 1932), 426.

123. See *Arbeiter-Zeitung*, March 14 and June 22, 1928. This form of censorship persisted throughout the period whenever the Chamber of Labor proposed programs that went beyond technical work issues to economic or social relations. In the two cases cited, the head of the Chamber of Commerce and the minister of education insisted that the subjects were not within the competence, as they narrowly defined it, of the Chamber of Labor. The Ravag administration hewed to this line, and the socialists on the committees denounced the censorship and threatened to take action.

124. Venus, "Sender Kanzel des Volkes," 251–52, 256–57. Catholic and conservative interests were compensated with programs on Easter and Christmas, national evenings, religious music, military subjects, and such events as the huge German Song Society Festival in 1928. The socialists did succeed in blocking the Ravag administration's demand for radio advertising. See "Die Wirtschaft bei der Ravag," *Arbeiter-Zeitung*, December 13, 1930.

125. Venus, "Sender Kanzel des Volkes," 249. These substantial investments were made at the expense of creating a real team of radio actors and musicians and of experimenting with the technical and artistic possibilities of radio as a mass medium.

126. *Reichspost*, September 29, 1929. The socialists blamed the growing Heimwehr influence on Ravag for the absence of any mention of Martin Andersen-Nexö's socialism when he gave a reading from his work. *Arbeiter-Zeitung*, November 17, 1929. Not content with its indirect influence on Ravag, the Heimwehr created its own club (Vaterländischer Radiohörer) in 1932 and thereby replaced moderate conservatives on the program subcommittee. The Nazi *Deutscher Volkhörerbund* was created at the same time. See Venus, "Sender Kanzel des Volkes," 260.

127. The Chancellors Johann Schober and Engelbert Dollfuss and President Wilhelm Miklas gave addresses, as well as the Heimwehr leader Guido Jakoncig. That privilege was denied to Mayor Seitz and other socialists politicians. Venus, "Sender Kanzel des Volkes," 255, 262.

128. In the municipal election of April 1932, the Austrian Nazis received over 200,000 votes and 17.4 percent of the mandates (as compared to 27,500 in 1930). See Rabinbach, *Crisis*, 89, and Frei, *Rotes Wien*, 59.

129. In a later reflection on the period, Ravag director Czeija maintained that the Schuschnigg government was able to put the worker rising of February 1934 down so quickly because he had two instruments: artillery and radio. Cited in Glaser, "Kulturleistung," 29.

130. Turning news broadcasts over to the national government controlled by their opponents remains inexplicable.

131. Venus, "Sender Kanzel des Volkes," 264–66. On September 11, 1933, Dollfuss declared Austria to be a corporate state; thereafter, the struggle could no longer be waged by such peripheral means as mass resignations from radio membership. But considering Vienna's preponderance in the radio audience until 1928, mass resignations then would have meant the economic ruin of Ravag, which was a business enterprise after all. That fact seemed to have been understood by the Ravag management; it was not acted upon by the socialists.

132. For the following, see "Die Hörerbefragung der Ravag," *Radio-Wien* 9:6 (1931/32): 2–4. Socialist demand for such a survey was motivated by the desire to have the distribution of listeners on record, so that the established preponderance of worker subscribers could be used to increase their representation on the program subcommittee. Listeners' attitudes and their evaluation of the programming were of less interest to the SDAP.

133. Operettas, one-act plays, geography, and recorded music likewise received high marks. Low marks were also received by organ concerts, readings by authors, literary lectures, and gymnastics.

134. A survey of radio listeners in Weimar Germany came to virtually the same conclusions. See von Saldern, "Ennobling Mass Culture," 18.

135. Age differences were selective: young men had greater interest in sports;

young women were negative on factual lectures; and the young of both sexes expressed a liking for jazz.

136. "Hörerbefragung," 5.

137. See for instance "Die Hörer wünschen mehr Heiterkeit," *Das klein Blatt*, October 10, 1932.

138. There were some exceptions, particularly in the Bildungszentrale, but they had little influence over the party's cultural decision makers. See for instance Fritz Rosenfeld, "Der Rundfunk und das gute Gewissen," *Bildungsarbeit*, 19:10 (Oct. 1932): 189–90.

139. Michel de Certeau offers a brilliant analysis of the relationship between production and consumption in which the *use* made of mass culture products by the consumer is viewed as a process of transformation—a kind of production in response. See *The Practice of Everyday Life*, trans. Steven F. Rendall (Berkeley, Ca., 1984), 31.

140. Unfortunately, there exists no real history of Austrian soccer. Three existing studies give little but a soccer fan's view of teams and players, with only occasional references to size of audience. See Leo Schidrowitz, *Geschichte des Fussballsportes in Österreich* (Vienna, 1951); Karl Langisch, *Geschichte des Österreichischen Fussballsports* (Vienna, 1965); and Karl Kastler, *Fussballsport in Österreich* (Linz, 1972).

141. Julius Deutsch, *Unter Roten Fahnen: Vom Rekord zum Massensport* (Vienna, 1931), 3–12. For the SDAP's position on commercial sports, see ch. 4.

142. Hendrik deMan, *Zur Psychology des Sozialismus* (Jena, 1927), 36–39.

143. See for instance Peter Friedmann, "Die Krise der Arbeitersportbewegung am Ende der Weimarer Republik," in Friedhelm Boll, ed., *Arbeiterkulturen zwischen Alltag und Politik: Beiträge zum europäischen Vergleich in der Zwischenkriegszeit* (Vienna, 1986), 235–40. For intraspectator agression, see R. Horak, W. Reiter, K. Stöcker, eds., *"Ein Spiel dauert länger als 90 Minuten": Fussball und Gewalt in Europa* (Hamburg, 1988).

144. An Austria-Italy match attracted 90,000 spectators who caused a dangerous landslide. Schidrowitz, *Geschichte Fussballsportes*, 125.

145. For the season 1932–33, official statistics list 446 professional championship games and 121 professional cup games. See *Handbuch der Gemeinde Wien* (Vienna, 1935), 201.

146. The idea that mass culture manipulates the consumer, imposing false needs and false consciousness on him, has been challenged only recently. Hans Magnus Enzensberger, for instance, suggests that the success of mass culture depends in part on its appeal to real needs. See "Constituents of a Theory of the Media," *The Consciousness Industry: On Literature, Politics, and the Media* (New York, 1974).

147. For an interesting account of how young workers creatively both survived and used their leisure time during the depression, see Safrian, "'Wir ham die Zeit.'"

148 It is interesting that the socialists paid hardly any attention to the oldest forms of spontaneous noncommercial leisure-time activities (rambling, swimming), which clearly, could not be thrown in the pot with "cheap capitalist distractions." These activities continued to make a considerable claim on the workers' free time, which the party demanded for itself. The party apparently chose to treat the subject with silence.

149. See Larry May, *Screening Out the Past: The Birth of Mass Culture and the Motion Picture Industry* (Chicago, 1980), 34–42; Sklar, *Movie-Made America*, ch. 2; and Roy Rosenzweig, *Eight Hours for What You Will: Workers and Leisure in Worcester, Massachusetts* (Cambridge, Mass., 1983), ch. 8.

Chapter 6

1. See Richard J. Evans, "Introduction: the Sociological Interpretation of German Labour History," in idem., *The German Working Class, 1888–1933: The Politics of Everyday Life* (London, 1982), 40–41. For a more general discussion of the worker family as receptor and resister, see the unpublished conference paper of Geoff Eley, "Some Thoughts on the History of the Family and Its Relation to the History of the Working Class," Second Meeting, SSRC Research Seminar Group on Modern German Social History, January 12–13, 1979.

2. See Joseph Ehmer, *Familienstruktur und Arbeitsorganisation im frühindustiellen Wien* (Vienna, 1980), 208–36.

3. See Joseph Ehmer, "Vaterlandslose Gesellen und respektable Familienväter: Entwicklungsformen der Arbeiterfamilie im internationalen Vergleich, 1850–1930," in Helmut Konrad, ed., *Die deutsche und die österreichische Arbeiterbewegung zur Zeit der Zweiten Internationale* (Vienna, 1982), 136–38.

4. See particularly Otto Bauer, *Mieterschutz, Volkskultur und Alkoholismus: Rede im Arbeiter-Abstinentenbund am 20. März 1928* (Vienna, 1929).

5. This development had been presumed by Marx and Engels and, more recently, by August Bebel, Klara Zetkin, and Lilly Braun.

6. Reinhard Sieder, "Behind the Lines: Working-Class Family Life in Wartime Vienna," in Richard Wall and Jay Winter, eds., *The Upheaval of War: Work and Welfare in Europe, 1914–1918* (Cambridge, 1988), 134.

7. For the origins and later meaning of the concept "ordentliche Familie," see Joseph Ehmer, "Familie und Klasse: Zur Entstehung der Arbeiterfamilie in Wien," in Michael Mitterauer and Reinhard Sieder, eds., *Historische Familienforschung* (Frankfurt, 1982).

8. See Joseph Ehmer, "Frauenarbeit und Arbeiterfamilie in Wien: Vom Vormärz bis 1934," *Geschichte und Gesellschaft* 7:3/4 (1981): 451, 470, tables 1 and 2. Almost half (41.3 percent) of all women were married.

9. The subject quite naturally received extensive and repeated coverage in the publications intended for women: *Die Frau, Die Unzufriedene, Die Mutter,* and *Einheit.* But it was also a major concern of *Die sozialistische Erziehung, Der Vertrauensmann, Bildungsarbeit, Das kleine Blatt, Der Kuckuck,* and *Der Kampf.* It is interesting to note that female white-collar employees, the majority of whom were unmarried, were subject to versions of the "triple burden" experienced by blue-collar married female workers. The burden of the white-collar employee took place in the environment of the family of origin where, as a daughter, she was expected to share in the housework and even look after children with other female family members. See Erna Appelt, *Von Landenmädchen, Schreibfräulein und Gouvernanten: Die weiblichen Angestellten Wiens zwischen 1900 und 1934* (Vienna, 1985), 169–78.

10. See Gottfried Pirhofer, "Politik am Körper: Fürsorge und Gesundheitswesen," *Ausstellungskatalog Zwischenkriegszeit—Wiener Kommunalpolitik* (Vienna, 1980), 69.

11. See Marianne Pollak, "Die Unnahbarkeit der Frau," *Der Kampf* 20:9 (Sept. 1927): 435–37.

12. See Marianne Pollak, *Frauenleben von gestern und heute* (Vienna, 1928), 23–24. The municipality even created a clinic to correct physical deformities (bellies, buttocks, breasts) without charge. See Pirhofer, "Politik am Körper," 69.

13. See James F. McMillan, *Housewife or Harlot: The Place of Women in French Soci-*

ety, 1870–1940 (New York, 1981), 166. The masculinization of female dress was also greatly influenced by such Parisian designers as Coco Chanel.

14. For Germany, see Atina Grossmann, "The New Woman and the Rationalization of Sexuality in Weimar Germany," in Ann Snitow et al., eds., *Power of Desire: The Politics of Sexuality* (New York, 1983), 156–57. For the United States, see John D'Emilio and Estelle B. Freedman, *Intimate Matters: A History of Sexuality in America* (New York, 1988), 233–35. For England, see Deirdre Beddoe, *Back to Home and Duty: Women Between the Wars, 1918–1939* (London, 1989), 22–24.

15. See Therese Schlesinger, *Die Frau im sozialdemokratischen Parteiprogramm* (Vienna, 1928), 5–9.

16. See Reinhard Sieder, "Hausarbeit oder die andere Seite der Lohnarbeit," *15. österreichischer Historikertag, Salzburg 1981* (Salzburg, 1984), 159–61, and Joseph Ehmer, "Frauenarbeit und Arbeiterfamilie," 459.

17. See Emmy Freundlich, "Zur Frage Einküchenhaus," *Die Frau* 34:7 (July 1, 1925): 5–6, and Marianne Pollak, "Wie kommt die berufstätige Frau zu ihrem Achtstundentag?," *Arbeit und Wirtschaft* 1 (Jan. 1, 1929): 44–46.

18. See Irena Hift-Schnierer, "Die neue Frau im neuen Haushalt," *Die Mutter* 1:12 (May 1925): 16–17.

19. Pollak, *Frauenleben*, 39–40.

20. See for instance "Hilf dir selbst," *Der Kuckuck* 2:4 (Jan.26, 1930).

21. Pollak, *Frauenleben*, 37–38.

22. See Therese Schlesinger, "Proletarisches Spiessbürgertum," *Der Jugendliche Arbeiter*, 23:3 (March 1924): 10–11; Marianne Pollak, "Beruf und Haushalt," *Handbuch der Frauenarbeit in Österreich* (Vienna, 1930), 413–19; and Otto Feliz Kanitz, "Vortrag auf dem 2. Kongress für Sozialismus und Individualpsychologie," *Die sozialistische Erziehung* 7:11 (Nov. 1927).

23. See Robert Danneberg, *Die neue Frau* (Vienna, 1924), 9. This pamphlet was announced (in its preface) as the first in a quarterly series intended for women and girls already in the Socialist party. The stated aim was an "exchange of views" between the party and its female members. It is difficult to imagine what the mechanism for such an exchange might have been.

24. And thus become a good friend to their children and a comrade to their husband. Pollak, *Frauenleben*, 45. For Danneberg, the model for such emotionalization was the bourgeois woman. See *Neue Frau*, 9.

25. Helene Bauer, "Ehe und soziale Schichtung," *Der Kampf* 20:7 (July 1927): 319–22.

26. See Neschy Fischer, "Ehe als soziales Problem, *Der Kampf* 20:8 (Aug. 1927): 387–89.

27. Anton Hanak's "Caring Mother" was representative of the motherhood genre. See Pirhofer, "Politik am Körper," 48, 69.

28. See for instance Fedora Auslaender, "Frauenarbeit und Rationalisierung," *Handbuch der Frauenarbeit*, 30.

29. See Martha Eckl, "Körperkultur und 'proletarische Weiblichkeit', 1918–1934: Eine Untersuchung am Beispiel der Frauenzeitschriften der Sozialdemokratischen Arbeiterpartei Deutsch/österreichs" (Diplomarbeit, Institut für Wirtschafts- und Sozialgeschichte University of Vienna, 1986), 54–55.

30. See Veronika Kaiser, "Österreichs Frauen, 1918–1938: Studien zu Alltag und Rollenverständnis in politischen Frauenblättern" (Ph.D. diss. University of Vienna, 1986), 87.

31. Pollak, *Frauenleben*, 24–25.

32. See Marie Deutsch-Kramer, "Die Befreiung der Frau durch den Sport," *Die Frau* 38:6 (June 1, 1929): 10–11.

33. See Dr. Wanda Reiss, "Die Pflege des weiblichen Körpers," *Die Mutter* 1:1 (Jan. 1925): 5.

34. *So leben wir . . . 1320 Industriearbeiterinnen berichten über ihr Leben* (Vienna, 1932). This study was based on a one-third return of 4,000 questionnaires distributed in 1931 by shop stewards at the workplace, supplemented by interviews and written communications from the workers. The sample was drawn from all the leading industrial sectors in which women were employed. But it represents workers whos condition was above the average: somewhat older, more secure at the workplace, not excessively dulled by misery, and already in contact with the working-class movement (party and/or trade union membership, or sympathy with the same). The general condition of Viennese working women, Leichter cautions, was considerably worse (3–4). 42.8% of the sample were single; 39% were married; 0.09% lived with a companion; 11.4% were widowed; and 5.9% were divorced. See also *Wie leben die Wiener Heimarbeiter?: Eine Erhebung über Arbeits- und Lebensverhältnisse von tausend Heimarbeitern* (Vienna, 1928). This survey was conducted in 1927 along the same lines as indicated above, save for the assistance of shop stewards; 94.91 % of the homeworkers were women.

35. See Ehmer, "Frauenarbeit," 470, table 1. As Ehmer acknowledges, the census figures left out a large sector of working-class women: those who considered themselves employed for less than full time. This category included tens of thousands of homeworkers.

36. Ibid., 472, table 6. The growth in factory workers was greatest in the newer industries, with women comprising 40% in the electrical industry, 50% in metalworking, and 80% in lightbulb production. Women continued to dominate in the traditional female textile and weaving industry. Domestics had decreased by 50% since 1910, whereas employees—mainly office workers and sales clerks—were the fastest growing occupational group (454).

37. See Joan W. Scott and Louise A. Tilly, "Woman's Work and the Family in Nineteenth-Century Europe," *Comparative Studies in Society and History* 17 (1975).

38. Leichter, *So leben wir*, 41–44.

39. Ibid., 78–79.

40. Ibid., 81–83. Only 14% of the women received some assistance from men with housework and childcare. The average workday of German female textile workers was equally long and filled with tensions brought on by multiple responsibilities and the lack of time to fulfill them. Though single women disposed of more free time after factory work, they were invariably forced to lend a hand with housework and child care in the family household. See Deutscher Textilarbeiterverband, *150 Berichte von Textilarbeiterinnen* (Berlin, 1930).

41. Leichter, *So leben wir*, 73–74.

42. Reinhard Sieder, "Housing Policy, Social Welfare and Family Life in 'Red Vienna,' 1919–1934," *Oral History: Journal of the Oral History Society* 13:2 (1985): 39.

43. See Gottfried Pirhofer and Reinhard Sieder, "Zur Konstitution der Arbeiterfamilie im Roten Wien: Familienpolitik, Kulturreform, Alltag und Ästhetik," in Michael Mitterauer and Reinhard Sieder, eds., *Historische Familienforschung* (Frankfurt, 1982), 342–43. For married couples living in crowded parental homes, see especially the oral history files of Frauen Schau, Win, Pre, and Fie available at the Institut für Wirtschafts- und Sozialgeschichte of the University of Vienna.

44. Leichter, *So leben wir,* 94–97. As a result, only 21.9% of the children under 6 went to kindergarten, and 18.1% of those under 14 made use of the after-school centers. Of the latter age group, 17% had no supervision whatever.

45. Ibid., 109–10.

46. Ehmer, "Frauenarbeit," 464–65.

47. Ibid., 461–62.

48. Ibid., 459–60. This conclusion is well demonstrated in an American study. See Ruth Schwartz Cowan, *The Ironies of Household Technology from the Open Hearth to the Microwave* (New York, 1983).

49. See Reinhard Sieder, "'Street Kids': The Socialization of Viennese Working-Class Children," typescript of paper delivered at the International Colloquim on Sociabilité of the Working Class, held in Paris in 1985, 21, and Robert Wegs, *Growing Up Working Class: Continuity and Change Among Viennese Youth, 1890–1938* (University Park, Pa, 1989), 140–42.

50. See Margarete Rada, *Das reifende Proletarier-Mädchen* (Vienna, 1931), 59–60, 82–84.

51. See Käthe Leichter, *Frauenarbeit und Arbeitterinnenschutz in Österreich* (Vienna, 1927), 58.

52. See Käthe Leichter, "Die Entwicklung der Frauenarbeit nach dem Krieg," *Handbuch der Frauenarbeit,* 40, 42, and Edith Riegler, *Frauenleitbild und Frauenarbeit in Österreich* (Vienna, 1976), 132. Leichter argues that low female wages rather than improved technology were the cornerstone of Austrian economic rationalization (34).

53. See for instance Käthe Leichter, "Vom Frauenberuf: Das Schwache Geschlecht bei der Arbeit," *Das kleine Blatt,* Oct. 19, 1927.

54. See Leichter, *Wie leben die Wiener Heimarbeiter?*, 11, 13, 19, 25, 37, 41, 45.

55. See Gabriele Czachay, "Die soziale Situation der Hausgehilfinnen Wiens in der Zwischenkriegszeit" (master thesis, University of Vienna, 1985), 143–48. Marianne Pollak's romantic novella in which a maid goes to vocational school, learns about the law protecting domestics, and finds love and marriage with a true comrade—all with the guidance of the SDAP—is far removed from reality. See *Aber schaun S', Fräul'n Marie!: Liebesgeschichte einer Hausgehilfin* (Vienna, 1932).

56. See Wilhelmine Moik, "Die Freien Gewerkschaften und die Frauen," *Handbuch der Frauenarbeit,* 581.

57. See Peter Stiefel, *Arbeitslosigkeit: Soziale, politische und wirtschftliche Auswirkungem am Beispiel Österreich* (Berlin, 1979), 200–202.

58. See "Frauenarbeit," *Arbeit und Wirtschaft* 7:15 (Aug. 1, 1929): 698, and "Doppelverdiener," *Die Arbeiterin* 7:4/5 (April–May 1930): 5.

59. See "Frauenarbeit," *Jahrbuch 1932 des Bundes der Freien Gewerkschaften Österreichs* (Vienna, 1933), 115.

60. "Frauenarbeit," *Arbeit und Wirtschaft,* 702.

61. At the trade union congress of 1931 the number of female delegates reached 11.3 percent. But female union membership was twice as high. See Heinz Renner, "Die Frau in den Freien Gewerkschaften Österreichs, 1901–1932: Statistische Materialien," International Conference of Labor Historians, *ITH Tagungsbericht 13* (Vienna, 1980), I: 322, 329.

62. *So leben wir,* 116, 122. But 73.3% of her sample were trade union members.

63. 41.2% of the husbands or life companions of these women were unemployed; 82.3% of the women supported others or at least themselves. Ibid., 13, 103, 107.

64. Ibid., 54. Leichter exaggerates the importance of the fact that 31.9% of the

single women said they would continue working in any case. She overlooks the fact that these women had as yet only limited household and childcare responsibilities.

65. See Marie Jahoda, Paul Lazersfeld, and Hans Zeisel, *Die Arbeitslosen von Mariethal: Ein soziographischer Versuch* (1933; Bonn, 1980), 91–92 and Ehmer, "Frauenarbeit," 466. In the female network of factory labor, information about birth control and abortion was traded freely. Ibid., 468–69.

66. Generally the women workers' lunch brought to the factory consisted of bread and vegetables often eaten unheated. For the generally high carbohydrate and fat content of working-class diets, see "Der Lebensstandard von Wiener Arbeiterfamilien im Lichte langfristiger Familienbudgetuntersuchungen," *Arbeit und Wirtschaft* 13:12 (Dec. 1959): supplement 8, 10. See also Roman Sandgruber, *Bittersüsse Genüsse: Kulturgeschichte der Genussmittel* (Vienna, 1986), 81, 182.

67. Leichter reports (*So leben wir*, 108–15) that 78.7 % spent evenings at home doing housework. Meetings were attended by a mere 4.4%. Entertainment outside the cinema was virtually unknown. Only the radio (aside from the press) offered a steady contact with the wider world, but only for 36.1% of the sample. Leichter makes too much of the young, unmarried women who were able to get out of the home. She neglects the fact that, to make this freedom possible, some other, usually older, woman in the household had to bear the full burden.

68. Leichter, "Entwicklung der Frauenarbeit," 38.

69. For the SDAP, see Helene Maimann, ed., *Die ersten 100 Jahre: Österreichische Sozialdemokratie, 1888–1988* (Vienna, 1988), 351. I have been unable to find reliable figures for Viennese trade union and Chamber of Workers and Employees functionaries. If one includes the lowest level of these, an estimate of several hundred might be realistic.

70. But it must be kept in mind that, here as well as in the SDAP, trade unions, and Chamber of Workers and Employees, women were grossly underrepresented. The one exception was the municipality's Department of Social Welfare, in which 2,884 women were employed. It was the only branch of the administration with a heavy concentration of female employees. See Anna Grünwald, "Die Frau in der Gemeindeverwaltung in der Gemeinde Wien," *Handbuch der Frauenarbeit*, 653.

71. This distance was deplored by individual socialists leaders. See Max Adler and Käthe Leichter in ch. 4.

72. The birthdates of six doyennes were as follows: Anna Boschek, 1874; Adelheid Popp, 1879; Emmy Freundlich, 1878; Therese Schlesinger, 1863; Gabriele Proft, 1879; Amalie Seidel, 1876. See Hermine Agnezy, "Die Frauenbewegung in der Sozialdemokratischen Partei von 1918 bis 1934" (Hausarbeit, Institut für Zeitgeschichte, University of Vienna, 1975), 92–94.

73. For a suggestive complaint about socialist female employers who exploited their domestics, see Helen Goller, "Klassenkampf im Haushalt," *Die Frau* 37:3 (March 1, 1928): 5.

74. In the sharpest attack on such wishful thinking that has come to light, Sophie Lazersfeld challenged the basis of a program for the sexual education of youth drafted by Therese Schlesinger and Dr. Paul Stein. No one is served, Lazersfeld argued, "if he is told how desirable things ought to be, but only if he is shown how he can get there." See "Zur Frage der sexuellen Aufklärung der Jugend," *Bildungsarbeit* 20:1 (Jan. 1933): 52.

75. For a useful introduction to sexuality as a historical problem, see Jeffrey Weeks, *Sex, Politics and Society: The Regulation of Sexuality Since 1800* (London, 1981). For a concise review of theoretical approaches, see Ellen Ross and Rayna

Rapp, "Sex and Society: A Research Note from Social History and Anthropology," *Comparative Studies in Society and History* 23 (Jan. 1981). For the problems of writing the history of sexuality in everyday life, see Dorothee Wierling, "Alltagsgeschichte und Geschichte der Geschlechterbeziehungen: Über historische und historiographische Verhältnisse," in Alf Lüdtke, ed., *Alltagsgeschichte: Zur Rekonstruktion historischer Erfahrungen und Lebensweisen* (Frankfurt, 1989).

76. Whereas I fully agree with Michel Foucault's insistence that the complexity of feelings and activities called sexuality can be understood in historical terms only as an integral aspect of human experience, I reject his position—implicit throughout his text—that sexuality cannot really be studied. See *The History of Sexuality, I: An Introduction* (New York, 1980).

77. For an interesting treatment of worker selfhood *(Eigensinn)* as an interplay of the private and political, see Alf Lüdtke, "The Historiography of Everyday Life: The Personal and Political," in Ralph Samuel and Gareth Stedman Jones, eds., *Culture, Ideology and Politics* (London, 1982).

78. "Leitsätze für sexuelle Aufklärung der Jugend," *Bildungsarbeit* 19 (1932), 234.

79. One is unpleasantly reminded of the use made of sexuality in antiutopian novels such as Aldous Huxley's *Brave New World* and George Orwell's *1984*.

80. See Anna Hauer, "Sexualität und Sexualmoral in Österreich um 1900: Theoretische und literarische Texte von Frauen," in *Die ungeschriebene Geschichte: Historische Frauenforschung* (Vienna, 1985), 143–47. For parallel expressions in Victorian England, see Judith Walkowitz, *Prostitution and Victorian Society: Women, Class, and the State* (Cambridge, 1980), ch. 4–7: For Germany, see Regina Schulte, *Speerbezirke: Tugendhaftigkeit und Prostitution* (Frankfurt, 1979), 11–56.

81. See for instance H. Montane, *Die Prostitution in Wien* (Hamburg/Vienna, 1925), and Karl F. Kocmata, *Die Prostitution in Wien: Streifbilder vom Jahrmarkt des Liebesleben* (Vienna, 1927). On occasion, socialist images of the horrors of prostitution gave way to sentimental and even romantic views. See "Die andere Welt: Unter Dirnen, Zuhälter und Verbrecher," *Das klein Blatt* 2:46 (April 15, 1928): 3–4.

82. See Alfred Pfoser, "Verstörte Männer und emanzipierte Frauen: Zur Sitten- und Literaturgeschichte der Ersten Republik," in Franz Kadrnoska, ed., *Aufbruch und Untergang: Österreichische Kultur zwischen 1918 and 1938* (Vienna, 1981), 206.

83. "Erhebung über Sexualmoral," *Studien über Autorität und Familie: Forschungsbericht aus dem Institut für Sozialforschung* (Paris, 1936), 279–80. Between 1930 and 1933 the number of prostitutes in Vienna under public control declined from 870 to 747; the number of women arrested for soliciting declined from 3,594 to 2,260. See *Handbuch der Gemeinde Wien, 1935* (Vienna, 1935), 62.

84. For instance, Erwin Wexberg, *Einführung in die Psychologie des Geschlechtslebens* (Leipzig, 1930), 118–19.

85. "Prostitution und Gesellschaftsordnung," *Die Arbeiterinnen-Zeitung* 22 (Nov. 18, 1919): 3–4.

86. *Glückliche und unglückliche Ehe? Ein Mahnwort an junge Ehe- und Brautleute* (Vienna, 1922), 5–9. His novels bore such titles as *Küsse die Leben werden; Die nicht Mütter werden dürfen* and *Am Kreuzweg der Liebe*. That Ferch was considered a sexual reformer in socialist circles gives some indication of the general conservatism on the sexual question in those ranks.

87. For instance, *St. Pöltner Diözesanblatt* 1 (1919).

88. Dr. Gertrud Ceranke, "Willst due heiraten?," *Die Uzufriedene* 6 (Aug. 7, 1926): 7. This journal ran advertisements for contraceptive devices; gave tips on

health, beauty, clothing, and cooking in an uncommercial fashion; and offered a column on "Women Speak from the Heart" and a personal column for marriage seekers. By 1933 it reached a circulation of 160,000 and was mentioned as the prefereed weekly of female industrial workers. See Leichter, *So leben wir,* 116.

89. Similar caution was expressed in the popular tract by Hans Hackmack, *Arbeiterjugend und die sexuelle Frage* (Berlin, 1922), 15–16.

90. "Erfahrungen und Probleme der Sexualberatungsstellen für Arbeiter und Angestellte in Wien," *Der Sozialistische Arzt* 5 (1929): 99.

91. See Karl Sablik, *Julius Tandler: Mediziner und Sozialreformer* (Vienna, 1983), 278–80.

92. "Wohnungsnot und Sexualreform," in Weltliga für Sexualreform, *Sexualnot und Sexualreform: Verhandlungen* (Vienna, 1931), 5–14.

93. Tandler's eugenicist views were reflected in his interventionist approach to public welfare. See ch.3.

94. Dr. Rudolf Dreikurs, "Wohnungsnot und Sexualreform," 39–41.

95. Dr. Siegfried Kraus, ibid., 41–42.

96. "Die sexualnot der Werktätigen Massen und die Schwierigkeiten der Sexualreform," ibid., 74–75, 80–83.

97. Otto Bauer rarely intervened in the discussion on sexuality. But leading figures of the party's cultural, educational, youth, and welfare programs acted as spokesmen. Most influential behind the scenes in determining the SDAP's position was its executive secretary, Robert Danneberg, who was responsible for the "political and moral purity" of the party. See Wolfgang Neugebauer, "Robert Danneberg (1885–1942): Eine biographische Skizze," *Archive: Jahrbuch des Vereins der Geschichte der Arbeiterbewegung* 1 (1985); 86–88.

98. An attempted abortion was punishable by a term of from six months to one year, and a successful abortion by from one to five years; midwives and physicians implicated were subject to the same terms. A law of leniency, however, was at the disposal of the judges to reduce or cancel prison sentences. See Dr. W. Gleisback, "Das Verbrechen gegen das keimende Leben im geltenden und künftigen Strafrechte," *Zeitschrift für Kinderschutz, Familien- und Berufsführsorge* 20:1 (Jan. 1928): 2.

99. See W. Latzko, *Wiener medizinische Wochenschrift* 26 (1924): 1387. In Germany the estimates were one million abortions in 1931 in a female population of 31.2 million; the average working-class woman was thought to have two or more abortions in her lifetime. See Atina Grossman, "Abortion and Economic Crisis: The 1931 Campaign against #218 in Germany," *New German Critique* 14 (Spring 1978): 121–22, 125.

100. See Karin Lehner, "Reformbestrebungen der Sozialdemokratie zum Paragraph 144 in Österreich in der 1. Republik," *Ungeschriebene Geschichte,* 298–99.

101. See Benno Wutti, "Die Stellung der Sozialdemokratischen Partei Österreichs zur Frauenfrage," (Ph.D. diss., University of Vienna, 1975), 102–11.

102. See "Die Schwangerschaftsunterbrechung: Eine Tagung der sozialdemokratischen Aerzte," *Arbeiter-Zeitung* 144 (May 25, 1924).

103. Julius Tandler, "Ehe und Bevölkerungspolitik," *Wiener medizinische Wochenschrift* 74 (1924).

104. See Lehner, "Reformbestrebungen," 302–3, and Sablik, *Tandler,* 281–82.

105. See Therese Schlesinger, *Die Frau im sozialdemokratischen Parteiprogramm* (Vienna, 1928). The pertinent paragraphs were reprinted in the journals aimed at women.

106. See *Mit uns zieht die neue Zeit: Arbeiterkultur in Österreich, 1918-1934* (Vienna, 1981), 225. An assembly of 2,500 delegates representing SDAP members and voters, meeting in Vienna on September 25, 1927, demanded the revisions of paragraph 144 adopted by the SDAP party congress at Linz. Lehner, "Reformbestrebungen," 186-88.

107. Otto Bauer was chairman; the commission included the notables Max Adler, Julius Deutsch, Robert Danneberg, Wilhelm Ellenbogen, Oskar Helmer, Karl Kautsky, Jr., Karl Renner, Paul Richter, and Karl Seitz. The sole woman was Adelheid Popp, who had made it clear at the prior women's conference that she supported the position of the socialist physicians out of party loyalty. See Lehner, "Reformbestrebungen," 146-47.

108. For the following, see *Frauenarbeit und Bevölkerungspolitik: Verhandlungen der sozialdemokratischen Frauenreichskonferenz, Oktober 29-30, 1926 in Linz* (Vienna, 1926), 15-50.

109. Leopoldine Glöckel summed up the position of the majority that women could not have the right to control their own bodies until they had been completely enlightened.

110. See Dr. Margarete Hilferding, *Geburtenregelung* (Vienna/Leipzig, 1926), 14-15, and Gertrud Ceranke, "Willst du heiraten?"

111. See for instance Herwig Hartner, *Erotik und Rasse: Eine Untersuchung über gesellschaftliche, sittliche und geschlechtliche Fragen* (Munich, 1925), 52-53, and Robert Hofstädter, *Arbeitende Frau: Ihre wirtschaftliche Lage, Gesundheit, Ehe und Mutterschaft* (Vienna, 1924).

112. For instance Adelheid Popp, "Geburtenregelung und Menschenökonomie," Weltliga, *Sexualnot*, 503.

113. "Zur Psychologie der Geschlechter," *Der Kampf* 18 (June 1925): 25-27.

114. See "Geburtenregelung und Kinderschutz," *Die Unzufriedene* 35 (Aug. 28, 1926): 1.

115. For instance Dr. Margarete Hilferding, "Probleme der Geburtenregelung," *Die Mutter* 1 (April 1925): 6.

116. The first of these was created in 1917. After 1924 they were spread throughout Vienna by the municipal council in response to Tandler's campaign against syphilis. The clinics gave advice but no treatment, so as not to conflict with private physicians. See Sablik, *Tandler*, 283.

117. See *Mit uns zieht die neue Zeit*, 226, 230-31. Attempts by *Die Unzufriedene* to sponsor consultation hours for the psychological needs of women or the showing of the antiabortion film *Cyankali* were valiant efforts along these lines.

118. See Eckl, "Körperkultur," 91-92.

119. See Karl Fallend, "Wilhelm Reich: Dozent der Psychoanalyse, Sexualberater und rebellischer Parteigenosse" (Ph.D. diss., University of Salzburg, 1987), 169-74; Reich, "Erfahrungen und Probleme," 98; and David Badella, *Wilhelm Reich* (Bern, 1981), 71-72.

120. For instance *Sexualerregung und Sexualbefriedigung* (Vienna, 1929), and *Geschlechtsreife, Enthaltsamkeit, Ehemoral* (Vienna, 1930). Both were published by the Münster Verlag and appeared in four or more printings. The former discussed the safety and use of condoms, pessaries, and antispermatic pills, recommended the best brands, and quoted the approximate price.

121. For the numerous sex manuals readily available to workers in Weimar Germany, see Grossman, "New Woman and Rationalization of Sexuality," 159-62.

These, however, were neither published nor distributed by the German Socialist party. Sexual enlightenment was undertaken by unorthodox or renegade socialists and communists frequently allied with bourgeois reformers.

122. See Andrea Schurian, "Der Agitationswert der Abtreibungsfrage in den sozialdemokratischen Medien der Ersten österreichischen Republik," (Ph.D. diss. University of Vienna, 1982), 252–53.

123. Grossman, "1931 Campaign," 128–32. Again, it was not the parties of the left but a broad coalition of "liberal and radical lawyers, doctors, and other intellectuals, Social Democrats, Communists, and thousands of women of all classes and many parties" which mounted the 1931 antiabortion campaign.

124. See Ferdinand Klostermann et al, *Kirche in Österreich, 1918–1965* (Vienna, 1965), 241–71, and Alfred Diamant, *Austrian Catholics and the First Republic: Democracy, Capitalism, and the Social Order, 1918–1934* (Princeton, N.J., 1960), 169–70.

125. See Wolfgang Maderthaner, "Die Schule der Freiheit—Otto Glöckel und der Wiener Schulreform," *Archiv: Mittteilungsblatt des Vereins für Geschichte der Arbeiterbewegung* 24 (July–Sept. 1984): 9–10.

126. All bishops' pastoral letters can be found in *Wiener Diözesanblatt* and/or *St. Pölten Diözesanblatt* by year of proclamation.

127. *Wiener Dözesanblatt*, July 10, 1926, 41–43; *St. Pölten Diözesanblatt* 4 (1926): 50–51.

128. See Alfred Pfoser, "Politik im Alltag: Zur Kulturgeschichte der Ersten Republik," *Zeitgeschichte im Unterricht* 5 (1978).

129. For the following, see Alfred Pfoser, "Der Wiener Reigen-Skandal: Sexualangst als politisches Syndrom der Ersten Republik," in Konrad and Maderthaner, eds., *Neuere Studien*, III.

130. The storming of the theater and other Catholic Action street violence led to the closing of the play by the municipal authorities on grounds of danger to public safety.

131. For the following, see Murray Hall, *Der Fall Bettauer* (Vienna, 1978), ch. 3, and Alfred Pfoser, *Literatur und Austromarxismus* (Vienna, 1980), 162–64, 194–97.

132. Ibid., 65–68.

133. For the seminal work on street socialization in Vienna, see Hans Safrian and Reinhard Sieder, "Gassenkinder, Strassenkämpfer: Zur politischen Sozialisation einer Arbeitergeneration in Wien 1900 bis 1938," in Lutz Niethammer and Alexander von Plato, eds., *Wir kriegen jetzt andere Zeiten* (Berlin, 1985). See also Wegs, *Growing up Working Class*, 68–74.

134. See Paul Wenger, "Kann die sexuelle Verwahrlosung der jugendlichen Mädchen wirksam bekämpft werden?," *Zeitschrift für Kinderschutz, Familien- und Berufsführsorge* 20 (1928): 175–76.

135. See Otto Felix Kanitz, *Das proletarische Kind in der bürgerlichen Gesellschaft* (Jena, 1925), 36–44, 72–76. Editor of *Die sozialistische Erziehung*, Kanitz was very influential among groups of young socialist teachers of the Schönbrunnerkreis, a power in the Kinderfreunde, leader of the Socialist Worker Youth from 1926 to 1934, and head of the SDAP propaganda bureau after 1931. See also Marianne Pollak, "Wer soll das Proletarierkind erziehen?," *Die sozialistische Erziehung*, 1:1 (1921): 17–19.

136. See Anton Tesarek, *Die österreichischen Kinderfreunde, 1908–1958* (Vienna, 1958).

137. The organizers admitted that 87 percent of Viennese children had not been weaned from the street. The number was probably larger. See *Die sozialistische Erzie-*

hung 6:3 (1926). In the centers the emphasis on washing and cleanliness was monomaniacal.

138. Anton Tesarek, *Das Buch der Roten Falken* (Vienna, 1926) 10–12, lists 15,117 members nationally of whom approximately 40 percent were in Vienna. See also Rosi Hirschegger, *Lasst die roten Fahnen weh'n: Die Geschichte der Roten Falken* (Innsbruck, 1987).

139. See Wolfgang Neugebauer, *Bauvolk der kommenden Welt: Geschichte der sozialistischen Jugendbewegung in Österreich* (Vienna, 1975), 113–218.

140. By 1931 this led to a rebellion against political constraints within the SAJ, expressed in a general critique of the passivity of the SDAP in the face of increased right-wing threats to the party. See Rabinbach, *Crisis*, ch. 3.

141. Neugebauer, *Bauvolk*, 138. The number is small when compared to the adults in the SDAP or the 14–21-year-olds in the Viennese population. The explanation offered by Neugebauer and others that the SAJ was a feeder organization which sent a large percentage into the party each year does not alter the low ratios mentioned above. Both the Rote Falken and the SAJ were coeducational, but girls constituted only 25 percent of either organization.

142. See Therese Schlesinger, *Wie will und soll das Proletariat ihre Kinder erziehen?* (Vienna, 1921), 2–4.

143. The church attacked the SDAP's youth organizations in general as subverters of parental authority and especially as corruptors of morals in their comingling of the sexes. See Gulick, *Austria*, 359–60, 609–10.

144. Virtually every party leader of consequence wrote glowingly on this subject. *Der jugendliche Arbeiter* and *Die sozialistische Erziehung* featured it regularly.

145. Commandment #9. Tesarek, *Rote Falken*, 8.

146. Comradeship was offered as a substitute for sexual drives. See Otto Felix Kanitz, *Kampf und Bildung* (Vienna, 1920), 22–23. For similar tendencies in the German SAJ, see Karen Hagemann, "Wir jungen Frauen fühlten uns wirklich gleichberechtigt," in Wolfgang Ruppert, ed., *Die Arbeiter: Lebensformen, Alltag und Kultur von der Frühindustrialisierung bis zum "Wirtschaftswunder"* (Munich, 1986), 77–78.

147. "Worte eines Proletariervaters," *Der jugendliche Arbeiter* 25 (July 1926): 108–9.

148. "Mädel von Heute—Zur Ehe," ibid., 29 (June 1930): 12–13.

149. *Gewerkschaft, Jugend und Kultur* (Vienna, 1928), 3–20.

150. "Unsere Arbeit," *Handbuch für die Tätigkeit in der sozialistischen Jugendbewegung* (Vienna, 1929), 166–69.

151. See Therese Schlesinger and Dr. Paul Stein, "Leitsätze für die sexuelle Aufklärung der Jugend," *Bildungsarbeit* 19 (Dec. 1932).

152. *Irrefahrten: Aus dem Tagebuch eines suchenden Mädels* (Vienna, 1929).

153. On the conflict and gradual blending of traditional values and changing circumstances, see Joan W. Scott and Louise A. Tilly, "Women's Work and the Family in Nineteenth-Century Europe," *Comparative Studies in Society and History* 17 (1975), 42–64.

154. *Ortswechsel: Die Geschichte meiner Jugend* (Frankfurt/Main, 1979)(, 94–96. Buttinger's case is particularly interesting because he was a graduate of the SDAP's prestigious *Arbeiterhochschule* and was the model young socialist the party's program sought to create. After the SDAP was outlawed in 1934, he became head of the underground party (renamed Revolutionäre Sozialisten Österreichs) in 1935.

155. Among the older workers the main transgressions were drunkenness and

wife beating. See Anson Rabinbach, "Politik und Pädagogik: Die österreichische sozialdemokratische Jugendbewegung, 1931–32," in Gerhard Ritter, ed., *Arbeiterkultur* (Königstein, 1979), 174–75.

156. See for instance Dr. Karl Kautsky, "Die Pflicht der Gesundheit," *Der jugendliche Arbeiter* 25 (July 1926): 106–8, and Gerda Brunn-Kautsky, "Proletariermädchen und Körperkultur," ibid., 23 (March 1924): 12–13.

157. Marie Jahoda-Lazersfeld pointed out that the party leaders had never resolved the problem of freedom and authority in the conduct of party life at the organizational level. On the sexual question, she found, the leaders retained a surprising degree of inhibition. See "Autorität und Erziehung in der Familie, Schule und Jugenbewegung Österreichs," *Studien über Autorität*, 720–21.

158. Reich, "Sexualnot," 87–92.

159. Reich, *Geschlechtsreife*, 122–24; *Sexualerregung*, 7–14; and "Sexualnot," 75, 80–83.

160. Ernst Fischer, *Krise der Jugend* (Vienna, 1931), especially 15, 17, 22–25, 33–38, 52–53.

161. This included the domination of men over women in the sexual realm and in general. For a thorough discussion of Fischer's views, see Anson Rabinbach, "Ernst Fischer and the Left Opposition in Austrian Social Democracy," (Ph.D. diss. University of Wisconsin, 1973), ch. 3.

162. See Fallend, "Reich," 256–75; Erich Wittmann, "Wilhelm Reich," *Wiener Tagebuch*, June 1985, and idem., "Reich in Wien," *Die Linke* 7 (April 9, 1986).

163. For the confrontation at the party congress, see Rabinbach, *Crisis*, 73–154. Shortly after the abortive rising of February 1934, Fischer joined the Communist party.

164. See Leichter, *So leben wir*, introduction, and Sophie Lazersfeld, *Wie die Frau den Mann erlebt* (Leipszig, 1931).

165. For the problem of evidence in reconstructing everyday life, see Alf Lüdtke, ed., *Alltagsgeschichte: Zur Rekonstruktion historischer Erfahrungen und Lebensweisen* (Frankfurt, 1989); Peter Borscheid, "Plädoyer für eine Geschichte des Alltäglichen," in Hans Teuteberg, ed., *Ehe, Liebe, Tod: Zum Wandel der Familien-, Geschlechts- und Generationsbesichtigung in der Neuzeit* (Münster, 1983); Helene Maimann, "Bemerkungen zu einer Geschichte des Arbeiteralltags," in Gerhard Botz et al., *Bewegung und Klasse: Studien zur österreichischen Arbeitergeschichte* Vienna, 1978); Elizabeth Roberts, *A Woman's Place: An Oral History of Working Class Women, 1890–1940* (Oxford, 1984); Hubert Ch. Ehalt, ed., *Geschichte von Unten* (Vienna, 1984); Robert Wheaton and Tamara K. Hareven, eds., *Family and Sexuality in French History* (Philadelphia, 1980); and Gerhard Botz and Joseph Weidenholzer, eds., *Mündliche Geschichte und Arbeiterbewegung* (Vienna, 1984).

166. The following relies on Reinhard Sieder, "'Vater derf i aufstehn?': Kindheitserfahrungen in Wiener Arbeiterfamilien um 1900," in Hubert Ch. Ehalt and Gernot Heiss, eds., *Glücklich ist wer vergisst . . . ? Das andere Wien um 1900* (Vienna, 1986). Here and elsewhere Sieder's work is based on some sixty extensive oral histories deposited on tape and in transcript at the Institute for Economic and Social History of the University of Vienna.

167. See Reinhard Sieder, "Housing Policy, Social Welfare, and Family Life," 42.

168. These conditions were attested to by the child psychologist and municipal councillor Joseph Friedjung, *Die geschlechtlich Aufklärung im Erziehungswerk* (Vienna, 1926), 7–8, and the psychologist and socialist youth functionary Hildegard Hetzer, *Kindheit und Armut* (Leipzig, 1929), 122.

169. See Sieder, "Vater," 62–72. A study of worker sexuality in late-nineteenth-century Germany also reports early sexual knowledge of the young through direct observation. See Robert Neumann, "Industrialization and Sexual Behavior: Some Aspects of Working-Class Life in Imperial Germany," in Robert J. Bezucha, ed., *Modern European Social History* (Lexington, Ky., 1972), 277–78.

170. *Das reifende Proletariermädchen* (Vienna, 1931). This study was based on personal observations, letters from students, questionnaires, and home visits, and was published in a monograph series of the Institut für pädagogische Psychologie. The following discussion is based on 67–80. For a surprisingly opposite reading and interpretation of the evidence on childhood encounters with sexuality, see Wegs, *Growing Up Working Class*, 125–27.

171. As a loyal member of the socialist pedagogical establishment, Rada concluded that only idealistic socialist organizations had the power to combat the negative influence of the home to which the girls were exposed (80).

172. Sieder, "Vater," 47.

173. *Kindheit und Armut*, passim.

174. But even the cared-for appeared to have knowledge at an early age about coitus and childbirth. See Alice Rühle-Gerstel, *Das Frauenproblem der Gegenwart: Eine psychologische Bilanz* (Leipzig, 1932), 413–19.

175. See Charlotte Bühler, *Kindheit und Jugend: Genese des Bewusstseins* (Leipzig, 1931).

176. See Safrian and Sieder, "Gassenkinder," passim, for the following discussion.

177. Alfred Adler had made a study of the rich texture and gender interaction of children's street play in which early sexual exploration and expression were apparent. See "Erotische Kinderspiele," in Friedrich S. Krauss, ed., *Anthrophyteia: Jahrbücher für folkloristische Erhebungen und Forschungen zur Entwicklungsgeschichte der geschlechtlichen Moral* (Leipzig, 1911), VIII, 257–58.

178. Sieder observes that this peer group freedom in the street had its daily end with the return of the father from work and the resumption of parental authority in the home. With the onset of work, the freedom outside the home of males increased, whereas that of females was diminished. See "Vater," 52–53, 60–61.

179. See Hans Safrian, "'Wir ham die Zeit der Orbeitslosigkeit schon richtig genossen auch': Ein Versuch zur (Über-) Lebensweise von Arbeitslosen in Wien zur Zeit der Weltwirtschaftskrise," in Botz and Weidenholzer, eds., *Materialien*, 293–331.

180. Among the court cases for abortion on file for Vienna, the average age was seventeen to eighteen years. See "Strafprozessakte zum Paragraph 144," Archiv der Stadt Wien (1921–32).

181. Buttinger, *Ortswechsel*, 125–27.

182. Reich, *Sexualerregung*, 46–69.

183. See Weltliga, *Sexualnot*, 114–15.

184. Safrian and Sieder, "Gassenkinder," 130.

185. The difficulty of attracting females to the youth organizations was a constant refrain in SDAP publications. See for instance, "Gibt es eine Mädlfrage in der Jugendorganisation?," *Die sozialistische Erziehung* 9 (April 1930): 90.

186. See J. Robert Wegs, "Working Class Respectability: The Viennese Experience," *Journal of Social History* 15 (Summer 1982): 630–31.

187. See Lazersfeld, "Autorität," 720–21.

188. See Friedrich Scheu, *Ein Band der Freundschaft: Schwarzwalder Kreis und Entstehung der Vereinigung Sozialistische Mittelschüler* (Vienna, 1985), 153–55.

189. See Jenny Strasser, "Manchmal hat die Polizei all festgenommen," in Fritz Keller, ed., *Lobau—die Nackerten von Wien* (Vienna, 1985), 64–65.
190. "Gassenkinder," 131.
191. See Maria Bayza, "Die schönste Art unglücklich zu sein," in Keller, *Lobau*, 69–70.
192. *Die sozialistische Erziehung* 1 (Jan. 1922).
193. See Henrietta Kotlan-Werner, *Otto Felix Kanitz und der Schönbrunner Kreis: Die Arbeitsgemeinschaft sozialistischer Erzieher, 1923–1934* (Vienna, 1982), 297–300.
194. See *Geschlechtliche Aufklärung*, 16–31.
195. See Ernst Glaser, *Im Umfeld des Austromarxismus: Ein Beitrag zur Geistesgeschichte des österreichischen Sozialismus* (Vienna, 1981), 273–333.
196. See *Die Sozialistische Erziehung* 7 (Nov.–Dec. 1927).
197. *Die Ehe von heute und morgen* (Munich, 1927), 66–69.
198. Gottfried Pirhofer and Reinhard Sieder, "Zur Konstitution der Arbeiterfamilie im Roten Wien: Familienpolitik, Kulturreform, Alltag und Ästhetik," in Mitterauer and Sieder, eds., *Historische Familienforschung*, 348. Premarital intercourse between courting couples was also common in Lancashire, England. See John R. Gillis, *For Better or Worse: British Marriages, 1600 to the Present* (New York, 1985), 235. For similar practices in France and Germany, see Daniel Bertaux and Isabelle Bertaux-Wiame, "Jugendarbeit bei freier Unterkunft und Verpflegung—Bäckerlehrlinge und Hausmädchen im Frankreich der Zwischenkriegszeit," in Botz and Weidenholzer, eds., *Mündliche Geschichte*, 266–70, and Carola Lipp, "Sexualität und Heirat," in Ruppert, *Arbeiter*, 193–94.
199. Pirhofer and Sieder, "Konstitution," 346–47.
200. See Eva Viethen, "Wiener Arbeiterinnen: Leben zwischen Familie, Lohnarbeit und politischen Engagement" (Ph.D. diss., University of Vienna, 1984), 309, 357.
201. For the very best of these, see Ehmer, "Frauenarbeit," 438–73.
202. See Elizabeth Maresch, *Ehefrau im Haushalt und Beruf: Eine statistische Darstellung für Wien auf Grund der Volkszählung vom 22. März 1934* (Vienna, 1938), 13, 36. Among working wives 49.44% had no children, 44.45% had one child, and 6.11% had two children.
203. The same explanation for France and England, beginning with the later nineteenth century, is given in Louise Tilly and Joan W. Scott, *Women, Work, and Family* (New York, 1978), 170–72. The authors observe that a decrease in family size actually led to an increase in the mother's responsibility for child care (210–11). For comparable conditions in Germany, see Neumann, "Industrialization," 289–91.
204. Ehmer, "Frauenarbeit," 451.
205. Applying condoms or diaphragms in a dark bedroom crowded with children and possibly other adults was surely no easy feat. Reich's instructions for this procedure are daunting. See *Sexualerregung*, 24–26. The average cost, according to Reich, of a package of three condoms was 1.5 to 3 Schillings. The average worker budget in 1932 allowed five Schillings or less for incidentals that included tobacco, beverages, and entertainment. If it was increased from time to time, it was at the expense of essentials such as food. See Fritz Klenner, *Die österreichischen Gewerkschaften* (Vienna, 1953), II, 893, and Bendikt Kautsky, *Die Haushaltstatistik der Wiener Arbeiterkammer, 1925–1934*, supplement of *International Review of Social History* 2 (1935): 245–46.
206. The same applies for France and England. See Etienne van de Walle, "Motivations and Technology in the Decline of French Fertility," in Wheaton and Har-

even, eds., *Family and Sexuality,* 147–52, and D. Gittins, *Fair Sex: Family Size and Structure, 1900–1939* (London, 1982), 169. For Germany, see Dr. Max Marcuse, *Der eheliche Präventivverkehr: Seine Verbreitung, Verursachung und Methodik* (Stuttgart, 1917), 168–72. Coitus interruptus was supplemented by various homely methods and devices—douching, cotton wads, postcoital urination—of negligible effectiveness.

207. Ehmer, "Frauenarbeit," 468–69.

208. In the period 1851–1920 seventy case files were deposited in the Vienna archives, including the notorious Mittermayer file. See Katharina Riese, *In wessen Garten wächst die Leibesfrucht?: Das Abreibungsverbot und andere Bevormundungen Gedanken über die Widersprüche im Zeugungsgeschäft* (Vienna, 1983), 49, 89–119.

209. The cases of Anna Ernst, Anna Sternak (1921); Rosalia Kaufmann, Anna Haselmayer, Marie Prudek, Barbara Zimmermann (1922); Margarethe Hatzl, Marie Schmidt, Anna Konrath, Anna Geist (1923); Sophie Rauch, Joseph Banauer (1929); Emma Goldmann (1930); Aloisia Bribitzer, Leopoldine Heinz (1931); Veronika Unger (1932). "Strafprozessakte zum Paragraph 144."

210. The judiciary in other countries (Czechoslovakia, Germany, Switzerland) was considering the decriminalization of abortion. See the report on the Swiss supreme court for instance, "Schweizer Richter gegen die Abtreibungsbestrafung," *Arbeiter-Zeitung,* June 7, 1927.

211. What the migraine was for the middle-class wife, work in the evenings, until her husband went to bed and fell asleep, was for the working-class wife: a means of avoiding intercourse. See Petra Helm, "Interviews mit Frauen über Sexualität und Hygiene," Institut für Wissenschaft und Kunst, *Oral History Projekte in Österreich* (Vienna, 1984), 66.

212. The following is based on Reinhard Sieder, "Die Geschichte der einfachen Leute—ein Thema für Geschichtswissenschaft und Unterricht," *Beiträge zu Historischen Sozialkunde* 14 (Jan.–March 1984), 27–31.

213. Cases of wives refusing intercourse for similar reasons are recorded in England. See Ellen Ross, "Fierce Questions and Taunts: Married Life in Working-Class London, 1870–1914," *Feminist Studies* 8 (Fall 1982): 595, and Roberts, *Woman's Place,* 95.

214. See Appelt, *Ladenmäden, Schreibfräulein und Gouvernanten,* 158–61. Appelt reports one case in which a young wife was forced to terminate her pregnancy by her mother-in-law because of the crowded conditions in the common dwelling (172).

215. *So leben wir,* 41–44, 73–74, 78–79, 81–83, 109–10.

216. See the pioneering empirical social science study of Marie Jahoda, Paul Lazersfeld, and Hans Zeisel, *Die Arbeitslosen von Marienthal: Ein soziographischer Versuch* (1933; Frankfurt, 1980), 83–112. A more recent study finds this loss of male affect primarily among settled heads of families. See Safrian, "Wir ham die Zeit," 316–20.

217. The SDAP killed the squatters/garden city movement, which was based on initiatives from below and on worker self-management, in 1923. See Klaus Novy, "Selbsthilfe als Reformbewegung: Der Kampf der Wiener Siedler nach dem 1. Weltkrieg," ARCH: Zeitschrift für Architekten, Sozialarbeiter und kommunalpolitische Gruppen 55 (March 1981). Initiatives from below and attempts at self-management were quickly nipped in the bud by the party bureaucracy even in the new municipal housing. See Alfred Frei, *Austromarxismus und Arbeiterkultur* (Berlin, 1984), 110–13.

218. Weeks, *Sexuality,* 57–59.

219. See Pirhofer, "Politik am Körper," 69.

220. See *Mieterschutz,* 8–9.

221. See Appelt, *Ladenmädchen,* 124–25; Reinhard Krammer, *Arbeitersport in Österreich* (Vienna, 1981), 181.

222. See Scheu, *Band der Freundschaft,* 127, 190.

223. See Kotlan-Werner, *Kanitz,* 72–79, 189, 296.

224. On Bauer and Adler, see Peter Lowenberg, *Decoding the Past: The Psychohistorical Approach* (New York, 1983), 194–96, 149.

225. It was published as *Du mariage* in 1907 and reprinted in 1937, despite the anticipated right-wing and anti-Semitic onslaught. See Jean Lacouture, *Léon Blum* (New York, 1982), 80–85.

226. See Hans Schafranek, "'Die Führung waren wir selber'—Militanz und Resignation im Februar 1934 am Beispiel Kaisermühlen," in Konrad and Maderthaner, eds., *Neuere Studien,* II: 439–70, and Hans Safrian, "Mobilisierte Basis ohne Waffen—Militanz und Resignation im Februar 1934 am Beispiel der Oberen und Unteren Leopoldstadt," ibid., II: 471–90. For the general paralysis of will, see Rabinbach, *Crisis,* ch. 4.

227. That the Christian Social party drew no distinction between politics and culture insofar as the socialists were concerned can be seen from the telling blow it struck against Vienna in 1932 by drastically reducing its tax apportionment from the national budget.

228. An imaginative Foucaultian study suggests the SDAP used sexuality as a means of disciplining, regimenting, and controlling the workers. See Doris Beyer, "Die Strategien des Lebens: Rassenhygiene und Wohlfahrtswesen—Zur Entstehung eines sozialdemokratischen Machtdispositivs in Österreich bis 1934" (Ph.D. diss., University of Vienna, 1986), 238–55.

Conclusion

1. See Julius Braunthal, *Die Arbeiterräte in Deutschösterreich* (Vienna, 1919); Otto Bauer, *Die österreichische Revolution* (Vienna, 1923); Rolf Reventlow, *Zwischen Aliierten und Bolschewiken: Arbeiterräte in Österreich, 1918 bis 1923* (Vienna, 1969); Hans Hautmann, *Die verlorene Räterepublik: Am Beispiel der Kommunistischen Partei Deutschösterreichs* (Vienna, 1971): and Helmut Gruber, *International Communism in the Era of Lenin* (Ithaca, N.Y., 1967), 191–217.

2. The KPÖ had about 10,000 members at the end of 1919, 4,300 of which were in Vienna; at the election for the Constituent Assembly it failed to win a mandate. Neither its membership nor electoral strength changed significantly throughout the period. See Herbert Steiner, *Die Kommunistische Partei Österreichs von 1918 bis 1933: Bibliographische Bermerkungen Meisenheim/Glan, 1968),* 24 and passim. The SDAP was able to dismiss the KPÖ's radical critiques as parrotings of the Communist International. The Austrian socialists' ability to claim the undisputed leadership of the whole working class was not shared by socialists in France and Germany, for instance, where strong communist parties offered radical critiques and programs in competing for worker affiliation and support. See for example Julian Jackson, *The Popular Front in France: Defending Democracy, 1934–38* (Cambridge, 1988), and Heinrich August Winkler, *Der Weg in die Katastrophe: Arbeiter und Arbeiterbewegung in der Weimarer Republik, 1930 bis 1933* (Berlin, 1987).

3. For Bauer's prognostication, see *Der Kampf um die Macht* (Vienna, 1924), 25, and the many daily leaders he wrote in *Die Arbeiter-Zeitung* from 1924 to 1927. In Vienna the SDAP gained 57–60% of the vote. Its proportion of the national vote

fluctuated between 36% and 42.3%; in mandates, from 37.7% to 43.6%. See Charles A. Gulick, *Austria: From Habsburg to Hitler* (Berkeley, Calif., 1948), I: 690, 792; II: 914–15.

4. In the national election of 1930 the SDAP received 1,517,251 votes and had a membership of 698,181. This meant that 53% of its votes came from others than organized socialists. In the Viennese municipal election of 1932 the SDAP received 683,295 votes and had a membership of 400,484; 41% of its votes came from outside the party.

5. See Helmut Gruber, *Léon Blum, French Socialism, and the Popular Front: A Case of Internal Contradiction* (Ithaca, N.Y., 1986), 11–12. See also Charles S. Maier, "The Weaknesses of the Socialist Strategy: A Comparative Perspective," in Anson Rabinbach, ed., *The Austrian Socialist Experiment: Social Democracy and Austromarxism, 1918–1934* (Boulder, Colo., 1985). Maier calls Bauer's notion of undertaking a "thoroughgoing socialist transformation [on the basis of a 51% majority] a disastrous type of programmatic concept" (249).

6. See Otto Bauer, "Das Gleichgewicht der Klassenkräfte," *Der Kampf* 17 (Jan. 1924): 57–67. For the critics, see Hans Kelsen, "Dr. Otto Bauers politische Theorie," ibid., 17 (Feb 1924); 50–56, and Otto Leichter, "Zum Problem der sozialen Gleichgewichtszustände," ibid., 17 (May 1924): 184.

7. Coalition discussions were resumed in the SDAP in the early 1930s, but disbelief in the good faith of the opposition and fear of abandoning red Vienna prevented serious consideration. Bauer stuck to his position on coalitions. See Labour and Socialist International, *After the German Catastrophe: The Decisions of the International Conference of the LSI in Paris, August 1933* (Zurich, 1933), 11–12.

8. The concept "republic" was unpalatable to Chancellor Seipel, who never spoke of the "Austrian republic" but only of the "Austrian state." See Ernst Hanisch, "Der politische Katholizismus als ideologischer Träger des 'Austrofaschismus,'" in E. Talos and W. Neugebauer, eds. *"Austrofaschismus": Beiträge über Politik, Ökonomie und Kultur, 1934–1938* (Vienna, 1984), 57.

9. On the question of whether or not there is a socialist art, see Brigitte Emig, *Die Veredelung des Arbeiters: Sozialdemokratie als Kulturbewegung* (Frankfurt, 1980), 278–86.

10. For utopian orientations of cultural movements, see Dieter Kramer, *Theorien zur historischen Arbeiterkultur* (Marburg, 1987), 239–43.

11. The literature on Gramsci's theories has reached avalanche proportions. For a ready access to "hegemony," see Quintin Hoare and Geoffrey N. Smith, eds., *Selections from the Prison Notebooks of Antonio Gramsci* (New York, 1971), 206–76. For the culturist interpretation, see Gwyn A. Williams, "The Concept of 'Egomonia' in the Thought of Antonio Gramsci: Some Notes on Interpretation," *Journal of the History of Ideas* 21.4 (Oct.–Dec. 1960); Christian Riechers, *Antonio Gramsci: Marxismus in Italien* (Frankfurt, 1970), 192–223; and James Joll, *Antonio Gramsci* (New York, 1977), 116–34.

12. For the latter, see Sheila Fitzpatrick, *Cultural Revolution in Russia, 1928–31* (Bloomington, Ind., 1978).

13. The massive exhibitions mounted in Vienna in the past fifteen years to celebrate the red Vienna of the First Republic have provided the municipal socialist administration with a heroic past and even, as cynics would have it, *Touristenzukerln* (bonbons for tourists). But these displays of past accomplishments also appear to embody a nostalgia for former aspirations rendered obsolete by a less idealistic present.

Index

N.B.: Numbers in italics refer to illustrations.

Abortion, 159–63, 179
  as birth control method, 176
  incidence of, 159, 246n
  justification for, 160
  SDAP's position on, 159–63, 179
  sentences for, 176–77, 246n
  trimester model and, 159–160
Adler, Alfred, 76
  ego psychology of, 112–13, 125, 175
Adler, Friedrich, 20, 26, 30, 31, 33, 107
  sexual affairs of, 178
Adler, Max, 26, 30, 31, 32
  Austromarxism and, 34–36
  confrontation with Renner, 42
  criticism of SDAP's cultural efforts, 91
  cultural revolution and, 39
  defensive-force position of, 40–41
  postwar role of, 33–34
  socialist culture and, 83–84, 86
Adler, Victor, 53, 107–8
  socialist culture and, 83
After-school centers *(Horte)*, 69, 166
Agrarian League, 42
Allianz (film distribution company), 133
All Quiet on the Western Front (film),
  controversy over, 235–36n
American Tragedy, An (film), 129
Anna Christie (film), 129
A nous la liberté ((film), 129
Anticlericalism, 28
Anti-Semitism, 25–27
  of Christian Social party, 72, 163
Apartment inspector, in public housing
  projects, 63
Apartments. *See* Public housing; Tenements
Arbeiterbank, 133

Arbeiterbund für Sport und Körperkultur in
  Österreich (ASKÖ), 103, 104, 106, 142
  membership of, 112, 141
Arbeiterheime, 117
Arbeiterhochschule, 90, 92
Arbeiterinnen-Zeitung, Die, 88
Arbeiterschulen, 92
Arbeitersymphoniekonzerte, 82
Arbeiter-Zeitung, Die, 37, 40, 41–43, 87, 96,
  97, 170
  on Austromarxism, 30
  circulation and readership of, 88, 89
  film reviews in, 131, 134
  on housing program, 52
  on pornography, 164
Architecture, 206–7n
  housing program and, 56
Aristocratic titles, 23
Art. *See also* Fine arts; *specific arts*
  socialist, definition of, 97
Art reproductions, 100–101
ASKÖ. *See* Arbeiterbund für Sport und
  Körperkultur in Österreich
Association for Birth Control, 161
Association for Social Science Education, 32
Association for Sports and Body Culture, 82
Atlantic (film), 130
Austerlitz, Friedrich, 26
Austerlitz Spricht, 89
Austria
  question of national viability of, 24–25
  republic declared in 1918, 13, 15
  republic of 1920, *14*
Austria-Hungary, Dual Monarchy of, *14*
Austrian Socialist party (SDAP)
  abortion and, 159–63, 179

257

Austrian Socialist party (SDAP) (*continued*)
  advantage in Constitutional Assembly, 181
  anti-Semitism and, 26–27
  attempt to control workers' power, 19–20
  attempt to force Seipel to make concessions, 41
  Austrian republic and, 13, 15, 24
  basis of attempt to develop proletarian counterculture, 5–11
  birth control and, 161, 162
  Catholic church's power and, 28–29
  coalition with Christian Socials, 16, 21
  cultural directors of, 85–86
  cultural organizations of, 81–82
  cultural project of. *See* Cultural project; Socialist party culture
  defensive-force position and, 40–41
  dichotomy between leaders and followers of, 7–8, 20, 54, 178
  educational reform program of, 77–80
  educational reforms of, 73–80
  election of 1919 to Constituent Assembly and, 21
  elections of 1927 and, 42
  elections of 1919 to 1930 and, 182–83
  emergence as mass party, 181
  failure to involve workers in housing plans, 52–53
  film and, 130–31
  intervention in cinema, 132–36
  Jews in, 26, 195n
  lack of preparation for role in establishing republic, 182–83
  lecture program of, 91–92
  mass resignations from, 140
  membership of, 10, 20, 81, 108, 220n, 223n, 255n
  "new woman" and, 147–55
  opponents of, 183–84
  opposition to coalition of Christian Socials and Pan-Germans, 29
  organization and decision-making bodies of, 53–54
  paid functionaries of, 53–54
  paternalism of, 8, 52–53, 90, 147, 162–63
  practical advice offered to women, 148, 150–51, 154–55
  program for children, 166–67
  proletkult adapted by, 109
  publications of, 87–91
  radio and, 136, 138, 139–41
  reforms of 1918–19 and, 21–22
  Schnitzler, Bettauer, and Baker sex scandals and, 179
  schools for leadership of, 90, 92
  settlers' movement and, 48
  sexuality and, 156–59
  social composition of, 20
  sports programs of, 102–7
  strategy of passivity of, 4
  theoretical orientation regarding proletarian culture, 86–87
  unsalaried cadres of, 54
  women's roles and, 150, 178
  worker festivals and, 107–12
  worker libraries and, 92–96
Austromarxism, 5, 29–44, 183, 197n, 198n
  culture and class struggle and, 5–11
  environmentalism and, 46
  family and, 147
  first appearance of, 32
  intellectual portrait of Austromarxists and, 30–32
  Jews and, 31
  two distinctive groups in, 30
Avenarius, Ferdinand, 86

Bach, David Joseph, 37, 85, 86
  Sozialdemokratische Kunststelle under, 96–102
Bahr, Hermann, 109
Baker, Josephine, 119, 164, 179
Balance of class forces, 37–41, 179
Balazs, Béla, 131
Bathhouses, municipal, 60, 66
Bauböck, Rainer, 192n, 202n, 203n, 204n, 206n, 207n
Bauer, Helene (Gumplowicz), 26, 150
Bauer, Otto, 5, 6, 20, 21, 23, 26, 30, 31, 36, 38, 43, 89, 90, 178
  arguments for separation of church and state, 29
  Austromarxism and, 30, 33, 37–41
  confrontation with Renner, 42
  customs union proposed by, 24
  defensive-force position of, 40–41
  explanation of 1919 elections to Constituent Assembly, 192n
  housing program and, 50, 54
  party hierarchy and, 7
  power of SDAP and, 182, 183
  response to anti-Semitism, 27
  on role of women, 178
  socialist culture and, 86
  socialization program of, 22
Bauer, Otto ("der kleine Bauer"), 28
Beethoven, Ludwig van, 99
Berlin
  educational reform in, 77
  public housing in, 60

Berlin Alexanderplatz (film), 129
Bettauer, Hugo, 95, 179
   Catholic church's campaign against, 164
Bettelheim, Bruno, 32
Beyer, Doris, 211–12n
Bichlmaier, Georg, 28
Bildung (civilization), 39
   Austromarxism and, 34
   education and, 73
   socialist commitment to, 35–36
Bildungsarbeit, Die, 89, 131–32
   film reviews in, 131
Bildungszentrale, 82, 86, 91, 92, 93, 100, 103, 111, 132, 134
Birth control, 46, 160
   abortion as, 176
   practiced by youth, 169
   SDAP's stand on, 161, 162
Birth rate, sexuality and, 176
Blaue Engel, Der (film), 129, 130, 131
Blum, Léon, 3, 178, 183
   response to anti-Semitism, 27
Böhm-Bawerk, Eugen, 32
Bolschwismus oder Sozialdemokratie? (Bauer), 89
Bolshevik party, 5
   SDAP rejection of elitism of, 7
Bolshevik Revolution, impact of, on Austrian workers, 19
Book clubs, middle-class, 219n
Books, 88
   inexpensive, 89
   sex manuals, 162, 247–48n
   value placed on, 87
   worker libraries and, 92–96, *94*
Boschek, Anna, 155
Botz, Gerhard, 188n, 190n, 191n, 201n, 204n, 225n, 227n, 228n, 230n, 231n, 237n, 250n, 252n
Boxing, 141
Brandstifter Europas, Die (film), 130
Braunthal, Julius, 26, 37, 89
Breitner, Hugo, 26, 37, 49, 133, 138
Bühler, Charlotte, 71, 76, 171–72
Bühler, Karl, 76
Building fund, 49
Building industry, housing program and, 55
Building methods, public housing and, 56–57
Bundeserziehungsanstalten, 76
Bunte Woche, 88
Burgtheater, 98
Buttinger, Joseph, 8, 168–69, 173–74
Bub und Mädl (Hodann), 162

Café Central, 33
Café Elektric (film), 130
Carltheater, 97
Carnap, Rudolf, 84
Catholic Action groups, 27–28, 164
   abortion and, 163
Catholic church
   abortion and, 163, 176
   anticlericalism and, 28
   educational reform and, 78–79
   marriage consultation clinics and, 68–69
   municipal crematorium and, 72
   power of, 27
   separation of church and state and, 27–29, 182
   separation of school and church and, 74
   sexuality and, 163–64
Catholic Day, 225n
Catholic radio club, 138, 139
Catholic youth, 24
Certeau, Michel de, 239n
CGT. *See* Confédération Générale du Travail
Chamber of Agriculture, 138
Chamber of Commerce, 138
Chamber of Workers and Employees, 21, 138
Charley's Aunt (film), 129
Chauvinism, ethnic, 25
Chess club, 103
Childbearing, exercise and, 151
Child care. *See also* Motherhood
   mothers' consultation clinics and, 69
   as women's responsibility, 152
Children
   after-school centers for, 69, 166
   aid to, 66
   illegitimate, 159
   infant layettes distributed to, 69, *70*
   kindergartens and, 66, 69, 152
   mothers' consultation clinics and, 69
   municipality's power to separate from parents, 68–69, 71–72
   public education and, 73–80
   recreational activities of, 121
   SDAP's program for, 166–67
   socialization of, *67,* 165, 171–72, 173
Childrens' diagnostic service *(Kinderübernahmestelle),* 71
Christian Social party, 4, 181
   abortion and, 159, 162, 163, 176
   anti-Semitism of, 26, 72
   attacks on Tandler's policies, 68–69
   coalition with SDAP, 16, 21
   commitment to monarchy, 13, 24
   confessional schools and, 215n

Christian Social party (*continued*)
  elections of 1927 and, 42
  health and welfare programs and, 68–69
  housing program and, 51
  housing requisitioning law and, 48
  1919 elections to Constituent Assembly and, 21
  pornography and, 164
  SDAP educational reform program and, 73–75, 78
  SDAP opposition to coalition of Pan-Germans with, 29
  sexuality and, 163–64
Church. *See* Catholic church
Cinema. *See* Films
Cinema conference, 130–31
Circuses, 118–19
Circus Kludsky, 118
Civilization. See *Bildung*
Class forces, balance of, 37–41, 179
Clothing
  utilitarian versus bourgeois, *149*
  of workers attending concerts, 99
Coitus interruptus, 169, 176
Combat gymnastics, 105–6
Comintern, 4
Commercial culture, 115
  leisure time and, 116–20
  problem of dealing with, 9
Communal facilities, 60–61, *62*, 72
Communist International, 4
Communist party, Austrian (KPÖ), 20, 181, 254n
Communist party, German (KPD), cultural efforts of, 81
Concierge, in public housing projects, 63
Condoms, 176
Confédération Générale du Travail (CGT), 232n
Confessional schools, 215n
Congress of Marxist Individual Psychology, 175
Conjugal rights, denial of, to prevent conception, 177
Conley, Andrew, 3
Constituent Assembly, 181
  1919 elections to, 21
Constitutional Assembly, SDAP seats in, 181
Contraception. *See* Birth control
Cooper, James Fenimore, 95
Cooperative societies, newspapers published by, 87
Cosmetics, SDAP discouragement of use of, 150

Council Movement, 33
Credé, Karl, 97
Crematorium, municipal, controversy over, 72
Croats, 25
Cultural project, 9. *See also* Socialist party culture
  shortcomings of, 184–85
  as substitute for politics, 183–84
Cultural revolution, Bauer's separtion of, from political revolution, 39
Cultural transformation, experiment in, 5–11, 180–86
Culture
  commercial, 9, 115, 116–20
  elite, rejection of and desire for, 83–87
  mass, 9
  noncommercial, 115
  popular, condemnation of, 123–26
  socialist. *See* Socialist party culture
Curtiz, Michael (Kertesz), 126
Czechs, 15, 18, 25, 194n
Czeija, Oskar, 136
Czeike, Felix, 190n, 202n, 203n, 206n, 207n, 211n, 212n, 213n, 229n

Dance, 119
Dancing, 120, 124–25
Danneberg, Robert, 26, 37, 133
  socialist culture and, 86
Danubian confederation, 15
Death rate, decline of, 66
Defensive-force position of SDAP, 40–41
deMan, Hendrik, 142
Demonstrations, *19*. *See also* Worker revolts; July 15, 1927, revolt
Department of Social Welfare, women employed in, 244n
Deutsch, Julius, 21, 26, 37, 111, 134
  paramilitary orientation of sports under, 105–6
  socialist culture and, 86
  on spectator sports, 142
Dietrich, Marlene, 125
Dollfuss, Engelbert, 3, 4, 187n
Domestic workers, 153
Drama, 97
Dreigroschenoper (Brecht and Weill), 97
Dreiser, Theodore, 95
Dreyfus (film), 130
Dreyfus Affair, 27
Dual Monarchy, *14*

Index 261

Dumas, Alexandre, 95
Düsseldorf, health and welfare programs in, 66

Economic legislation, 21–22
Economy
 inflation and, 22–23
 of postwar Austria, 10
Education, 214n. *See also* Schools
 cultural revolution and, 39
 public, 73–80
 religious, 74, 78–79, 182, 215n
 on sexuality, 175
Ego psychology. *See* Individual psychology
Ehmer, Joseph, 117, 191n, 202n, 240n, 242n
Einküchenhaus Heimhof, 51–52
Eisenstein, Sergie, 109
Ekstein, Therese. *See* Schlesinger, Therese
Election, 182–83. *See also* Suffrage
 campaign films and, 133
 Constituent Assembly, of 199, 21
 of 1927, 42
 of 1932, 238n
 parliamentary, of 1920, 21
 socialist vote and, 192n
Electoral law of 1918, 21
Elite culture, rejection of and desire for, 83–87
Ellenbogen, Wilhelm, 26
Emil und die Detektive (film), 129
Engels, Friedrich, 34–35
England
 public education in, 77
 public housing in, 64
Environmentalism, municipal reform and, 46
Ethnic groups. *See also specific groups*
 German Austrian chauvinism and, 25
 Viennese ethnic enclaves and, 18
 in Viennese population, 15, 18, 25, 194n
Eugenics
 abortion and, 160–61
 Tandler on, 68
Exercise
 during International Worker Olympics, *106*
 for women, 151

Factory councils, 21–22, 33
Family, 146–79. *See also* Children
 institutions set up to assist, 68–69
 municipal inspections of, 69
 nuclear, 146, 147
 orderly. *See* Ordentliche Arbeiterfamilie
 public housing and, 63
 sexuality and, 155–78
 socialist attempt to strengthen and shape, 146–47
 women and. *See* Women
Faust (Goethe), 86
Feasts, 100, 107–12, 225–26n
Federal educational institutes, 76
Ferch, Johann, 157
Ferris wheel, 117
Festivals, 100, 107–12, *110*, 225–26n
Feyder, Jacques, 126
Figaro, Le, 4
Films
 audiences at, 127–28
 as democratic art, 126–35
 election campaign, 133
 movie theaters and, 120, 127
 reviews of, 131, 132, 134, 234–35n
 socialist policy on, 132
Finanzkapital (Hilferding), 35
Fin-de-siècle Vienna (Schorske), 12
Fine arts, 100–101
Finker, Aegidius, 209n
Fischer, Ernst, 111
 criticism of socialist youth policy, 169–70
Flamme, Die, 28, 72
Folk dances, 124–25
Food
 lunchtime, 148, 150, 228n, 244n
 shortage of, 65
Forster, Friedrich Wilhelm, 178
France
 educational reform in, 79
 socialist response to anti-Semitism in, 27
Frankel, Victor, 174
Frankfurt
 educational reform in, 77
 public housing in, 57, 60, 64
Frankfurter Küche, 60
Frau, Die, 88, 90
Free Trade Unions, 20
Frei, Alfred G., 192n, 201n, 203n, 209n, 220n, 228n, 233n
Freie Schule, 74
Freie Volksbühne, 82
Frensham Heights, 77
Freud, Anna, 76
Freudlose Gasse, Die (Pabst), 13
Freud, Sigmund, 226–27n
Freundlich, Emmy, 26, 160
Friedjung, Joseph, 175

Friends of Nature, 103
Fuel, shortage of, 65
Furniture, *61*

Ganghofer, Ludwig, 95
Garbo, Greta, 126
Garçonne, La (Margueritte), 148
Garçonne/flapper figure of "new woman," 148, 150
Garden plots *(Schrebergärten)*, 123, 230n
Gas (film), 129
Gastgeb, Hans, 104
Gasthäuser, 18, 116–17, 120, 228n
  books as alternative to, 93
  worker festivals as alternative to, 108
Gefährdete Mädchen (film), 130
Gender roles
  conflicts created by work and, 153
  of women, 150, 178
German Socialist party, cultural efforts of, 81
German Song Society Festival, 118
Germany
  educational reform in, 77
  public housing in, 57, 60, 64
Glaser, Ernst, 231n, 236n
Glöckel, Otto, 37, 87
  educational reforms and, 74–76, 77, 78, 79–80
Golem, Der (film), 130
Gramsci, Antonio, 8, 255n
Grand Hotel (film), 129
Groh, Dieter, 39
Gruber, Helmut, 188n, 189n, 191n, 192n, 205n, 226n, 255n
Grünberg, Carl, 32
Gulick, Charles, 30
Gumplowicz, Helene. *See* Bauer, Helene

Hadow Commission, 77
Hahn, Otto, 84
Hamber, Edmund, 133, 134
Hamber, Philip, 133
Hamburg
  educational reform in, 77
  public housing in, 60, 64
Hanak, Anton, 69, 100
Hartmann, Ludo, 32
Hauptmann von Köpenick, Der (film), 129
Health and welfare programs, 65–73
  coercive nature of, 71
  in Düsseldorf, 66
  population politics and, 66, 68
  social workers and, 69–72

Health care
  incorporated into school system, 75
  prophylactic, 66
  socializing children for, *67*
Heimwehr, 40, 187n
  Ravag and, 139, 238n
Hetzer, Hildegard, 172, 173
Heurigen, 117, 120
Hiking, 121
Hilferding, Margarete, 211n, 247n
Hilferding, Rudolf, 30, 31, 32, 33, 35
Hirsch, Johann, 131
Hodann, Max, 162
Hofmannsthal, Hugo von, 109
Hohe Warte, 142
Homelessness, 47, 65
Homeworkers, 153
Horte. *See* After-school centers
Hospitals
  municipal, 69
  overcrowding in, 65
Housework, rationalization of, 148, 150–52
Housing. *See also* Housing shortage; Public housing; Tenements
  overcrowding in, 151–52, 154, 158, 159, 171
Housing requisitioning law (Wohnungsanforderungsgesetz), 48
Housing shortage, 45–46
  and emergency decrees of 1918–19, 48
  rent control and, 47
Housing tax, 49, 203n
Huckleberry Finn (film), 129
Hugenberg, Alfred, 131
Humanité, L', 3–4
Hunchback of Notre Dame, The (film), 129
Hungarian Soviet Republic, fall of, 21

IFTU. *See* Trade Union International
Illnesses
  tuberculosis and, 16, 65, 66, 211n
  venereal disease and, 16, 65, 157
Illustrierte Kronen-Zeitung, 88
Independent Association of Socialist Students and Academicians, 32
Individual psychology, 76, 125
  influence of, on SDAP's cultural program, 112–13
  sexuality and, 175
Industry
  SDAP's failure to nationalize, 182
  Viennese, 15
Infant layettes, distribution of, 69, *70*

Index 263

Infant mortality, decline of, 66
Inflation, rent control and, 22–23
International Worker Olympics, *106, 107,*
  109–10, 139, 183
  second, 107
Italians, 25

Jahoda, Marie, 6, 113
Jews, 25. *See also* Anti-Semitism
  Austromarxist, 31
  concentration camps and, 195n
  prominent, 195n
  in SDAP, 26, 195n
  self-hatred among socialist leaders and,
    26–27
  value placed on books by, 87
  in Vienna, 15, 16, 18, 25–27, 195n
John, Michael, 191n, 192n, 193n, 203n,
  204n, 206n, 208n, 222n
Jugendliche Arbeiter, Der, 131–32
Jules Ferry Laws, 79
July 15, 1927, revolt, 10, 41–43, 108, 183
  radio and, 138
  as turning point in history of First
    Republic, 42–43
Jungen, Die, 74

Kabinett des Doktor Caligari, Das (film), 130
Kaffeehaus, 33
Kampf, Der, 33, 88, 131–32
Kampf um die Macht, Der (Bauer), 90
Kanitz, Otto Felix, 37, 111, 167–68, 175
  views on sexuality, 178
Kant, Immanuel, 36
  Max Adler's attempt to integrate ideas of,
    with Marxist ideas, 34–35, 36
Karl-Marx-Hof, *61,* 64, 65
Karl-Seitz-Hof, 65
Katholischer Lehrerbund für Österreich, 79
Kautsky, Benedikt, 228n, 252n
Kautsky, Karl, 158, 160
Kelsen, Hans, 40
Kiba. *See* Kino-Betriebsgesellschaft m.b.H.
Kinderfreunde, 166
Kindergartens, 152
  attempts to produce orderly family, 69
  increase in number of, 66
Kinderübernahmestelle (children's
  diagnostic service), 71
Kino-Betriebsgesellschaft m.b.H. (Kiba),
  133–34, 235n
Kitsch
  art and, 100

SDAP's struggle against, 95–96
  socialist culture and, 85
Klassenkampf—Marxistische Blätter, Der, 34
Kleine Blatt, Das, 88, 89, 90, 142
  film reviews in, 131
  readership of, 89
Kleine Volkszeitung, Die, 88
Kollwitz, Käthe, 184
Kolowrat, Alexander, 126
Kommunisten und Sozialdemokraten
  (Braunthal), 89
Kongressbad, *122*
Korda, Alexander, 126
Kortner, Fritz, 126
KPD. *See* Communist party, German
KPÖ. *See* Communist party, Austrian
Kracauer, Siegfried, 190n, 234n
Kraus, Karl, 97, 218n
Kreisky, Bruno, 188n
"Kreta," 55
Kronen-Zeitung, readership of, 89
Kuckuck, Der, 88, 142, 148
Kulemann, Peter, 192n, 219n, 220n, 228n,
  233n, 235n
Kulturkampf, 135
  waged by Catholic church against SDAP,
    29
Kunschak, Leopold, 25
  Jewish exclusionary law prepared by, 26
Kunststelle. *See* Sozialdemokratische
  Kunststelle
Kunst und Volk, 96, 100
Kunstwart, 86

Labor and Socialist International, 7, 33,
  107
Labor legislation
  of 1918–19, 21–22
  workday and, 115
Lang, Fritz, 126, 130–31
Langbehn, Julius, 86
Langewiesche, Dieter, 87, 216n
Language spoken at home, 194n
Laski, Harold, 3
Laundry, communal, *62*
  supervisor of, 63
Lazersfeld, Paul, 140
Lazersfeld, Sophie, 170, 175
Leaders, socialist
  dichotomy between followers and, 7–8,
    20, 54, 178
  schools for, 90, 92
Lebensunfähigkeit, 24–25
Lectures, 91–92

Legislation
  labor, 21–22, 115
  rent-control, 22–23
  separation of school and church and, 74
Leichter, Käthe, 90, 154, 155, 170, 177
Leichter, Otto, 40, 97–98, 100
Leicht-Varieté, 119
Leisure time, 114–45
  activities enjoyed by workers during, 112
  available to workers, 115–16, 150, 230n
  cinema and, 126–35
  commercial culture and, 116–20
  condemnation of popular culture and, 123–26
  noncommercial activities and, 120–23
  radio and, 135–41
  spectator sports and, 141–42
  unemployment and, 116
  use of, 124
  of women, 150, 230n
Leser, Norbert, 188n, 197n, 200n
Letzte Mann, Der (film), 129
Levi, Paul, 34
Libraries, worker, 92–96, *94*
Lichtheim, George, 30–31
Linz Program, 30, 40, 183
Lobau, 121, *122*
Lobmeyerhof, 49–50
Loewenberg, Peter, 198n, 200n
London, Jack, 95, 184
Lorre, Peter, 126
Lueger, Karl, 25
Lukacs, Paul, 126
Lunch, 228n, 244n
  Sunday, SDAP's practical suggestions for, 148, 150
Luxury taxes, 49, 203n

McGrath, William J., 189n, 190n, 224n
Mach, Ernst, 32, 36
  Max Adler's attempt to integrate with Marxist ideas, 34–35, 36
Maderthaner, Wolfgang, 196n, 213–14n, 222–23n, 248n
Magic Mountain, The (Mann), 129
Mann, Heinrich, 131
Mann, Thomas, 129
Margueritte, Victor, 148
Marius (film), 129
Mark, Karl, 97
Marriage. *See also* Family; Women
  religious versus secular ceremonies and, 196n

  sexual intercourse and cohabitation before, 175–76
  sexuality and, 157, 177, 178
Marriage consultation clinics, 68–69, 158
Marx, Karl, 31
Marxism, 36. *See also* Austromarxism
  Max Adler's attempt to integrate Kantian and Machian ideas with, 34–35, 36
Mass culture, 115
  problem of dealing with, 9, 141–45
May, Ernst, 65
May, Karl, 184
  SDAP's attack on, 95–96
Menger, Carl, 32
Metropolis (film), 129
Middle class, 18–19
  book clubs and, 219n
Middle schools, Glöckel's proposal to establish, 75, 76
Military terms, 24
Ministry of Education, 77
Mistinguett, 119
Morberger, Gerta, 168
Motherhood. *See also* Child care
  burden imposed by, 152
  emphasis on role of, 150
Mothers' consultation clinics, 69
Movie theaters, 120, 127
Municipal bathhouses, 60, 66
Municipal Council
  housing program and, 51
  infant layettes distributed by, 69, *70*
Municipal socialism, 45–80
  public education and, 73–80
  public health and social welfare and, 65–73
  public housing and, 46–65
Music, 98–100
  Schrammeln quartets and, 116

National Conference of SDAP Women, 159
National Women's Conference, 160
Nazi party, Austrian
  attack on Ravag headquarters, 139
  election of 1932 and, 238n
Neill, A. S., 77
"Neue Menschen," *165*
  Bildung and, 73
  condemnation of popular culture and, 123–26
  cultural efforts and, 82
  film as barrier to creating, 135
  population politics and, 72–73

Index 265

resistance to attempts to transform workers into, 112
as SDAP goal, 143
socialist culture and. *See* Socialist party culture
Neue Menschen (Adler), 39
"Neue Sachlichkeit," 130
Neurath, Otto, 52
  criticism of Max Adler's position on socialist culture, 84–85
News broadcasts, Ravag and, 136
Newspapers
  published by cooperative societies, 87
  published by SDAP, 87–91
  published by trade unions, 87, 90, 218n
  worker uprising of 1934 and, 4–5
New woman, SDAP goal of, 147–55
New York Times, 4
Nobelheurigen, 117
Noncommercial culture, 115
Nuclear family, 146, 147
  public housing and, 63

Oberst Redl (film), 130
Oela (film distribution company), 133
Olympia Arena, 118
Opera, 97, 98
Ordentliche Arbeiterfamilie (orderly worker family), 147
  public health and social welfare and, 65–73
  public housing and, 45–65
  women's role in, 178
Österreichische Lehrerbund, 79
Österreichische Radio-Verkehrs A.G. *See* Ravag

Pabst, G. W., 13, 126
Palace of Justice, burning of, 10, 108, 183
Pan-German party, 181
  Anschluss and, 15, 24
  anti-Semitism of, 26
  commitment to monarchy, 13
  1919 elections to Constituent Assembly and, 21
  1927 elections and, 42
  SDAP opposition to coalition of Christian Socials with, 29
Paragraph 144, abortion and, 159–61, 162, 176–77, 179
Paramilitary sports, 105–6
Parliament, 1920 elections to, 21

Parteischulen, 92
Paternalism of SDAP, 8, 52–53, 90, 147, 162–63
People's clubhouses, 91
"People's palaces," *51*, 55, *57*, 58, *58*, *61*, 64. *See also* Public housing
Pernerstorfer, Engelbert, 83
Personal hygiene for women, emphasis on, 150
Peukert, Detlev J. K., 230n, 233n
Pfabigan, Alfred, 197–200n
Pfoser, Alfred, 197n, 218–24n, 226n, 248n
Pfrimer, Walter, 110
Pictograms, 84
Piffl, Gustav (cardinal), 28, 139
Pirhofer, Gottfried, 209n, 210n, 212n, 240n, 242n, 252n, 253n
Point system for allocation of public housing, 61–62
Political forces, 181–84
Political revolution, Bauer's separation of, from cultural revolution, 39
Pollak, Marianne, 155, 168
Popp, Adelheid, 155, 159, 160
Popular culture, condemnation of, 123–26
Population politics
  health and welfare programs and, 66, 68, 72–73
  sexuality and, 157–64
Pornography
  Catholic church and, 164
  SDAP's fight against, 163
Poverty, municipal intervention in family and, 71
Prater, 117–18, 119, 123, 141
Praxis, Die, 124
Pregnancy
  abortion and. *See* Abortion
  exercise and, 151
  before marriage, 159
  SDAP's advice to women regarding, 154–55
Premarital sex, 159, 175–76
Preminger, Otto, 126
Press. *See* Books; Newspapers; Publishing; *specific newspapers*
Price of consumer articles and services, 229n
Professionalized housework, 52
Professor Unrat (H. Mann), 131
Proft, Gabriele, 155, 160
Proletarian Riviera, 121, *122*
Proletkult, adaptation of, by SDAP, 109
Prometheus company, 134

Prostitution, 16
  socialist obsession with dangers of, 157
Provisional Assembly, 13, 15
Public education, 73–80. *See also* Schools
  development of new curricula and teaching methods for, 75–76
  in England, 77
  SDAP reform program for, 77–80
  in United States, 77
  in Weimar Republic, 77
  workers' views on, 80
Public health. *See* Health and welfare programs
Public housing, 6, 46–65, *51, 57, 58, 61,* 72, 206–7n
  architecture and, 56
  in Berlin, 60
  communal facilities and, 60–61, *62*
  construction materials and methods and, 56–57
  effect of small size of apartments in, 62–63
  in England, 64
  in France, 64
  in Frankfurt, 57, 60, 64
  in Hamburg, 60, 64
  lack of planning for, 50–53
  "red fortress" theory and, 210n
  regimentation imposed by management of, 63–64
  shortcomings of housing built by, 58, 60
  tenants of, 61–62
  workers' lack of control over, 63–64
Public sanitation, breakdown of, 65
Public Welfare Office, 68
Publishing. *See also* Books; Newspapers; *specific publications*
  Austromarxist, 32–33

Rabinbach, Anson, 36, 197n, 199n, 202n, 218n, 225n
Rada, Margarete, 171, 173
Radio, 135–41, *137*
  enthusiasm for, 136–37
  listeners' program preferences and, 140
  "neutral" programming for, 136, 138
  price of receivers and, 237n
Radio clubs, 82, 138, 139, 140
Ravag (Österreichische Radio-Verkehrs A.G.), 136–41
  Heimwehr influence on, 238n
  struggle over control of, 136–140
Readership
  of books, 92–96
  of socialist publications, 87–91

Red Falcons (Rote Falken), 24, 109, 121, 166, 167
  sexuality and, 174
Reich, Wilhelm, 158, 159, 161, 162
  criticism of socialist youth policy, 169, 170
Reichspost, Die, 41
Reigen, Der (Schnitzler), 163–64
Reinhardt, Max, 109
Religious instruction, 74, 78–79, 182, 196n, 215n
Remarque, Erich Maria, 95
Renner, Karl, 8, 15, 21, 30, 31, 33, 36, 108, 183, 199n
  confrontation with Max Adler, 42
  opposition to defensive-force position, 40–41
  postwar role of, 33
  socialist culture and, 86
Rent control, 22–23, 203n
  housing shortage and, 45, 47
Renz circus, 118
Resurrection (film), 129
Reumann, Jakob, 20, 37, 163–64, 202n
Reumannhoff, *51, 57*
Revolutionäre Sozialdemokraten, 170
Riesenrad, 117
Rintelen, Anton, 136
Rituals of old order, 23–24
Rosenfeld, Fritz, 131, 132, 134, 234–35n
Rosenfeld, Kurt, 34
Rote Falken. *See* Red Falcons
Russell, Bertrand, 77

Sacco und Vanzetti (film), 129
Safrian, Hans, 175
SAJ. *See* Socialist Worker Youth
Saldern, Adelheid von, 204n, 207–8n, 209n, 220n, 222n, 236n, 238n
Sandleitenhof, *58*
  communal laundry at, *62*
  worker library at, *94*
Sascha-Film, 126
Schlesinger, Therese (Ekstein), 26, 155, 160, 161
Schlick, Moritz, 84
Schmalzbrot, 71
Schmitz, Richard, 77
Schnitzler, Arthur, 163, 179
Schober, Johannes, 41–43
Schools. *See also* Education; Public education
  confessional, 215n
  health care incorporated into, 75

middle, Glöckel's proposal to establish, 75, 76
for SDAP leadership, 90, 92
separation from church, 74
work, 76
Schorske, Carl, 12, 189n, 190n
Schrammeln quartets, 116
Schrebergärten. *See* Garden plots
Schütte-Lihotzky, Margarete, 60
Schutzbund, 41–43, 87, 103, 105–6, 166
SDAP. *See* Austrian Socialist party
Seidewitz, Max, 34
Seipel, Ignaz, 136
  annexation with Germany, Czechoslovakia, or Italy proposed by, 24–25
  anti-Semitism of, 26
  Catholic church and, 27, 28
  censorship and, 164
  controversy over municipal crematorium and, 72
  refusal to compromise following July 15, 1927, 41–42
Seitz, Karl, 37, 133, 138
  controversy over municipal crematorium and, 72
  educational reforms and, 74
  housing program and, 50
Settlers' movement, 48
Sex education in socialist youth organizations, 175
Sex manuals, 162, 247–48n
Sexual abstinence, socialist promulgation of, 158
Sexual consultation clinics,161–62
Sexual division of labor in household, 150
Sexuality, 155–78
  Catholic church's position on, 163–64
  as distraction from party, 156–57
  orderly family and, 156
  overcrowding in housing and, 171
  place accorded to, by SDAP, 156–57
  population politics and, 157–64
  realities of everyday life and, 170–78
  repression of, 159
  sexual precocity and, 172
  youth and, 165–70, 174–75
Sexual promiscuity, SDAP condemnation of, 158–59
Sieder, Gerhard, 172, 175
Sinclair, Upton, 95
Singing societies, 100
Sklavenkönigin, Die (film), 130
Slezak, Walter, 126
Slums, Viennese, 55

Soccer, 123
  as spectator sport, 141–42, *143*
Soccer association, 103
Social and Economic Museum, 84
Social Democratic Workers' party. *See* Austrian Socialist party
Socialist art, definition of, 97
Socialist art center, 82
Socialist cultural center, 82
Socialist Party culture, 81–113
  concept of, 216n
  elite culture both rejected and desired and, 83–87
  lectures and, 91–92
  music, theater, and fine arts and, 96–102
  perspective guiding program for, 82
  printed word and, 87–91, 92–96
  questions about content of, 184–85
  sports and festivals and, 102–13
Socialist Performance Group, 100
Socialist Sports International, 107
Socialist symbols, 226n
Socialist Worker Youth (SAJ), 103, 109, 111, 124, 166, *167*, 169
  condemnation of popular culture and, 125
  membership of, 231n
  sexuality and, 174–75
Socialization of children, *67*, 165, 171–72, 173
Socialization Commission, 22
Social welfare. *See* Health and welfare programs
Social workers
  home visits carried out by, 66, 69, 71
  negative perceptions of, 71–72
  professional training of, 69–70
Sodom und Gemorrha (film), 130
Soldiers' councils, *18*
Sozialdemokrat, Der, 87
Sozialdemokratische Kunststelle, 82, 83–87
  music, theater, and fine arts and, 96–102
  worker festivals and, 108, 109
Sozialistische Bildungszentrale, 82, 86, 91, 92, 93, 100, 103, 111, 132, 134
Sozialistische Erziehung, 124
Sozialistischer Lehrerverband, 79
Spectator sports, 141–42, *143*
Sports
  bourgeois, 104
  paramilitary orientation of, 105–6
  politicization of, 104–5
  spectator, 141–42, *143*
  women in, 106
Sports programs, 102–7

Sprechchor, 109
Squatters, 48
Starhemberg, Ernest Rüdiger von, 187n
State opera. *See* Opera
Stern, Joseph Luitpold, 37, 85–86, 168
  attack on Karl May, 95–96
  worker festivals and, 108
  worker libraries and, 93
Stiftungshof, 49–50
Strasser, Georg, 139
Strauss, Richard, 109
Strikes
  national, of transportation and information services, 41–43
  in 1918, 19
Ströbel, Heinrich, 34
Subcultures, working-class, 8–9, 115, 184
Suffrage
  in 1919, 6
  electoral law of 1918 and, 21
Summerhill, 77
Swimming, 121–22, *122*
  nude, 230

Tandler, Julius, 26, 37, 46
  abortion and, 158, 159, 160
  attitude toward women, 178
  Christian Social attacks on policies of, 68–69
  marriage consultation clinics and, 158
  proposal to distribute infant layettes and, 69
  resistance to anti-Semitism, 72
  social welfare and, 65–66, 68
Taxes
  federal apportionment of, 193n
  health and welfare programs and, 65
  housing, 49, 203n
  luxury, 49, 203n
  Viennese power of local taxation and, 22
Teachers
  educational reforms and, 79–80
  training of, 76
Temps, Le, 4
Tenements (*Zinskasernen*), 18, *47*, 55, 206n
  rent control and, 22–23
  sharing of bedrooms in, 171
Tesarek, Anton, 37, 104
Theater, 97
  Varieté, 118, 119–20
Tiller Girls, 119
Times (London), 4
Titles, aristocratic, 23
Toller, Ernst, 97

Trade Union International (IFTU), 187n
Trade unions
  lack of support of women by, 153–54
  membership of, 10, 81, 232n
  newspapers published by, 87, 88–89, 90, 218n
  women in, 88, 243n, 244n
Trade Union Youth, 103
Trash. *See* Kitsch
Traven, B., 95
Trimester model, abortion and, 159–60
Tröpferlbäder, 60
Trotsky, Leon, 198n
Tuberculosis, 16, 66, 211n
  breakdown of public sanitation and, 65
Twenty-Four Hours (film), 129
Two-and-a-half International, 33

UFA, 133
Unemployment, 6
  burden placed on women by, 177
  extent of, 189n
  leisure time and, 116
  postwar, 10, 191n
Uniforms, 24
Union Against Forced Motherhood, 157
Unions. *See* Trade unions
United States, public education in, 77
University of Vienna, 32–33
Unzufriedene, Die, 88, 148, 158
  readership of, 89
  women's issues and, 161

Vacation, statutory, 21
van de Velde, Hendrik, 162
Varieté theater, 118, 119–20
Veidt, Conrad, 126
Venereal disease, 16, 157
  breakdown of public sanitation and, 65
Verne, Jules, 95
Vertrauensmann, Der, 87, 89
Vertrauensmänner, 54
Viebig, Klara, 95
Vienna
  Austromarxism in, 32
  ethnic enclaves in, 18
  housing shortage in, 45–46
  landscape of, 15
  made a province of Austria, 22
  middle class of, 18–19
  movie theaters in, 127
  municipal socialism in. *See* Municipal socialism

in 1919–21, 13–29
in 1929, *59*
population of, 15, 16, 25
postwar hardships in, 16
public housing in, 46–65
slums in, 55
social tension in, 15–16
worker concentration in, in 1921, *17*
worker revolt of July 15, 1927, in. *See* July 15, 1927, revolt
workers in, 16, 18
Vienna Circle, 84
Viennae, Robert, 126
Vienna Union, 33
Visual statistics, 84
Vita studio, 133
Volksheime, 91
Volks-Kino-Verband, 134
Volkswehr, *18*, 19
Volks-Zeitung, readership of, 89
Vollkomene Ehe, Die (van de Velde), 162

Wages
 real, 193–94n
 of women, 148, 153
Wagner, Richard, 85, 100
 socialist culture and, 86
Walter, Gabriele, 52
Wassermann, Jakob, 95
Weber, Anton, 37
Webern, Anton von, 98, 99
Wegs, J. Robert, 189n, 203n, 206n, 208n, 211n, 212n, 215n, 220n, 229n
Weidenholzer, Josef, 91, 216n
Weimar Republic
 educational reform in, 77
 public housing in, 57, 60, 64
Welfare workers. *See* Social workers
Wiener Dreigroschenbücher, 89
Wiener Messe, 118
Wilder, Billy, 126
Winter, Jay, 211n
Wirtshaus. *See Gasthäuser*
Wohnungsanforderungsgesetz (housing requisitioning law), 48
Wolff, Friedrich, 97
Women, 147–55
 abortion and. *See* Abortion
 anticlericalism and, 28
 as comrade and friend, 150
 denial of sexual intercourse by, as birth control method, 177
 as domestic workers, 153
 everyday life of, 151–52

exercise for, 151
as factory workers, *152*
as homeworkers, 153
leisure time of, 150, 230n
lunches consumed by, 228n, 244n
new, 147–55
postwar expulsion of, from industry, 191n
professional housework and, 52
publications aimed at, 88
rationalization of housework and, 148, 150–52
reading habits of, 89
reasons for working and, 154
role of, 150, 152, 178
as SDAP members, 20
second-class status of, in sports, 106
socialist view of, 148
in trade unions, 88, 243n, 244n
trade unions' lack of support for, 153–54
triple burden of, 150, 152, 177
wages of, 148, 153
work as burden for, 150, 152–53
workday of, 151
workweek of, 192n
Workday
 eight-hour, 115
 of women, 151
Worker chess club, 103
Worker cultural organizations, 217n
Worker festivals, 100, 107–12, *110*, 225–26n
Worker libraries, 92–96, *94*
 books borrowed from, 95–96
 patronage of, 94–95
Worker Olympics, 109–10, 139, 183
 second, 107
Worker Radio Club, 103
Worker revolts
 of July 15, 1927. *See* July 15, 1927, revolt
 of 1934, 3–4
Workers
 allocation of public housing to, 61–62
 benefiting from educational reforms, 80
 concentration of, in Vienna in 1921, *17*
 demonstrations of, *19*. *See also* Worker revolts; July 15, 1927, revolt
 family of. *See* Family; Ordentliche Arbeiterfamilie
 female, *152*. *See also* Women
 impact of Bolshevik Revolution on, 19
 lack of control of, over public housing, 63–64
 lack of involvement of, in planning for housing program, 52–53

Workers (*continued*)
　leisure time of. *See* Leisure time
　neighborhoods and, 18, 47–48
　patronage of worker libraries by, 94–95
　in postwar Vienna, 16, 18
　protests in 1919–20, 18–19
　reading habits of, 89
　response of, to attempts to produce orderly families, 71–72
　taste of. *See* Kitsch; Socialist Party culture
　unemployment and. *See* Unemployment
　views of, on public education, 80
　wages of. *See* Wages
Workers' college, 90
Workers' councils, 33, 181
"Workers' Hour" (radio program), 138–39
Worker soccer association, 103
Workers' Radio Club, 138, 140
Worker symphony concerts, 98–99
Work schools, 76
Workweek, forty-eight-hour, 115, 192n, 228n
World League for Sexual Reform, 158

Zentral circus, 118
Zinnemann, Fred, 126
Zinskasernen. See Tenements
Zirkus Gleich, 118
Zirkus Krone, 118, 119
Zola, Emile, 95
Zweig, Stefan, 74, 95